The Work System Method

Connecting People, Processes and IT for Business Results

Steven Alter

Work System Press
Larkspur, CA

Printed on acid-free paper in the United States of America and the United Kingdom by Lightning Source, Inc.

Cover designed by Melinda Brown

Cataloging-in-Publications Data
Alter, Steven.
The Work System Method: Connecting People, Processes, and IT for Business Results/ Steven Alter
p. cm.
Includes bibliographical references and index.
ISBN 0-9778497-0-8
1. Information Technology – Management 2. Reengineering – Management
3. Organizational change 4. Industrial management

HD30.2 .A483 2006
Library of Congress Control Number 2006902277

FIRST EDITION, version 1.2, October 2006

Contents

Preface

Current business organizations cannot operate without IT, but many IT initiatives fail to meet expectations and many IT-enabled systems satisfy neither employees nor customers. The following problems are disappointingly common:

- new systems that are supposed to improve performance but never meet expectations
- ineffective analysis and design projects that absorb time and effort but never produce consensus about what is to be done and why
- ineffective communication between business and IT professionals
- software implementation that proceeds despite disagreement about how the software is expected to improve work practices and provide benefits.

I wrote this book because I believe that many applications of IT would be more successful if business and IT professionals had an organized but non-technical approach for communicating about how current work systems operate and how they can be improved with or without changing technology. This book synthesizes ideas from many areas into a coherent, yet flexible approach that any business professional can use directly. I believe it is unique among the many current books that promise business and technical professionals insight about organizations, competition, technology, and system development. Although many of these books address important topics in compelling ways, most of them ignore this book's central issue for managers and business professionals:

> *How can I be more effective in evaluating systems and thinking about how to improve them, whether or not IT plays a major role, and whether or not the system is totally within my organization or links to other organizations?*

In other words, this book differs from the many books that explain how technology has made the world a better or worse place, how new technologies will shape competition in the future, how executives should allocate investment dollars, or how to thrive in the chaotic times ahead. Anyone who reads those books already knows that the pace of change is relentless and that technology is both a source of opportunity and a glutton for resources and attention. This book presents ideas that can help in making decisions for which IT may seem part of the solution, but may also become part of the problem.

Goal and audience. This book's primary goal is to provide a rigorous, but non-technical approach that any manager or business professional can apply for visualizing and analyzing system-related opportunities and problems. While this is not rocket science, just communicating more effectively about systems would help in seizing missed opportunities, minimizing wasted effort, and attaining business results.

Unfortunately, common tools and methods for IT professionals do not support this goal adequately. Graphical documentation tools are certainly useful for explaining the detailed structure of a proposed or actual system, but these tools omit many key topics related to work system characteristics, participants, customers, and the surrounding environment. Carefully scripted meetings facilitated by internal or external consultants have been used widely, but except in very complex situations, a meeting of the minds shouldn't require engaging a consultant. And regardless of how expertly IT professionals do their technical work, they too would often benefit from a better way to think and communicate from a business viewpoint.

This book is written for business professionals who want to participate more effectively in building, using, and managing systems that apply IT, and who may wonder how to create practical meaning from the technology-related sales pitches, jargon, and speculation they often hear. Even if they have a coherent view of what to do about applying IT, they often find it difficult to explain that view to their business and IT colleagues. This book addresses these concerns by providing an organized approach for thinking and communicating realistically about how systems operate in organizations and how they can be improved.

The book's secondary audience is IT professionals who want to work more effectively with business professionals when building or maintaining systems. The mere fact that they are IT professionals often typecasts them as technology advocates even when they realize that a more balanced, business-oriented approach would lead to greater clarity, greater trust, and better results.

It is important to say what this book is not. First, it is not another book about reengineering and does not assume that you need to change your company to compete in the 21st century. Although it uses some ideas that are associated with total quality management, it is not another book about total quality, statistical process control, or Six Sigma. Rather, it is about establishing a basic understanding of a system. That type of understanding is required before a more detailed analysis is performed. Similarly, it is not about software or software development even though it uses some of the terms used in software development projects. Many established tools and procedures help software engineers generate the extensive documentation needed to produce high quality operational software. The ideas in this book address neither that type of work nor that level of detail, but they might help software engineers develop software that better reflects the needs of the business.

Basic Ideas. This book's central concept is the work system. All businesses and organizations consist of multiple work systems that perform essential functions such as hiring employees, producing products, finding customers, selling to customers, providing customer service, and planning for the future. Almost all important systems in today's organizations use IT extensively, but there is no reason to believe that IT will be the source of all improvements. Even if a particular system uses IT extensively, from a business viewpoint IT is not the headline. Rather, the headline is about how well the work is done and how it provides value for internal or external customers.

The work system method is organized around the Work System Framework™, which summarizes the components that should be considered in even a basic understanding of a work system. Using this framework to identify the elements of a particular system helps in establishing mutual understanding of the system's scope and operation. A deeper analysis examines these elements and their interactions. The work system method provides an organized, but flexible structure for analyzing a system from a business viewpoint, identifying possible changes, and then justifying a design recommendation.

Organization. This book is organized in layers. Some readers probably want just the first layer, a basic understanding of the approach. Other readers want to understand the work system method in enough depth to apply it when analyzing a work system.

With the exception of establishing a rigorous definition of work system, this book is written using familiar business terms. It chapters use a format that is designed for skimming. They contain tables, bullet lists, and bolded subheadings that should help in finding what you need.

Part I consists of seven chapters that introduce the work system approach and summarize how to analyze a work system from a business professional's viewpoint.

- Chapters 1 and 2 introduce basic ideas about work systems, including the definition of work system, the Work System Framework, and the work system snapshot that is used to summarize a work system on one page.
- Chapter 3 presents an overview of the work system method, which is designed to help business professionals clarify system-related issues, identify possible directions for change, and produce and justify a recommendation. This chapter emphasizes the logical flow and alternative paths for using the work system method.

- Chapters 4, 5, and 6 present clarifications and specific topics related to the three main steps in the work system method. Chapter 4 discusses issues related to deciding a work system's scope and resolving ambiguities about the work system's elements. Chapter 5 uses the Work System Framework to organize numerous topics that provide hints for identifying issues and possible improvements during the analysis process. Chapter 6 covers aspects of explaining and justifying a recommendation.

- Chapter 7 uses the work system life cycle model to explain how work systems evolve over time. This iterative model is quite different from the "system development life cycle" models that are often proposed for guiding software projects.

Part II looks at each element of the Work System Framework in more depth. The main purpose is to identify a range of issues that should be considered when analyzing a work system. All of these issues are important in a large number of work systems, although many of them may not be important in the particular one you need to think about. Accordingly, these chapters are organized to make it as easy as possible to find topics that are important in a particular situation.

- Chapter 8 uses an example to show how the top levels of the work system method can be applied to organize the analysis of a work system. The example concerns a regional bank's system for approving large commercial loans. It combines aspects of a number of real world situations, thereby illustrating the potential relevance of topics identified in tables in Chapter 5 and explained in Chapters 9 through 14. Many of those topics are relevant to a work system's operation and success but are downplayed or ignored in techno-centric analysis methods for IT professionals. The example is presented as a discussion document, including comments by a management reviewer.

- Chapters 9 through 14 explore concepts used for describing and evaluating each work system element in particular situations. Sections within the chapters cover performance indicators, strategies, tradeoffs, and related issues. These chapters include:

  Chapter 9: Customers and Products & Services
  Chapter 10: Work Practices
  Chapter 11: Participants
  Chapter 12: Information
  Chapter 13: Technology and Infrastructure
  Chapter 14: Environment and Strategies

- Chapter 15 closes the book by looking at work system ideas in a broader context. It shows how these ideas can be used to identify omissions in IT- and system-success stories, to interpret IT- and system-related jargon, and to understand information system categories that are changing continually.

- The Appendix completes the example that was introduced in Chapter 8 by showing how it might be described in a work system questionnaire. It includes checklists related to performance indicators, work system strategies, risk factors and stumbling blocks, work system principles, and possibilities for change. It also includes questions and templates for justifying the recommendation.

Background and Motivation. This book is based on a combination of experience as vice president of a software firm and as a business professor teaching MBA and EMBA students at the University of San Francisco. In both the business world and the academic world I found that people had difficulty explaining how systems work, how they should be improved, and why. Part of this was an unfortunate tendency to focus on what computers do and to de-emphasize what people do. Another part was the lack of an organized method and vocabulary for thinking and communicating about systems. The explosion of hype about ebusiness and the subsequent dot-com bust may have shifted the discussion a bit, but the underlying problem persists.

I first recognized the need for greater understanding of computerized systems thirty years ago during interviews of decision support system users as part of my Ph.D. research at MIT's Sloan School of Management. The need for greater understanding of systems shifted from an academic interest to a pocketbook issue when I served as vice-president of Consilium, a successful manufacturing software firm (acquired by Applied Materials in 1998). Working with customers in a variety of management roles related to consulting, customer service, training, documentation, and product design convinced me that many business professionals need a simple, yet organized approach for thinking about systems without getting swamped in details. Such an approach would have helped our customers gain greater benefits from our software and consulting, and would have helped us serve them more effectively across our entire relationship.

This book is one of the results of a multi-year research effort aimed at developing a systems analysis method that business professionals can use for themselves without relying on consultants or IT professionals to help them get started. A number of articles based on this research were published in the last six years in the *Communications of AIS*, an online journal of the Association for Information Systems.

The effort started in 1992 with a series of presentations supporting a book tour for the first edition of my information systems textbook, whose four editions were used by several hundred thousand business students in universities around the world. The topics in the original presentations were the basis for a working paper that MBA and EMBA students used to analyze computerized systems in their own businesses.

Around 1997 I suddenly realized that I, the professor and textbook author, had been confused about what system the students should be analyzing. Unless they are focusing on software or hardware details, business professionals thinking about information systems should not start by describing or analyzing the information system or the technology it uses. Instead, they should start by identifying the work system and summarizing its performance gaps, opportunities, and goals for improvement. Their analysis should focus on improving work system performance, not on fixing information systems. The necessary changes in the information system would emerge from the analysis, as would other work system changes separate from the information system but necessary before information system improvements could have the desired impact.

With each succeeding semester and each succeeding cycle of papers by employed MBA and EMBA students, I tried to identify which confusions and omissions were the students' fault and which were mine because I had not expressed the ideas completely or clearly enough. The original working paper evolved into a workbook and then into an analysis outline that became more effective over many iterations of use in situations that these typical business people faced at work every day.

Acknowledgements. I would like to acknowledge the direct and indirect help of a large number of colleagues in the information system field. Many of the ideas in this book either describe or address issues revealed in their practical work, research, conference presentations, and discussions.

Among many colleagues who have contributed in a variety of ways, I want to thank two in particular. Around ten years ago John King pointed out the inconsistent use of the term *system* in a draft article I showed him. That single observation helped me recognize the need to develop a rigorous but useful way to talk about IT-enabled systems. Paul Gray, Founding Editor of the *Communications of AIS*, created a unique and innovative forum in which researchers could discuss and debate long-standing issues and new ideas. Publishing a series of articles in *CAIS* helped me reach interim closure on a number of topics that are integrated into the ideas in this book.

I also want to thank my wife Linda for her patience, encouragement, and willingness and ability to include systems among the many topics for our morning walks.

This book is dedicated to many hundreds of MBA and EMBA students whose successes and occasional confusions were essential in developing the ideas that are presented here. I greatly appreciate the many lessons they taught me.

Steven Alter
April 2006

Chapter 1: Why Are So Many Systems Such a Mess?

- A System Gone Awry
- Common Disappointments
- Seven Common Temptations
- Overcoming the Seven Temptations

Today's technology has had a huge impact on business and society. Current business practices would be impossible without IT. A seemingly endless stream of innovations brings images of boundless opportunity and change.

The progress of IT over the last five decades is an incredible accomplishment, yet IT is often under-utilized and misused. Its adoption has too many unanticipated consequences and sometimes causes many new problems. Unreliability and internal complexity often inhibit change instead of fostering it, converting a dream of flexibility into a maze of electronic concrete.

A System Gone Awry

Consider the following story[1] reported in *CIO Magazine* in 2005. "Global Giant," a global telecommunications company, installed a new "CRM system"[2] for customer-facing activities such as entering orders and providing customer service. From a purely IT viewpoint, the project went well. Even though the software vendor made significant modifications to its software package, the project was completed on time and within budget. However, the project was allowed to proceed without full agreement by Global Giant's sales, marketing, customer service, and channel management departments about how the new CRM system would affect their work practices. "They argued they couldn't realistically assess the business impact of a major system they'd never used before. They would work out their differences in the rollout."

"Unfortunately, no one—except IT—used the CRM the way it was supposed to be used, including the customers and the channels." Contrary to expectations, certain sales people figured out how to offer better prices and terms to their customers. Other customers complained and received rebates. The new system was supposed to help customer service representatives with cross-selling and up-selling, but it complicated their jobs so much that customer satisfaction suffered. Some customers would not use the new web-based order entry capabilities and phoned instead, causing inbound calls to surge. By the end of the story, a corporate innovation went down in flames, interdepartmental relationships turned sour, key customers complained, and the sales vice president was fired.

How could a seemingly successful project at a major company turn into an unmitigated disaster? According to the story's lead-in, "Getting people to use a new system correctly is much harder than getting it up and running." That partial explanation sounds reasonable, but it ignores a basic distinction about the nature of the system. Ask yourself which best describes the system in this story:

- The system is the software acquired from the software vendor, as modified and installed on the Global Giant's computers.

- The current system is the way Global Giant currently performs its customer-facing work, such as entering orders and providing customer service. The intended system is the new way Global Giant will do that work in the future.

- The system can be described in either of the two ways, depending on who happens to be talking and what they are talking about.

The people in the story acted as though they used the first definition. According to the story, the IT group acquired and installed the software; the sales, marketing, customer service, and channel management departments believed they would figure out how to use it once it was available. In other words, they believed system is a computerized tool that people use.

Things might have happened differently if the people in the story used the second definition and acted as though the system is the way Global Giant performs specific work. From that viewpoint, modification of the software was only part of a project that was never completed. The full project was not about acquiring and installing software. Rather, it was about changing the way work is performed. It may seem a bit crazy for an organization to make a major software investment based on the assumption it will figure out how to use the software. It seems even crazier for an organization to invest in major changes in work systems without knowing what the changes will be.

On the other hand, what would have happened if the people in the story chose the third possibility and believed that the system is either the software or the work system that uses the software, depending on who is talking and what they are saying? Accepting two different definitions for the same thing might seem crazy as well, except that it happens frequently. When you hear someone in your organization say, "we improved our sales system" how do you know whether they are referring to the software used in sales or the system of doing sales?

Part of a larger problem. The Global Giant story illustrates a larger problem, the lack of a practical, organized method that business professionals can use for thinking about business systems (not just software) from a business viewpoint. Viewing a system as an IT-based tool that people use is limited, techno-centric, and ultimately misleading. This view also puts business professionals at a disadvantage due to the many technical aspects of IT-based tools that they cannot appreciate fully because they lack the required technical background and interest.

This book provides a practical, but in-depth method for thinking about systems in organizations whether or not IT is involved. The method assumes that a system includes work practices, human participants, information, and technology, and that it exists to produce products and services for internal or external customers. Such a system is called a work system.[3] That idea is the basis of the work system method. The remainder of this chapter says more about issues that the work system method addresses. The second chapter begins the discussion of work system concepts.

Common Disappointments

The Global Giant story is a dramatic example of a widespread problem. Too many systems in organizations are disappointments or failures. Despite good intentions and hard work, the planned improvements and benefits are often elusive. Despite all the talk about process excellence, total quality, and agility, stories such as the following are commonplace in our personal lives:

- After switching to my local phone company's new service plan, it took three months and over four hours on the phone to correct the incorrect bills that we received.

- When I called a pharmacy to ask whether my prescription was ready, the pharmacist said that my insurance policy probably would not cover that medicine, but he couldn't be sure because the computer system linking the pharmacy to insurance companies had been down most of the day. The insurance company's call center agent gave the same answer. I was to leave on a trip the next morning and had to decide whether to pay the full price and try to obtain reimbursement later.

- My wife's former employer implemented ERP, a software package supporting business operations across multiple departments by means of an integrated database. Two weeks after the "go live" date, she was still unable to generate a simple invoice for an important client. Ironically, her employer was in the business of providing ERP training.

- After my wife's purse was stolen, her bank gave her a new account number and switched her account balance into the new account. When she deposited a check from my account into her account to pay bills, the bank put a hold on the funds until the check cleared because hers was a new account. She complained that she had been a customer for 14 years and that we had done this type of transfer before, but the teller said that the computer system would not let her remove that hold because it was a new account.

Similar stories from the everyday corporate world include:

- "Our customer service people are going crazy. They claim that the computer system is so complicated that they can't avoid making a lot of mistakes."

- "We installed great technology for sharing information but people are mostly using it as a personal tool and very little sharing is happening."

- "We installed a new enterprise system but almost everyone insists on receiving the same information they always had. What was the point?"

- "They called it sales force automation, but it doesn't help me and forces me to do clerical work so headquarters can monitor what I do."

- "The software sounded great, but after we bought it we realized we would have to make major changes in our processes."

Problems with IT-related projects and systems sometimes become visible to customers and stockholders despite typical efforts to avoid adverse publicity:

- Foxmeyer Drug Corporation, a successful wholesale pharmaceutical distributor, went bankrupt after a failed implementation of an enterprise software system that was supposed to save money and foster growth.[4]

- Cigna intended to create an integrated system for enrollment, eligibility, and claims processing. Ideally, the new system would improve customer service through an upgrade that would consolidate customer bills, process medical claims rapidly and efficiently, and provide customer service reps a single unified view of clients. Flawed implementation of the upgrade caused customer service problems that led to customer defections.[5]

- ChoicePoint, an information broker with an enormous database of personal information about almost every American adult, announced in 2005 that personal information about 145,000 people had been downloaded by identity thieves who could use the information for credit card fraud and other purposes. The impostors applied for access to the data by pretending to be legitimate business people.[6]

- Hershey's annual report for 1999 mentioned, "well-publicized problems associated with the implementation of the final phase of our enterprise-wide information system." Hershey's customers suffered order fulfillment delays and incomplete shipments, and Hershey lost some market share to major competitors during its key Halloween and December seasons.[7]

- Nike encountered difficulty implementing new supply chain software in 2001, resulting in an estimated $100 million in lost sales, a 20% drop in its stock price, and a flurry of class-action lawsuits. CEO Phil Knight was quoted as saying, "This is what we get for $400 million?"[8]

- Cisco Systems "wrote off $2.5 billion of inventory in 2001 when its forecasting systems told it to keep pumping out switches and routers even as many of its customers were going belly-up. The company said the errors were largely due to customers grossly inflating orders, fearing they might get caught without supply."[9]

Widely publicized stories such as these are unusual because no company wants to publicize bad news, but smaller examples of similar problems are surprisingly common. Industry surveys and estimates by leading consulting companies typically describe appallingly high levels of disappointment in IT-related projects. For example, even with major improvements over the last 10 years, the Standish Group's 2004 report on its biennial survey of thousands of IT projects reported that 51% missed their schedule, went over budget, or did not produce the intended functionality. Another 15% were total failures.[10] Other researchers produce somewhat different numbers depending on the types of questions they ask and the samples they look at, but the preponderance of the evidence says that the batting average for IT success is far lower than it should be. Business and government organizations would dissolve into chaos if most of their activities encountered the level of ineffective effort and disappointment that is common with computerized systems.

Seven Common Temptations

Among the many reasons for IT-related disappointments is a set of powerful temptations that somehow beguile business and IT professionals. These temptations lead to unclear communication, unrealistic expectations, and disappointing results.

Most organizations need an effective way to recognize these temptations and avoid their consequences. This book shows how thinking of IT-reliant systems as work systems, rather than IT systems, addresses every one of these temptations and is an effective method for understanding and communicating about systems.

Temptation #1: Viewing technology as the system. Have you every wondered why systems that involve IT are often called IT systems? For example, the title of a brief case in *Harvard Business Review* was "The IT System that Couldn't Deliver," even though the issue in the case was that a new sales system had not been implemented.[11]

The tendency to see technology as the central element in systems, or even as the entire system, reflects a techno-centric view of systems. That type of view typically focuses on whether the technology operates to specification and typically downplays human and organizational aspects of the situation. It often seems to assume that technology is directly responsible for success and for mistakes and errors. A more balanced view recognizes that people and work practices are essential ingredients in IT success stories, and that supposed "computer glitches" often involve sloppy work practices and human error. A more balanced view is less concerned with the tools a system uses and more concerned with what a system does and what it produces. Thus, even though a system may not be able to operate without IT, the headline should emphasize what the system does and what it produces.

Systems in organizations are always much more than the software that is being used. Defining the system as the software is a diversion from business issues and often leads to vendor-driven discussions of software features and purported benefits. The software is part of a system, just as lungs are a part of a human being. We all need lungs and feel better if our lungs are operating well, but that doesn't mean that every discussion of human endeavors needs to focus on lungs.

Ask yourself whether these temptations affect systems in your organization.

Temptation #2: Assuming technology is a magic bullet. Technology is often discussed as though it were a magic bullet,[12] a "solution," in the parlance of the computer industry. Contrary to what is often implied and sometimes said explicitly, IT is not a magic bullet that can provide all the information you need, make people smarter, transform organizations, or provide competitive advantage. Despite its incredible speed and power, IT cannot solve informational, personal, or organizational problems independent of larger systems that it supports. IT cannot supply all of your information unless your world operates like a totally predictable machine. IT cannot make you smarter even though it can provide access to information that may be useful. IT cannot transform anything other than data even though it can be used as one of the tools in a conscious organizational change effort. When the topic is a system in an organization, there is no such thing as a magical "IT solution," regardless of what the sales pitch said.

Unrealistic optimism about user enthusiasm for new technical capabilities is a common result of techno-centrism and techno-hype. Project teams indulging in this temptation pay little attention to the human and organizational aspects of systems. They may assume that since a particular technology was used successfully in one department it will probably be successful in another where the conditions are different. They may assume that procedures will capture all possible situations and that procedures will be followed. They may provide elaborate methods for sharing information despite the organization's culture of hoarding information. They may create elaborate computer security systems despite the organization's culture of supervising everything loosely.

Temptation #3: Abdicating responsibility for systems. Businesses operate through systems of doing work. That work might be manufacturing a product, finding potential customers, producing monthly accounting statements, or performing any other organized activity. The use of IT does not change the responsibility for organizing and managing the work. Too many managers and business professionals let their eyes glaze over when the discussion turns to IT and how it will be used. They know they need to be involved in IT-related projects to make sure the requirements are correct, but they still act as though systems are someone else's responsibility if IT is involved. IT professionals create computer-related tools and keep them running, but line managers are responsible for the work systems that define the organization's operation.

Managers sometimes abdicate their system-related responsibilities by confusing systems with IT, acting as though systems are someone else's work, and granting IT professionals excessive power to decide and define how systems will operate in business organizations. Sometimes they do this by allowing themselves to be misled through techno-hype and techno-centrism; sometimes they deceive themselves. In other cases, they are unwilling to get involved in detailed analysis, giving excuses such as "my eyes glaze over at those meetings about systems" or "as a manager I don't mess with that kind of stuff and I let the techies design the systems."

Abdication of system-related responsibilities creates problems. First, deep knowledge about IT and IT-related systems analysis techniques does not guarantee insight about how an organization should operate. IT experts provide essential help in developing IT-based tools, but they may not have much insight into personal and organizational issues that determine how things are really accomplished. Since computerized information systems often determine how work can or cannot be done, granting IT experts the freedom to define information systems increases the chances that these systems will miss the mark.

The other side of this question is technology hubris, the belief by some IT professionals that their knowledge of programming and systems analysis techniques qualifies them as experts in how work should be done in the organizations they support. Based on this belief they sometimes feel empowered to follow their own instincts and to ignore the concerns of the business professionals they are trying to support. Furthermore, since the IT staff typically has no authority over changing work practices in other departments, even a well conceived set of tools may not be used well unless managers in user departments are on board and committed.

Temptation #4: Avoiding performance measurement. A surprising avoidance of clear specifics about performance is one of the most striking phenomena I observed across many years of reading assignments written by part-time MBA and EMBA students about systems in their firms. After numerous repetitions of "performance is not good but would be better with these changes," I became increasingly explicit about asking students to identify key performance indicators and provide actual values or estimates of metrics in these areas. Even with increasing attention to measures of performance in the course, I still found employed MBA and EMBA students reluctant to engage this topic.

Willingness to manage without using goals, baselines, or measures of performance makes it difficult to say how well a system is operating and how much improvement is needed. The most egregious form of measurement avoidance occurs when no measures of performance are identified or tracked. A lesser form occurs when managers focus on one or two performance indicators while ignoring many others that are pertinent. A typical example is factories and service operations that measure productivity carefully but do not measure the consistency of the work or quality of the product as perceived by the customer. The same phenomenon occurs in the management of information systems when the technical efficiency of information storage hardware is monitored carefully but the error rate of the information itself is not measured.

Measurement avoidance has an important effect on projects involving IT because it goes hand-in-hand with unclear project goals. Instead of identifying current and desired levels of work system performance, project leaders focus on providing particular software capabilities and achieving indirect benefits such as better information and better decision making. The new capabilities and indirect benefits may be quite important, but without measurement it will be difficult to decide whether the investment was worth the effort.

Temptation #5: Accepting superficial analysis. Superficial analysis often starts at the beginning of a project when managers are unwilling or unable to participate fully in defining the goals of the project and exploring how the proposed changes will really accomplish those goals. The most extreme cases of superficial analysis can be called "management by slogan." The manager identifies a broad goal, such as cutting response time or providing the best information possible, but never gets involved with what needs to happen in order to accomplish the goal. Even when managers are willing and able to be involved, a techno-centric emphasis sometimes focuses on changes related to IT applications and downplays changes related to work practices and people. In many cases, less IT-intensive but equally valid approaches are not fully considered, such as business process changes and better staffing, training, and on-going support.

When IT applications are involved, superficial analysis sometimes takes a bizarre turn by trying to attain closure on microscopic details before resolving big picture issues. The temptation to leap into details prematurely stems from two types of motives. The first is pressure to keep the project on schedule. The second is a reluctance to raise difficult political and practical issues about how work will be done, who will be in charge, and who will gain or lose power. A way to avoid these issues is to plunge into IT-related details that are essential for producing computer programs, whether or not those programs actually address the main problems. In combination, these two forces result in system development projects that ignore major business and organizational issues even though they tenaciously clarify the smallest details about anything a computer touches.

Another aspect of superficial analysis is confusion about the difference between documentation and analysis. Documentation of IT-based systems is an organized description of the system's structure and details. Documentation is essential for creating well-crafted technical systems, but by itself is not an analysis. In contrast, an analysis is about defining a problem carefully, gathering information, identifying alternatives, and deciding what to do. A look at the documentation of most IT-based systems would reveal substantial emphasis on what should happen in the computer and comparatively little emphasis on what should happen in the organization.

Temptation #6: Accepting one-dimensional thinking. Each of the following book titles (minus the subtitle) focuses on one dimension of a much more complex situation.

- Digital Enterprise
- IT Doesn't Matter
- IT Doesn't Matter, Business Processes Matter
- The Process Edge
- The Information Payoff
- People Come First

- Customers Come First
- Customers Come Second
- Customer Focused Organizations
- Designing Effective Organizations

The authors of these books unquestionably recognize that many other things matter, but the main title of each book puts one topic in the foreground and other parts of the puzzle in the distant background. Focusing on one main topic is an effective and frequently essential strategy for writing a business book, but one-dimensional thinking is totally inadequate for understanding systems in organizations. These systems consist of work practices AND people AND information AND technology. They exist to produce products and services for customers, and they do that within an environment and using available infrastructure. The analysis of a system in an organization needs to take all of those elements into account.

One-dimensional analysis of a system often assumes that the system exists in a vacuum and will operate according to the designer's wishes regardless of incentives, skills, culture, history, inventiveness, workarounds, changes in the environment, and other factors affecting the way work is done. These context-related issues make it risky to assume that better technology, better information, and better interfaces will result in better business results. To build computerized capabilities that are on the mark, designers and managers need to validate both the likely benefits of the proposed changes and the organization's commitment to implement them.

Temptation #7: Assuming desired changes will implement themselves. Innovations do not implement themselves. To the contrary, changes in significant systems require careful planning, extensive communication, and effective responses to problems that emerge. Even well designed systems may encounter implementation difficulties if the planning for the implementation process is inadequate and if implementation resources are unavailable. At minimum, the conversion from the previous way of doing work to the new way may involve a great deal of extra effort by the people doing the work. This extra burden should be recognized and planned in advance so that it does not undermine the new system.

Unrealistic assumptions about implementation sometimes leads to resistance and to a variety of interpretations of why resistance is or isn't legitimate. Project teams may believe that the software capabilities they have produced are a given and that the people who use those capabilities are guilty of resisting change if they have difficulty learning how to use the software or if they criticize it directly. They sometimes take an arrogant stance of assuming they are the bearers of the truth and that anyone who does not accept this truth is at best misinformed, and at worst, lazy or stupid. Many observers have noted this type of rhetoric in the early reengineering movement, whose proponents said things like "carry the wounded but shoot the stragglers."[13] More recently, the comic strip Dilbert has mined this topic endlessly by showing Dilbert's boss or an HR consultant ignoring sensible comments and squelching any motivation the employees might have. Resistance sometimes stems from little more than personal interest, but in many other cases resistance occurs because the resistors genuinely believe the intended changes reflect design flaws that need to be addressed.

Overcoming the Seven Temptations

As with many seductive forces, yielding to any of the seven temptations is far from fatal, but may have negative impacts. You can probably demonstrate their importance by looking at your own organization and asking whether these temptations affect the way systems operate. Think about the meetings you and your colleagues attended to discuss systems. Was there any difficulty in focusing on the real issues? Was there a tendency to become entangled in computer details instead of focusing on how the real work was done and how well that work was done? Was there a lot of frustration about incomplete understanding and communication? Was there a feeling that the business issues were not adequately expressed? Was there an urgency to define computer system features so that project schedules could be met regardless of whether there was genuine agreement about how those features would affect the organization's operations and results? Was there enough discussion of how the changes would affect the people in the situation? These questions may hit a nerve because a surprisingly large percentage of the expense and effort devoted to using information technology fails to produce the expected benefits.

The ideas presented throughout this book will help you identify each temptation and respond to it. The next chapter shows how focusing on work systems automatically counteracts temptations to view technology as a system and to assume technology is a magic bullet. By making it easier for managers to understand systems, the work system approach reduces the temptation to abdicate responsibility for systems. Attention to performance indicators and metrics counteracts the temptation to avoid measurement. Antidotes to superficial analysis and unrealistic assumptions about system details and implementation start with an organized method for assuring that system changes are designed with a work system in mind.

The title of the next chapter summarizes the overall approach. Thinking of a system as a work system, not an IT system, is a step toward recognizing all seven temptations and reducing their impact. The next chapter uses brief examples to illustrate a work system approach for thinking about systems in organizations. The subsequent chapter summarizes the work system method.

Chapter 2: Work Systems - The Source of Business Results

Most business and IT professionals could benefit from a practical, organized way to think and communicate about systems in organizations. With this type of approach they would be more likely to spot opportunities to exploit technology and more able to explain why poorly conceived system projects probably won't succeed. The work system method is based on the idea of "work system," which is introduced here through four disparate examples.

Examples of Work Systems

The following examples of work systems are in different industries, use different technologies, and face different problems or improvement opportunities. The tasks in these work systems include approval of commercial loans, identification and qualification of sales prospects, use of an ecommerce web site, and development of software. When these disparate situations are stripped down to the basics, they can be described in terms of identical categories that form the basis of the Work System Framework.

Work system #1: How a bank approves commercial loans. A large bank's executives believe that its current methods for approving commercial loans have resulted in a substandard loan portfolio. They are also under pressure to increase the bank's productivity and profitability. The work system for approving loan applications from new clients starts when a loan officer works with a new prospect to identify financing needs. The loan officer helps the client compile a loan application including financial history and projections. A credit analyst prepares a "loan write-up" summarizing the applicant's financial history, providing projections explaining sources of funds for loan payments, and discussing market conditions and the applicant's reputation. Each loan is ranked for riskiness based on history and projections. Senior credit officers approve or deny loans of less than $400,000; a loan committee or executive loan committee approves larger loans. The loan officer informs the loan applicant of the decision. A loan administration clerk produces loan documents for approved loans. (This example will be used in several places to illustrate the work system method.)

Work system #2: How a software vendor tries to find and qualify sales prospects. A software vendor sells human resources software to small and medium sized enterprises. It receives initial expressions of interest through inquiries from magazine ads, web advertising, and other sources. In addition to cold calling using industry phone lists, a specialized sales group contacts leads from other sources and asks questions to qualify them as potential clients who might be interested in buying or using the software. A separate outside sales force contacts qualified prospects, discusses what the software can do, and tries to negotiate a purchase or usage deal. The company's managers are concerned that the sales process is inefficient, that it misses many good leads, and that the prospect lists received by the outside sales group contain too many unqualified prospects.

Work system #3: How consumers buy gifts using an ecommerce web site. A manufacturer of informal clothing for teenagers and younger adults has a web site that has not produced the anticipated level of sales. Both surveys and logs of web site usage reveal that customers who know exactly what they want quickly find the product on the web site, put it in their electronic shopping cart, and make the purchase. On the other hand, customers who are not sure what they want, such as parents wishing to buy gifts for their children, often find it awkward to use the site, make a low percentage of purchases relative to the number of web site visits, and have a high rate of after purchase returns. The company's strategy calls for extension of existing sales channels. Its managers want to do something that will improve the level of sales on the web site and will improve the overall experience of customers who are not sure what they want.

Work system #4: How an IT group develops software. The IT group in a major manufacturing company buys commercial application software whenever possible, but also produces home-grown software when needs are not met by software from external vendors. Managers of the IT group are displeased with many of the IT group's software projects, which often miss schedule deadlines, go over budget, and fail to produce what their internal customers want. Software developers often complain that users can't say exactly what they want and often change their minds after a lot of work has been done. Users complain that the programmers tend to be arrogant and unresponsive. Several years ago the company bought computer aided software engineering (CASE) software, but its use has been uneven. Some enthusiasts think it is very helpful, but other programmers think it interferes with their creativity. The IT group's managers believe that failure to attain greater success within several years could result in outsourcing much of the group's work.

Commonalities

The four cases have many commonalities. In each system, people are doing work (approving loans, qualifying sales prospects, buying gifts, or developing software) in a somewhat systematic way to produce a result that is of importance to someone else. In each situation they are using information and technology. Each situation calls for multiple performance indicators related to the time and effort that goes into the work, the quality of the work, and the quality of the results. In each situation, aspects of the surrounding environment have an important impact on the possibilities for change. The following commonalities apply to all four cases and are at the core of the work system method:

Focusing on work, not just IT. Each of these systems uses IT, but exists to perform a particular type of work. The concept of work has nothing to do with paying people to be present at a particular place during a particular part of the day. Work is the application of resources such as people, equipment, time, effort, and money to generate products and services for internal or external customers.

A closer examination of the work within each system would reveal one or more business processes that are followed, at least to some extent. However, the work practices in each case also involve more than an idealized set of business process steps. The loan approval system depends on the loan officer's ability to communicate and persuade. The prospect qualification system relies on the communication ability of phone agents. Different users of the ecommerce web site might use it in different ways and in different sequences for several different purposes. The software development system seems to have several different sets of work practices depending on who is doing the work.

Work systems. From a business viewpoint, systems in organizations are best understood as work systems, regardless of whether they use IT extensively. A work system is a system in which human participants and/or machines perform work using information, technology, and other resources to produce products and/or services for internal or external customers.

Problems or opportunities. Each story is about a work system that has problems or opportunities. Because life is very busy, it is unlikely that anyone would go to the trouble of thinking about a work system in depth without being motivated by a problem or opportunity.

Identification of the system. The precise scope of each work system is not obvious in advance and is determined based on the problem or opportunity that is being explored. In each of these situations, the analysis should start with more clarification about the work system's boundaries. For example, does the work system for selling through the web site involve anything other than the mechanics of the web site? Does it involve larger questions such as how the company organizes and presents its product lines to the market in general? Similarly, is the work system for producing qualified prospects mostly about what happens when the internal sales force receives the initial leads, or does it start with the way the initial leads are generated? Thus, work system is a mental construct rather than a physical thing. Different people looking at the same situation usually define the relevant work system somewhat differently.

The basis of the work system method is apparent in commonalities shared by four seemingly disparate systems.

Participants and customers. It is useful to think separately about the people who are participants in the work system and the people who are the work system's customers. In three of the examples, the participants are employees or contractors for the company that owns the work system. In the ecommerce example, however, the customer is a work system participant who performs self-service activities. Although participants in all four work systems are users of computers or software, being a participant in the work system is much more important to them than being a user of technology. In all four examples, the efficiency of the work and the quality of the outcome depends on the participants. Appropriate technology and well-designed processes certainly help, but uninterested or unwilling participants will not produce good results. In the software development example, the attitude that CASE software restricts creativity was one of several obstacles to successful implementation.

Information. Each work system involves computerized databases and other information not stored in databases. For example, transaction-related information in the ecommerce web site is stored in a highly structured database, but other relevant information is from advertisements, personal preferences, or other sources. The loan approval example uses information from the applicant's financial statements and from publicly available databases. Other important information includes the content of loan documents and the reputations of the loan officer and loan applicants.

Technologies. Each work system uses technology extensively. Those technologies range from telephones and personal computers through web sites and complex decision support software. The loan approval example includes a use of software that helps work system participants make the decision. In the ecommerce example, the work of deciding what to purchase and then making the purchase occurs through a web site. In the software development example, the programming and documentation work uses computers and software tools extensively, but important aspects of the requirements analysis and user collaboration may not involve technology at all.

Products and services produced for customers. Each work system produces things for internal or external customers. The loan approval system produces an approval decision for the loan applicant, but the loan officer, whose bonus depends on the decision, is also a customer. The sales person who will try to make the sale is an internal customer of the prospect qualification system. The ecommerce system helps external customers decide what to buy and executes the purchase transaction. The customers of the IT group are the people within the company for whom the software is being built. To some extent the IT group is also a customer of the software development system because it will have to maintain the software over time.

Environment. Each work system operates within an environment that affects its operation. The managers of the loan approval system, prospect qualification system, and software development system are all under pressure to improve their operational results. The environment surrounding the ecommerce web site includes an industry-wide expectation that web sites will be attractive and effective.

Infrastructure. Although shared infrastructure was not mentioned directly in the brief descriptions of the examples, each work system relies on human, technical, and information infrastructures. For instance, the software development system relies on the firm's human infrastructure to maintain the work environment, on its technical infrastructure to provide computers and networks, and on its information infrastructure to provide internal company information that may be relevant.

Strategies. Each work system has an operational strategy. These strategies include making loan approval decisions by committee, using several different channels for finding sales prospects, selling using a web site that assumes the customer knows what to buy, and allowing individual programmers to decide on their own programming methods. Whether or not articulated as a strategy with a capital S, each of these represents a strategy because the main activities might have been performed in fundamentally different ways. In addition, each work system is to be improved within a corporation that also has strategies. An attempted improvement that conflicts with the corporation's strategies is unlikely to succeed.

Relationships with other systems. Systems in organizations never exist in a vacuum. Most work systems receive inputs from other systems and produce products and services that are used by other work systems. For example, the work system for finding qualified sales prospects produces lists of qualified leads that are used by a separate sales force that actually tries to sell the software. Similarly, the customer's use of the ecommerce web site produces sales transactions fulfilled by other work systems that find, package, and ship the merchandise. Additional impacts on other systems are less direct. For example, the resources and attention devoted to one work system might have been devoted to other work systems.

Work system architecture and performance. The analysis of each of these systems requires attention to both architecture and performance. A work system's architecture is a description of how it is organized and how it operates. Architecture can be described at various degrees of detail under the headings of work practices, participants, information, and technology. Architecture is assumed to be relatively constant until it is changed. A work system's performance is a description of how well the work system operates during a particular time interval. Work system performance has many facets and can be measured using a variety of performance indicators and related metrics. In many situations, improving a work system's performance requires changing its architecture.

The Work System Framework

A work system is a system in which human participants and/or machines perform work using information, technology, and other resources to produce products and/or services for internal or external customers. Businesses operate through work systems. Typical business organizations contain work systems that procure materials from suppliers, produce products, deliver products to customers, find customers, create financial reports, hire employees, coordinate work across departments, and perform many other functions.

The work system method is organized around the Work System Framework™ (Figure 2.1), a graphical representation of the elements that are included in even a basic understanding of a work system's scope and operation. Just agreeing on the identity and scope of these 9 elements can eliminate fundamental confusions in many situations.

The Work System Framework provides an outline for describing the system being studied, identifying problems and opportunities, describing possible changes, and tracing how those changes might affect other parts of the work system. The arrows within the framework indicate that the various elements of a work system should be in balance. The first four elements are the basic components that actually perform the work. These include work practices, participants, information, and technologies.

The other five elements are included in even a basic understanding of a work system:

- Products and services the work system produces
- Customers for those products and services
- Environment that surrounds the work system
- Infrastructure shared with other work systems
- Strategies used by the work system and the organization.

Products & services and customers are not part of a work system, but are included in the framework because systems exist in organizations in order to produce products and services for internal or external customers. Environment and infrastructure are included because a work system's success often

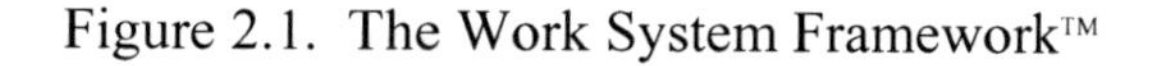
Figure 2.1. The Work System Framework™

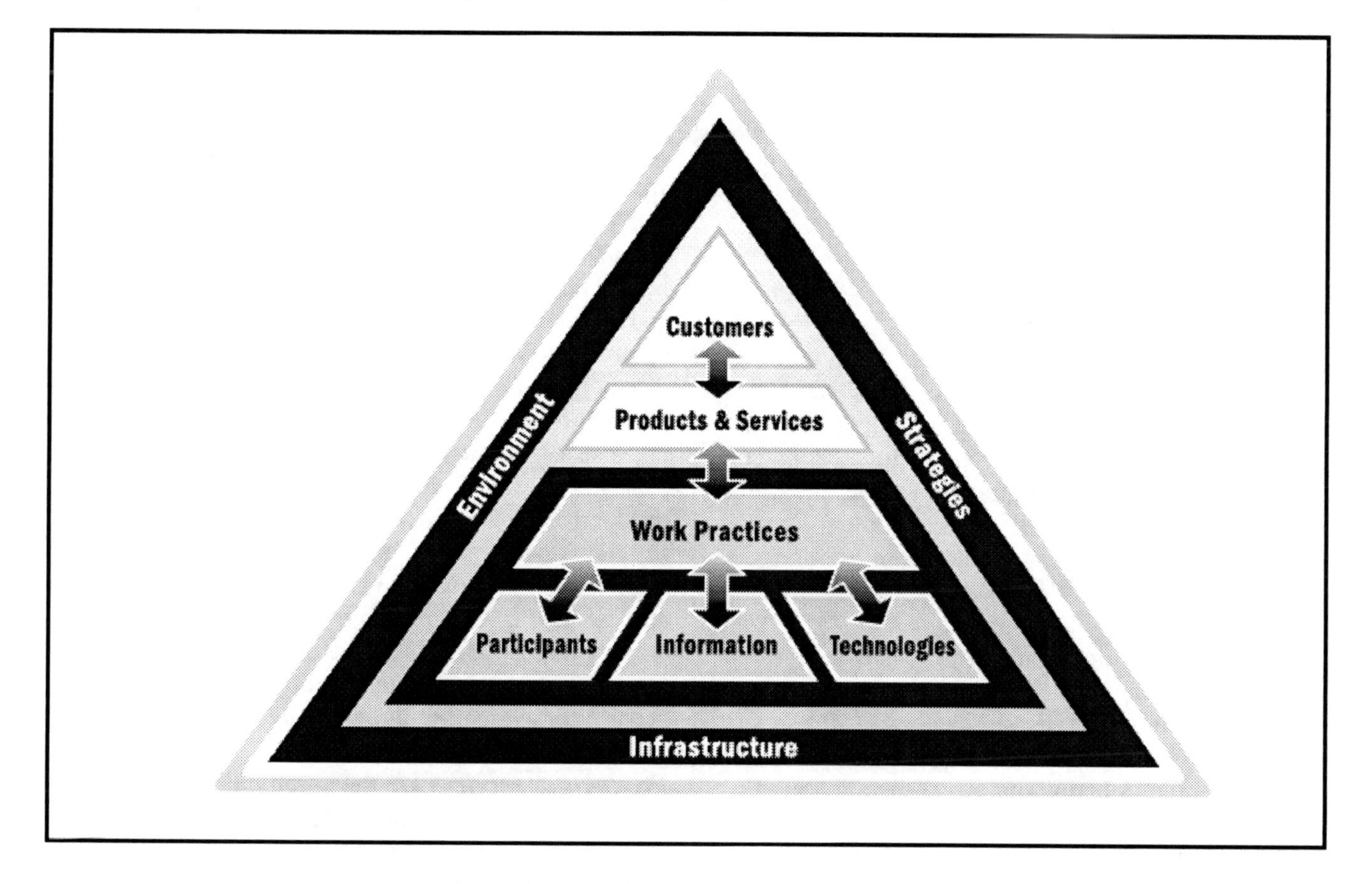

depends on its fit with the surrounding environment and on its use of available infrastructure that is shared with other work systems. Strategies are included in the framework as a reminder that work systems have strategies and that those strategies should be aligned with the organization's strategies.

The Work System Framework makes no assumptions about whether or not IT is used. It simply reserves a location for whatever technology is used. This is appropriate because any particular work system might not use IT, or might use IT only in a minor way. Furthermore, the framework does not create an artificial separation between the work system that produces products and services for the customer and the information system that often overlaps with the work system.

The Work System Framework also says nothing about how long a system will exist. Some work systems exist and produce their outputs over extended time spans. Others are projects, temporary systems designed to produce a particular output and then dissolve. In other words, the same framework can be used to summarize work systems in general, information systems, and projects. The concept of work system therefore provides a consistent starting point for thinking about work systems in general, about information systems that support work systems, and about projects that build or modify work systems.

The Work System Framework is easy to understand as an abstraction and has many uses in practice. At the beginning of an analysis it can help people agree about the scope of the system, and hence, what the analysis should include or exclude. Later, it is a useful reference point for keeping the analysis on target and recognizing whether the initial definition of the system and problem proves inadequate in relation to the realities that are uncovered.

The individual ideas underlying the Work System Framework are not revolutionary,[14] but the framework combines these ideas to provide an organizing perspective that is quite different from the way people in many organizations talk and think about computerized systems. You can verify this in your organization by listening to informal conversations or formal presentations about these systems, or by looking at written system proposals or documentation. You will probably find the informal conversations focusing on isolated details and frustrations related to information system use and maintenance. The formal conversations, proposals, and documentation will probably say a great deal about how the computerized parts of a system should operate. Aside from diagrams representing idealized work flows, however, rather little will typically be said about how people produce the work system's products with the help of technology, how well the various components of the work system are expected to operate, and how infrastructure and surrounding environment affect the work system.

Elements of a Work System

The nine elements of the Work System Framework were identified, and will now be defined. Chapter 4 will identify additional distinctions that have proven useful in applying the Work System Framework to understand systems in organizations.

Customers are the people who receive, use, or benefit directly from products and services that a work system produces. In most cases they can experience or perceive the quality of those products and services. Customers include external customers and internal customers. External customers receive and use the economic products and/or services that a firm produces. Firms exist to produce those products and services. Internal customers are employees or contractors who receive and use a work system's products and/or services while performing work in other parts of the same firm's value chain. A basic premise of quality management is that a system's customers are typically best able to evaluate the products and services it produces. Customer satisfaction is often linked to the entire customer experience, starting from determining requirements and acquiring the products or services. In many work systems, different groups of customers may receive different products and may have different criteria for evaluating those products.

Products & services are the combination of physical things, information, and services that the work system produces for its various customers. A work system's products and services may take various forms, including physical products, information products, services, intangibles such as enjoyment and peace of mind, and social products such as arrangements, agreements, and organizations.

Work practices include all of the activities within the work system. These activities may combine information processing, communication, decision making, coordination, thinking, and physical actions. In some work systems, work practices can be defined tightly as highly structured business processes; in other work systems, some or all of the activities may be relatively unstructured. The Work System Framework uses the term *work practices* instead of the more familiar term *business process* for two reasons. First, in many work systems some of the activities are unstructured, and coordination occurs through improvisation rather than pre-specified rules. In these situations, the term business process is misleading because it implies a set of steps that are highly structured and are related in a predictable way. Second, there is no reason to assume that business process is always the primary perspective for thinking about work practices. Other useful perspectives include decision making, communication, coordination, control, and information processing.

Participants are people who perform the work. Some participants may use computers and IT extensively, whereas others may use little or no technology. When analyzing a work system, the more encompassing role of work system participant is more important than the more limited role of technology user,[15] whether or not particular participants happen to be technology users.

Being a work system participant is more important than being a computer user.

Information includes codified and non-codified information used and created as participants perform their work. Typical codified information is the pre-defined information used in tracking packages, entering orders, and performing repetitive financial transactions. In each case, each data item must be defined precisely, and the information is usually processed using explicit rules. Typical uncodified information includes computerized or handwritten documents, verbal agreements, and formal or informal conversations. Information may or may not be computerized. Information not related to the work system is not directly relevant, making the common distinction between data and information secondary when describing or analyzing a work system.

Knowledge can be viewed as a special case of information. Explicit knowledge is recorded in documents, images, rules, and other forms. Tacit knowledge exists in people's heads and is not explicit. Maintenance of both explicit and tacit knowledge is an important management challenge.

Technologies are tools that help people work more efficiently. Some technologies, such as search engines, cell phones, spreadsheet software, and automobiles, are general-purpose because they can be applied in a wide range of business situations. Other technologies are tailored to specific situations. Examples include a spreadsheet model for calculating mortgage interest and a software package for designing kitchens. Technologies tailored to specific business situations usually involve a combination of general-purpose tools and specialized techniques, such as mortgage calculation formulas. The separation between tools and techniques is worth considering because it is often possible to use a different general-purpose tool (e.g., a better laptop) without changing the technique. Similarly, it is possible to change the technique (e.g., moving to a better mortgage calculation method) while using the same laptop or other tool. In other situations, different techniques require more powerful general-purpose tools (e.g., early PCs were too slow to support voice recognition software).

Environment includes the organizational, cultural, competitive, technical, and regulatory environment within which the work system operates. Factors in the environment affect system performance even though the system does not rely on them directly in order to operate. The organization's general norms of behavior are part of the culture in the environment that surrounds the work system, whereas behavioral norms and expectations about specific activities within the work system are considered part of its work practices.

Infrastructure includes human, information, and technical resources that the work system relies on even though these resources are managed outside of it and are shared with other work systems. Human infrastructure is the people and organizations that supply services shared by different work systems. For example, training organizations, internal consultants, and human resources departments are typically considered part of human infrastructure. Information infrastructure is information shared across various work systems, such as mutually accessible databases and other enterprise-wide information. Technical infrastructure includes the Internet, corporate computer networks, database management software, and other technologies shared by multiple work systems and often hidden or invisible to work system participants.

Strategies consist of the guiding rationale and high-level choices within which a work system, organization, or firm is designed and operates. Strategies at the department and enterprise level may help in explaining why the work system operates as it does and whether it is operating properly. Although sometimes not articulated clearly, high-level choices about a system can often be inferred by considering plausible alternatives that were not chosen. For example, a system designed based on an assembly line rationale is not using a case manager rationale, in which one individual performs multiple steps, usually with the help of computerized information and models.[16] Similarly, a system designed to produce based on mass customization is not using a commodity-like, "one size fits all" strategy. A work system's strategy should be aligned with the strategy of the organization and firm that it serves. For example, a work system designed to produce the highest quality products might not fit in an organization operating under a cost minimization strategy.

Work System Snapshot

The Work System Framework can help in summarizing virtually any system in an organization. It can be applied to produce a tabular one-page summary called a **work system snapshot**. This is a highly summarized but balanced view of a work system that a business professional can produce quickly or can understand quickly when it is presented. A work system snapshot uses six central elements to summarize what a system is and what it produces. At this level of summarization, distinctions between technology and technical infrastructure are unimportant.

The sample work system snapshot in Figure 2.2, summarizes the loan approval system mentioned earlier. Any summary of this type is a simplification of reality that attempts to identify the main features of the system and ignores many distinctions that are important when analyzing the system in depth. Even with no additional explanation, Figure 2.2 gives a general impression of what work system is being discussed. This type of summary could also be used effectively in an initial discussion with colleagues, especially since it would help in recognizing inconsistencies between different views of system content and operation. A glance at Figure 2.2 might reveal areas that seem unclear or questionable. Someone familiar with the situation might find details that are no longer correct or were never totally correct. Use of a work system snapshot makes it more likely that everyone in a conversation will be thinking about the same system and not just imagining that everyone else shares the same view of the system's scope.

A work system snapshot can help people verify that they agree on a work system's scope and purpose.

The identification of work system elements in the work system snapshot might lead to a variety of discussion points. Part of the discussion would be about defining the scope of the system that was being analyzed:

- What else does the loan approval system produce? For example, does it produce any management information for tracking the bank's current and future revenues and evaluating the quality of its current and future loan portfolio?

- Do we really want to talk about how the loan officer finds prospects, or is it better to think of the loan approval system as starting with a known prospect?
- Can we treat the production of loan documents as part of a separate system so that we can focus on how to make the approval/denial decisions more effectively?

Another part of the discussion would be about how well the loan approval system operates and what are some of the possibilities for improvement:

- How appropriate are the approval or denial decisions that have been made? What is the frequency of mistakes?
- Which of the delays in the system are due to the clerical work of assembling the loan applications, and which are due to loan committee schedules?

Figure 2.2. Work system snapshot for a loan approval system for loans to new clients

Customers	Products & Services
• Loan applicant • Loan officer • Bank's Risk Management Department and top management • Federal Deposit Insurance Corporation (FDIC) (a secondary customer)	• Loan application • Loan write-up • Approval or denial of the loan application • Explanation of the decision • Loan documents

Work Practices (Major Activities or Processes)
• Loan officer identifies businesses that might need a commercial loan. • Loan officer and client discuss the client's financing needs and discuss possible terms of the proposed loan. • Loan officer helps client compile a loan application including financial history and projections. • Loan officer and senior credit officer meet to verify that the loan application has no glaring flaws. • Credit analyst prepares a "loan write-up" summarizing the applicant's financial history, providing projections explaining sources of funds for loan payments, and discussing market conditions and applicant's reputation. Each loan is ranked for riskiness based on history and projections. Real estate loans all require an appraisal by a licensed appraiser. (This task is outsourced to an appraisal company.) • Loan officer presents the loan write-up to a senior credit officer or loan committee. • Senior credit officers approve or deny loans of less than $400,000; a loan committee or executive loan committee approves larger loans. • Loan officers may appeal a loan denial or an approval with extremely stringent loan covenants. Depending on the size of the loan, the appeal may go to a committee of senior credit officers, or to a loan committee other than the one that made the original decision. • Loan officer informs loan applicant of the decision. • Loan administration clerk produces loan documents for an approved loan that the client accepts.

Participants	Information	Technologies
• Loan officer • Loan applicant • Credit analyst • Senior credit officer • Loan committee and executive loan committee • Loan administration clerk • Real estate appraiser	• Applicant's financial statements for last three years • Applicant's financial and market projections • Loan application • Loan write-up • Explanation of decision • Loan documents	• Spreadsheet for consolidating information • Loan evaluation model • MS Word template • Internet • Telephones

- Could some type of preliminary screening by a person other than the loan officer save time and effort for everyone?

- What types of biases are introduced by the fact that the loan officer is both a participant and a customer (whose bonus is affected)?

- Does the information about the applicant provide an adequate picture of the risk of the proposed loan?

- How effective is the loan evaluation model? How often does it mislead the decision makers?

- Could the loan officers use the loan evaluation model directly to help them analyze the loan application, thereby reducing the number of situations in which a senior credit officer or loan committee rejects an application that should not have been submitted?

Although analysis of these questions would probably involve a detailed examination of different aspects of the situation, the work system snapshot would always be available to maintain the big picture. And the snapshot itself might change if the analysis revealed that the initial scope of the system was wrong, or that the system should really be subdivided into several separate systems.

The work system snapshot is a deceptively simple tool. For example, one team of Executive MBA students analyzing a system in one of their organizations complained that the team had argued for four hours about the scope of the system they were trying to analyze. This was much more effort than they had expected in the first step of the project, but their experience helped them realize why it is so important to define both the work system and the problem or opportunity before launching into technical details. They had seen expensive system projects that generated disappointing results because the technical work occurred without genuine agreement about how the new work system should operate. A comparatively small work system analysis early in the project might have dispelled big picture confusions that led to unnecessary technical features that were never used.

One Lens Fits Almost All

The work system concept is like a common denominator for many of the types of systems discussed in information system practice and research. Operational information systems, projects, supply chains, and ecommerce web sites can all be viewed as special cases of work systems.

- An **information system** is a work system whose work practices are devoted to processing information, i.e., capturing, transmitting, storing, retrieving, manipulating, and displaying information. Some information systems exist to produce information products for internal or external customers. Others exist to support work systems that produce physical products and services for internal or external customers. Increasing reliance on computerized information systems has led to greater overlap between work systems and the information systems that support them. This increasing degree of overlap makes it even more important to avoid confusion about the problem or opportunity that an analysis addresses, about which system has that problem, and about what is to remain constant.

- A **project** is a work system designed to produce a product and then go out of existence. Each major project phase or subproject might also be viewed as a separate work system with its own work practices, participants, and products.

- A **value chain** can be viewed as a large work system that crosses several functional areas of business and whose participants typically reside in different departments. Similarly, each major step in a value chain can be viewed as a work system.

- A **supply chain** is an interorganizational work system devoted to procuring materials and other inputs required to produce a firm's economic products and services. The firm and specific suppliers are participants in work practices that use specific information and technology to create, monitor, and fulfill orders.

- An **ecommerce web site** can be viewed as a self-service work system in which a customer uses the seller's web site in a process of matching requirements to product offerings and then making the purchase. By focusing attention on the customer's work practices and desired outcome, a work system view helps in recognizing why an attractive interface does not assure a web site's success.

The fact that all of these systems can be viewed as work systems implies that the same basic concepts and analysis ideas apply for all of these cases. For example, it is possible to use work system elements to summarize any work system, including any information system, project, supply chain, or ecommerce web site. Being able to start from the same big picture ideas and models makes it easier to think about a broad range of situations even though each special case may have its own specialized terminology for specific topics.

The work system method exploits these commonalities in many ways. For example, it uses a set of commonsense principles (presented in Chapter 5) that apply for every work system. These principles are part of a straightforward, organized approach for evaluating how well each aspect of a system is operating now and whether any particular change in the system would have beneficial or adverse effects in other parts of the system. The work system method also provides lists of common performance indicators, strategies, stumbling blocks, and possibilities for change that apply across most work systems.

Themes Expressed by the Work System Framework

The content and form of the Work System Framework express a number of themes that motivate the work system method.

Work system, not IT system. Systems in organizations involve much more than IT. Even a cursory understanding of a system in an organization starts by summarizing the work system itself, and its work practices, participants, information, and technologies. Recognition of five other elements (products and services, customers, environment, infrastructure, and strategies) is required for a meaningful discussion of whether the work system is producing what customers want and of the consequences of any particular change.

Work system, not just business process. Some people describe the work system approach as a "process approach" to systems in contrast to a technical approach. This characterization is partly correct. Yes, the work system approach focuses more on the work than on the technology that is used, but the Work System Framework is designed to show that a pure process view is also inadequate.

The framework contains *work practices* rather than business process because work involves much more than just the steps in an idealized business process. The steps in a business process can be described independent of the participants, information, and technology. However, different participants with different skills or motivation might perform the same business process steps differently. People performing the work also may invent workarounds for their own convenience or to adjust for exceptions or conditions not anticipated when the process was designed. The business process results might also be affected by variability related to information or technology. The discussion of the work system method will explain that business process is one of several different perspectives for looking at work practices, just as an organization chart and personal impacts on participants are different perspectives for attending to the people who participate in a work system.

Alignment between the elements. The Work System Framework's double-headed arrows express the need for internal alignment. They also convey the path through which a change in one element might affect another. For example, the arrows linking the work practices to participants, information, and technology say that a change in the work practices might call for a change in any of these elements and vice versa. Thus, changes in work practices may affect the participants by changing the quality of their work life. Conversely, the replacement of work system participants by people with different skills and motivation might affect how well the work is done and, hence, the results.

Participants, not users. The framework refers to *participants* rather than *users* because it focuses on how a work system operates rather than on how people use computers while doing work. The use of computers is certainly important, but it usually shouldn't be the headline when trying to understand a work system in an organization. Part of the common confusion about systems in organizations involves an under-emphasis on how system participants do their work and an over-emphasis on how they use computers and on what the computers do.

Products and services, not outputs. The terms products and services are used instead of "outputs" because that term brings too many mechanistic and computer-related connotations, especially when services and intangibles are involved. Computer programs produce outputs; work systems produce products and services for their internal and external customers.

Better information may not matter. Better information leads to better work system results only if the new information helps participants perform work more efficiently or effectively. In other words, access to a huge amount of seemingly relevant information may have no impact whatsoever.

Better technology may not matter. As with better information, better technology leads to better work system results only if the new technology helps participants perform work more efficiently or effectively. For example, writing a book using the very latest desktop computer instead of a two-year old computer would have little or no impact on the quality of the result.

Internal versus external performance. The managers of the work system want to promote both internal work system efficiency and customer satisfaction. The customers are primarily concerned with whether the products and services they receive meet their goals and expectations related to cost, quality, responsiveness, and other facets of product performance. Thus, evaluation of any work system should consider both internal and external views.

Work Systems, Not IT

This overview of four examples and the main ideas in the Work System Framework demonstrates that a work system approach provides an organized way to think about how systems in organizations operate, and how well they operate. The ideas are logical and straightforward. Other than the term work system, most of the ideas are familiar in the everyday work environment.

Applying a work system approach for understanding or analyzing an IT-reliant system helps in focusing on the quality of work and the satisfaction of customers. Technology assumes its rightful place as an essential part of the work system, but certainly not the headline. Seeing IT from this vantage point leads to a number of simple conclusions:

- IT success is about how IT is used in work systems. IT success is really about work system success.
- IT has a significant impact only when it is used in a work system.
- IT success is not the sole responsibility of IT professionals. They have to make sure that IT is set up properly and works technically, but IT success is largely the responsibility of line managers and work system participants.
- Reducing confusion and uninformed expectations about what IT can and cannot do is a powerful way to boost the likelihood of IT success.

A useful analysis of a system should produce recommendations that are practical to implement as part of a work system's evolution over time. For explanatory purposes it is easier to continue discussing the analysis of a work system. Chapter 7 will return to implementation issues by explaining how work systems change over time.

Chapter 3: Overview of the Work System Method

- Three Levels for Using the Work System Method
- Level One
- Level Two
- Level Three
- Different Uses of the Work System Method
- Conclusion - the Whole, the Parts, and the Specifics

The work system method (WSM) is designed to help business professionals understand systems in their organizations. It is especially valuable early in a system-related project when people identify the problem, think about alternative courses of action, and decide how to proceed. As a project unfolds, it is also useful for keeping the project on track by reinforcing the goal of improving the work system rather than just producing software on time and within budget. Unlike some methods that require specific steps performed in a specific order, the work system method is designed to be quite flexible. It provides usable guidelines and analysis concepts while at the same time permitting the analysis to occur in whatever order and at whatever level of detail is appropriate for the task at hand.

WSM outlines steps a business professional can use to clarify a system-related issue, identify possible directions for change, and produce and justify a recommendation. The goal is to create an organized, rigorous problem solving approach with enough freedom to encourage focusing on important issues and downplaying minor ones. To avoid becoming a straightjacket it encourages examination of a work system in as much depth as is warranted by a particular situation. Instead of emphasizing the kinds of rigorously consistent details required for creating a debugged, maintainable computer system, it focuses on business issues leading to a recommendation about desired work system changes.

WSM is organized around a typical problem solving process of defining a problem, gathering and analyzing relevant data, identifying alternatives, and selecting a preferred alternative. WSM is divided into three major steps that apply general problem solving to typical systems in organizations:

- SP - Identify the System and Problem: Identify the work system that has the problems or opportunities that launched the analysis. The size and scope of the work system depends on the purpose of the analysis.
- AP - Analyze the system and identify Possibilities: Understand current issues and find possibilities for improving the work system.
- RJ - Recommend and Justify changes: Specify proposed changes and sanity-check the recommendation.

The abbreviations SP, AP, and RJ minimize repetitive verbiage when referring to these steps.

An idealized problem solving process might call for a prescribed sequence of steps such as completing analysis of the current situation before trying to define possibilities for improvement. However, in reality people often redefine the problem and think about possibilities while they do the analysis. They may even start with a recommendation and work backward to see whether an analysis could support it. WSM's goal is to help people understand systems and recognize the completeness of their understanding, regardless of the order they use for thinking about the situation.

Awareness of the three steps is valuable because many problems result from inadequacies in each step. In too many cases, organizations purchase software without a clear understanding of which work systems have the problems or opportunities that are being addressed. The common result is a "solution" that doesn't fit the situation or that causes many new problems. In too many cases, little or no real analysis is done and managers simply assume that software designed for a specific type of problem, such as tracking sales calls, will be effective. Too often, the recommendation focuses on the software that will be built or acquired, but not on how the work system will operate after the software is available. Finally, too many system improvement efforts are not adequately evaluated and justified. Too many projects plunge ahead before creating a shared understanding of the extent and value of the intended changes, the new problems that might arise, and the feasibility of implementing changes whose rationale and impacts are unclear.

Before proceeding it is worth noting that other methods and techniques address some of these issues. For example the "rich pictures" produced by soft systems methodology[17] can be used in establishing the scope of the system that is being analyzed. Influence diagrams[18] can be used to map beliefs about important relationships. A wide range of techniques under the general heading of Six Sigma are used to gather and analyze statistical data about process performance and causes of problems. Our goal is to explain the work system method rather than to compare it to many other methods and techniques that address some of the same issues, but in different ways and with different goals.

Three Levels for Using the Work System Method

WSM is based on the assumption that a single, totally structured analysis method is not appropriate for all situations because the specifics of a situation determine the nature of the understanding and analysis that is required. In some situations, a manager simply wants to ask questions to make sure someone else has done a thoughtful analysis. At other times, a manager may want to establish a personal understanding of a situation before discussing it with someone else. When collaborating with IT professionals, managers can use WSM to clarify and communicate their own understanding of the work system and to make sure that the IT professionals are fully aware of business issues and goals that software improvements should address. From the other side, IT professionals can use WSM at various levels of detail to confirm that they understand the business professionals who are the customers for their work.

Recognizing the varied goals of WSM applications, WSM can be used at three levels of detail and depth. The user's goals and the need to communicate and negotiate with others determine the level to use in any particular situation.

- Level One: Be sure to remember the three main steps when thinking about a system in an organization: SP (system and problem), AP (analysis and possibilities), and RJ (recommendation and justification).

- Level Two: Within each main step, answer specific questions that are typically important.

- Level Three: Use checklists, templates, and diagrams to identify and consider issues that are easy to overlook.

Table 3.1 illustrates WSM's structure by showing how the three steps in Level One expand into specific questions in Level Two and use checklists, templates, and diagrams in Level Three.

Table 3.1. Three levels of the work system method

	First step in WSM	*Second step in WSM*	*Third step in WSM*
Headings in Level One	**SP**: Identification of the work system that has the problems or opportunities.	**AP**: Analysis of current issues and identification of possibilities for improvement.	**RJ**: Recommendation and its justification.
Questions in Level Two	**SP1** through **SP5**: Five questions about the system and problem.	**AP1** through **AP10**: Ten questions related to analysis and possibilities.	**RJ1** through **RJ10**: Ten questions related to the recommendation and its justification.
Topics and guidelines in Level Three	Checklists, Templates, & Diagrams	Checklists, Templates, & Diagrams	Checklists, Templates, & Diagrams

This chapter summarizes the three steps at Level One, identifies the 25 questions at Level Two, and describes a few of the issues that are introduced in the checklists in Level Three. It also explains some of the different ways you can use WSM. Subsequent chapters fill in guidelines and ideas needed for an in-depth analysis. Chapter 4 focuses on defining the system and the problem (the SP step). Chapter 5 introduces many ideas in the Level Three checklists in the AP step. Chapter 6 provides brief explanations of the RJ questions. Chapter 8 and the Appendix provide an example designed to illustrate how the levels can be used independently or in combination.

Level One

The most basic application of WSM encourages the user to think about the situation in work system terms, but provides minimal guidance other than saying that each of the three main steps (SP, AP, and RJ) should be included. For example, assume that several people are speaking in general about whether a particular CRM (customer relationship management) software package might be beneficial. Using WSM at Level One, they would not focus initially on the features and purported benefits of the CRM software. Instead, they would discuss:

- SP: What work system(s) are we talking about? From a business viewpoint, what are the problems and opportunities related to the work system(s)?
- AP: What are the shortcomings of each part of the work system(s) and what are the possibilities for eliminating or minimizing these shortcomings? Adopting CRM software might be one possible change, but there are probably others.
- RJ: What changes in the work system(s) do we recommend and how could we justify those changes?

Merely using these questions to stay focused on the work system(s) instead of plunging into software details and features would probably make the initial discussion more productive. The discussion would treat the CRM software as part of something larger. It would be clearer that addressing the business problems or opportunities requires many changes beyond just adopting CRM software.

Level One can be used in other ways as well. An individual can use it as a simple discipline for organizing a quick personal summary of a familiar situation. A manager can use it to organize a ten-minute summary of a proposed system change, thereby increasing the likelihood that colleagues would understand the purpose and scope of the proposed change. From the other side, someone reading a proposal or listening to a presentation can use the three headings in Level One to attain clarity about what system and problem are being discussed, whether any real analysis has occurred, and whether the recommendation is justified.

Level One is a minimal first cut that highlights three headings (SP, AP, and RJ) but provides no other guidance. For example, the headings provide no help in identifying topics that should be considered and concepts that are effective in exploring those topics. A seemingly reasonable proposal created and presented at Level One is likely to overlook many foreseeable difficulties that would emerge from a deeper, more detailed business analysis.

Level Two

As illustrated in Table 3.1, Level Two provides a set of important questions for each of the three steps in the Level One analysis. Because the steps in Level One are abbreviated SP, AP, and RJ, specific questions in Level Two are identified using abbreviations such as SP3 and AP8.

Every question in Level Two is relevant to almost any analysis of a system in an organization, whether or not IT plays a major role. Level Two can be used as a checklist to organize one's own analysis or to review someone else's analysis to make sure that major aspects of a current and proposed system are being considered. Level Two can be used to make sure that an analysis has not over-emphasized one aspect of a work system while ignoring other important aspects. If any of the Level Two questions has been ignored, the analysis is probably incomplete and may overlook important issues that are easily identified.

System and Problems in Level Two. The WSM step of identifying the work system that has the problems or opportunities is subdivided into five questions that provide a starting point for analyzing a work system in an organized way.

- SP1: Identify the work system that is being analyzed.
- SP2: Identify the problems or opportunities related to this work system.
- SP3: Identify factors that contribute to the problems or opportunities.
- SP4: Identify constraints that limit the feasible range of recommendations.
- SP5: Summarize the work system using a work system snapshot.

Taken together, SP1 and SP2 imply that the scope of the work system is not known in advance, but rather, depends on the problems or opportunities that are being pursued. To expedite the analysis effort and focus the recommendations, the work system should be the smallest work system that has the problems or opportunities that are being analyzed. The initial identification of the work system should be viewed as no more than a starting point that is definitely not cast in concrete. Insights as the analysis unfolds often reveal that the initial definition is inadequate and that a better definition would encompass unanticipated problems, opportunities, or issues.

Questions SP3 and SP4 are included because initial beliefs about contributing factors and constraints help in framing the analysis. They also may help in surfacing assumptions that may not be shared and in recognizing additional opportunities and issues that emerge during the analysis. For purposes of analyzing a system, contributing factors are relevant issues that might provide insight and might affect the analysis or recommendation even though they are not part of the system. For example, contributing factors might include recent turnover in key personnel, recent changes elsewhere in the organization, the frequent absence of a key decision maker due to other responsibilities, or month-end spurts in output from another work system that cause overtime in this work system.

Constraints are limitations that are assumed to be unchangeable within the time period covered by the analysis. It is worthwhile to identify constraints at the beginning of an analysis to avoid making recommendations that are impractical. For example, there are many times when it would be nice to replace some of the work system participants with people who are more skilled or motivated. If there is no possibility of doing that, the retention of the current participants should be treated as a constraint that precludes recommendations that would remove those individuals. Other typical constraints include budgets, capacity limitations, technology limitations, union rules, employment regulations, privacy regulations, and customer requirements for documentation of processes.

Question SP5 calls for clarifying the scope of the work system using a work system snapshot such as the one presented in Figure 2.2. A work system snapshot is a single page that identifies:

- the work system's customers, products and services, and participants
- the major steps or activities
- the most important information and technologies in the work system.

Even when there is initial agreement about the work system snapshot, looking at the situation in more depth as the analysis unfolds often results in revising the initial view of the work system's scope. The initial work system snapshot may reveal that several different work systems are involved and that it will be simpler to analyze each work system separately. Figure 4.3 in Chapter 4 provides guidelines for work system snapshots. Figures 4.4 through 4.9 provide additional examples of work system snapshots for a diverse set of situations.

Steps and Levels keep the analysis organized while it goes as deep as is needed.

Analysis and Possibilities in Level Two. Understanding current issues and finding possibilities for improvement is subdivided into ten questions, one for each element in the Work System Framework and one for the work system as a whole:

- AP1: Who are the customers and what are their concerns related to the work system?
- AP2: How good are the products and services produced by the work system?
- AP3: How good are the work practices inside the work system?
- AP4. How well are the roles, knowledge, and interests of work system participants matched to the work system's design and goals?
- AP5: How might better information or knowledge help?
- AP6: How might better technology help?
- AP7: How well does the work system fit the surrounding environment?
- AP8: How well does the work system use the available infrastructure?
- AP9: How appropriate is the work system's strategy?
- AP10: How well does the work system operate as a whole and in relation to other work systems?

Although some of these AP questions are inevitably more important in than others in specific situations, inclusion of questions about all nine elements of the Work System Framework increases the likelihood that analysis and design efforts will start with a reasonably balanced view of the work system and the range of possible improvements. In particular, this approach should avoid the common error of assuming that the system consists of little more than the software and computerized information.

Recommendation and Justification in Level Two. The goal of performing the analysis and identifying possibilities is to provide the basis for a recommendation for action. In addition to making sure the recommendation is stated clearly, the RJ questions at Level Two probe whether the recommendation and justification are coherent and compelling enough to explain to others:

- RJ1: What are the recommended changes to the work system?
- RJ2: How does the preferred alternative compare to other alternatives?
- RJ3: How does the recommended system compare to an ideal system in this area?
- RJ4: How well do the recommended changes address the original problems and opportunities?
- RJ5: What new problems or costs might be caused by the recommended changes?
- RJ6: How well does the proposed work system conform to work system principles?
- RJ7: How can the recommendations be implemented?
- RJ8: How might perspectives or interests of different stakeholders influence the project's success?
- RJ9: Are the recommended changes justified in terms of costs, benefits, and risks?
- RJ10: Which important assumptions within the analysis and justification are most questionable?

The RJ questions at Level Two serve several purposes. First, they identify issues that should be considered before pursuing a recommendation. Second, they make sure the recommendation is justified from a variety of economic, organizational, and practical perspectives, not just financial estimates such as internal return on investment or net present value. Third, they can help in recognizing topics that need a deeper discussion and/or additional expertise. Although none of these questions are intrinsically technical, several require input from IT professionals. For example, it is rarely feasible to produce a cost-benefit justification (RJ9) without help from someone who understands the technical issues and technical resources required.

WSM's justification questions help in explaining the rationale and completeness of a proposal, but they are not designed for comparison of disparate proposals competing for limited investment funds. WSM is used to analyze a current work system, describe a proposed work system, and justify a proposal. WSM assumes that budget allocation decisions occur at a different time. Its main contribution to a budget allocation process is in supporting the analysis of the systems that are being considered.

Level Three

The 25 questions in Level Two provide an organized approach for pursuing each of the three major WSM steps. Level Three identifies specific topics that are often worth considering when answering Level Two questions. Although each question in Level Two is almost always relevant, some of the Level Three topics for the Level Two questions may not be relevant to a particular situation.

As an example, consider the Level Two question AP3, "How good are the work practices inside the work system?" In a particular situation, it might be easy to produce a quick answer that the work is being done inefficiently, that it takes too long, and that the error rate is too high. That type of statement mimics question SP2 (identify the problems or opportunities), but does not go into enough specifics to determine whether any particular set of proposed changes would actually generate better results. A genuine understanding of the quality of the work practices would look more deeply at how the work is done and how well it is done. It would include consideration of a variety of possibilities for improvement.

The major topics and concerns identified in Level Three are basically hints about topics that are frequently important but that might not be important in any particular situation. Using Level Three involves deciding which topics are important and which can be ignored in the particular situation. Examples of Level Three topics related to AP3 include, among others:

- Roles and division of labor
- Relevant functions not performed by the work system

- Problems built into the current business process or functions
- Effect of strategies built into work practices
- Evaluation criteria for work practices
- Problems involving phases of decision making
- Problems involving communication.

Each of these topics provides a direction for looking at work practices. Each direction brings a set of everyday concepts, images, and examples that may be useful. For instance, problems built into current work practices might include any of the following:

- Unnecessary hand-offs or authorizations
- Steps that don't add value
- Unnecessary constraints
- Low value variations
- Responsibility without authority
- Inadequate scheduling of work
- Large fluctuations in workload
- Inadequate or excess capacity.

This type of list might help in identifying or categorizing problems that otherwise might be overlooked. At minimum, a glance at such a list provides greater assurance that widely applicable issues have been considered.

As you can imagine, lists of topics for each of 25 questions can be unwieldy to use. Experimentation with various alternatives led to an approach of using checklists, templates, and diagrams to make Level Three efficient for the user. The checklists and templates organize concepts and knowledge in an easily accessible form that allows a user to identify topics that are relevant and ignore other topics without devoting effort to them.

Checklists within themes. Chapter 5 presents five themes that can be used for identifying issues and possible improvements. The topics within these themes are presented in Figures 5.1 through 5.5. The extended example in the Appendix shows how those topics can be included in checklists organized around the themes:

- Work system principles
- Work system performance indicators (the work system scorecard)
- Strategic issues for a work system
- Possibilities for improvement
- Work system risk factors and stumbling blocks.

The principles checklist may trigger realizations that the original problem statement ignored important issues, or that certain possible changes would cause other problems. The scorecard encourages consideration of a range of performance indicators. The strategies checklist encourages consideration of big picture strategies that go beyond fixing details and local symptoms. The possibilities checklist helps in remembering that changes in one part of a work system often must be accompanied by related changes in other parts of the system. The risk factors and stumbling blocks checklist provides reminders of things that often go wrong.

Each of these checklists contains topics related to all of the work system elements. For example, the work system scorecard lists common performance indicators for each work system element. It also provides space for entering current and desired values of specific metrics for the performance indicators that are relevant to a specific work system.

Checklists for specific work system elements. The same topics can be organized in a different direction by using a separate checklist for each work system element, thereby drilling down within each element instead of following a theme across all of the elements. To illustrate this type of checklist, the last section of Chapter 5 presents a separate table of topics for each work system element. These tables serve as reminders of frequently relevant topics for each of the AP questions in Level Two. For example, a glance at the table for question AP3 (about work practices) identifies relevant work system principles, performance indicators, strategies, stumbling blocks, and possibilities for change.

Templates. The work system snapshot is an example of a template. It is an organized, tabular format for displaying information that is useful in answering a specific question, in this case, SP5. The Appendix presents a number of other templates that are useful for answering other questions.

Diagrams. A range of diagrams and visual display methods from various disciplines may be useful in analyzing a work system. For example, Chapter 10 (Work Practices) shows three types of diagrams that can be used in analyzing work practices. These include flow charts, swimlane diagrams, and data flow diagrams. The tables at the end of Chapter 5 mention some of the common types of diagrams that can be used to think about other work system elements. Many websites and books on data analysis and Six Sigma methods explain other types of diagrams that are not covered in this book but are useful for analyzing work systems when appropriate data is available. Examples include histograms, scatter plots, control charts, Pareto charts, and fishbone diagrams.

All of Level Three is designed to help in identifying issues and topics that might be overlooked. The analyst can decide whether or not to use any or all of the checklists, templates, and diagrams. Typically it is unnecessary to use all of the checklists because topics on one are often reminders of topics on other checklists.

WSM is designed based on the assumption that providing an organized structure with considerable depth is an effective way to help managers and business professionals pursue whatever amount of detail they want to pursue in a particular situation. Because WSM is designed to support a business professional's analysis, even Level Three does not approach the amount of detail or technical content that programmers and other IT professionals must analyze and document while producing computerized information systems. In addition, although Level Three might seem quite thorough, it focuses on a single work system and therefore skims over some of the details and complexities of interactions between work systems. In some cases it is worthwhile to address those issues by viewing several interacting work systems as a single, larger work system.

Different Uses of the Work System Method

After seeing all 25 questions in Level Two, you might wonder whether it is important for business professionals to look at systems in so much depth. The answer is a resounding *maybe*. If a particular situation calls for nothing more than a quick once-over, even the questions in Level Two may not be necessary. But when it is important to understand a significant situation, it is unlikely that any of the Level Two questions about the analysis and possibilities can be skipped with impunity. And if it is important to participate in decisions about a system-related project, all of the questions about the recommendation and its justification are important.

The organization of WSM into Level One steps, Level Two questions for each step, and Level Three checklists and templates is basically a device to help in recognizing different depths of analysis and deciding how to apply work system ideas in practical situations. The goal is to combine guidance with freedom in using a set of ideas that apply to any real world system but often are not applied in a complete or organized way.

In a well-managed organization, analyzing situations and justifying recommendations should be totally familiar even if the idea of focusing on work systems is unfamiliar. In such an organization, something similar to Level One should be part of the culture, even if different terms are used. Level Two fills in reminders about typical questions under each of the three steps in Level One. For individuals with substantial systems analysis experience, just the list of Level Two questions could suffice because that experience provides personal checklists and rules of thumb.

Most business professionals do not have substantial systems analysis experience, however. If it is important to understand or discuss a system in some depth before making or agreeing to a proposal, most business professionals would probably benefit from the Level Three reminders about typical issues and topics related to the Level Two questions. Experience to date implies that the questions at Level Two provide an organized way to go beyond the superficial, and that the checklists, templates, and diagrams at Level Three make it easier to consider each question in reasonable depth.

Shown next are some of the ways the work system method can be used at different levels of depth for different purposes. The first six types of use are posed as an individual's analysis, although teams can use the same approaches. The seventh and eighth invite analysts and IT professionals to customize parts of WSM when performing interviews or creating questionnaires to gather diverse viewpoints about a proposed project.

WSM's structure provides reminders about questions and issues that should be considered when a system improvement project begins.

1. Ask several basic questions.

Level 1	SP	AP	RJ
Level 2	SP1 - SP5	AP1 - AP10	RJ1 - RJ10
Level 3	Checklists Templates Diagrams	Checklists Templates Diagrams	Checklists Templates Diagrams

Assume that someone has proposed that software should be developed or that an existing system should be improved. The most basic review of the situation simply asks several basic questions to check whether all three Level One steps have been considered. Just looking these three steps provides three important reminders:

- ***SP reminder***: The work system, not the software or IT, is the primary unit of analysis from a business viewpoint.
- ***AP reminder:*** Before deciding what to do, it is a good idea to analyze the situation and consider different possibilities related to the work system, not just the technology.
- ***RJ reminder:*** Recommendations should be justified.

Unfortunately, the three general headings of Level One provide very little guidance about how to define the system, analyze it, or justify a recommendation.

2. Show care in defining the work system and problem.

Level 1	SP	AP	RJ
Level 2	SP1 - SP5	AP1 - AP10	RJ1 - RJ10
Level 3	Checklists Templates Diagrams	Checklists Templates Diagrams	Checklists Templates Diagrams

In many cases of system-related disappointments, merely defining the work system clearly probably would have helped firms minimize confusions and unrealistic expectations that led to costly project delays and rework. The term *system* often takes on different meanings in the same conversation. For example, "our manufacturing system isn't working properly" might refer to the way manufacturing is being performed, and later might refer to software that is being used. Defining the work system and problem using questions SP1 through SP5 adds clarity when distinguishing between the desired work system improvements and the information system changes that are necessary in order to realize those improvements.

3. Make sure the analysis is balanced and considers each element of the work system.

Level 1	SP	AP	RJ
Level 2	SP1 - SP5	AP1 - AP10	RJ1 - RJ10
Level 3	Checklists Templates Diagrams	Checklists Templates Diagrams	Checklists Templates Diagrams

The ten questions AP1 through AP10 look at each work system element and at the work system as a whole. Considering each question makes it less likely that information and technology will be over-emphasized and that customers and participants will be under-emphasized.

4. Use checklists to identify topics and issues for the analysis.

Level 1	SP	AP	RJ
Level 2	SP1 - SP5	AP1 - AP10	RJ1 - RJ10
Level 3	Checklists Templates Diagrams	Checklists Templates Diagrams	Checklists Templates Diagrams

Checklists and templates in Level Three provide a convenient way to identify topics and issues that should be considered when answering the Level Two analysis questions. As mentioned earlier, some checklists organize Level Three topics within themes such as work system principles, performance indicators, strategies, possibilities for change, and stumbling blocks. Other checklists are organized by work system element. WSM templates provide organized, tabular formats for displaying information that is useful in answering specific Level Two questions. Various diagrams can also be used, some of which are introduced in Chapters 10 and 12.

5. Examine the recommendation and make sure it is justified from a number of viewpoints.

Level 1	SP	AP	RJ
Level 2	SP1 - SP5	AP1 - AP10	RJ1 - RJ10
Level 3	Checklists Templates Diagrams	Checklists Templates Diagrams	Checklists Templates Diagrams

The ten questions about the recommendation and justification look at the recommendation from a variety of viewpoints, all of which are frequently useful. Attention to different facets of the recommendation may highlight issues that would not be identified with a less thorough approach. For example, these questions might help in realizing that a recommendation will cause new problems, that it may not solve the original problem, or that it may be very far from the ideal approach to the current problem. It is much better to recognize these situations than to pretend they don't exist.

The work system method is designed for flexibility. It organizes ideas that should be applied in whatever way makes sense in the user's situation.

6. Use recommendation checklists and templates to sanity-check whether the recommendation is justified.

Level 1	SP	AP	RJ
Level 2	SP1 - SP5	AP1 - AP10	RJ1 - RJ10
Level 3	Checklists Templates Diagrams	Checklists Templates Diagrams	Checklists Templates Diagrams

As illustrated in the example in the Appendix, checklists and templates for the RJ questions organize a large number of topics in a form that makes them readily accessible. For example, for question RJ7, "How can the recommendations be implemented?" the templates cover project ownership and management, a tentative project plan, and resource requirements. Many organizations have more complex templates for these topics, but these basic templates are at least a reminder of topics should that be considered in any well-justified recommendation.

7. Select questions and topics to use in an outline for interviewing work system participants and stakeholders.

Level 1	SP	AP	RJ
Level 2	SP1 - SP5	AP1 - AP10	RJ1 - RJ10
Level 3	Checklists Templates Diagrams	Checklists Templates Diagrams	Checklists Templates Diagrams

Analysts and IT professionals may find it useful to select particular questions and topics as the basis for interviews. For example, a newly hired IT professional had a mandate to improve software methods in a product engineering group. He created an interview questionnaire to help him learn about his new firm's product engineering system. The questionnaire started with "describe your role in the system" and included questions such as AP5 and AP6, "Would better information (or better technology) help?" He was intrigued that everyone seemed to have views about many topics but no one had strong suggestions for question RJ3, "What would be an ideal system?"

Effective use of WSM requires conscious decisions about what to consider and what to omit.

8. Create a questionnaire to gather information for a project.

Level 1	SP	AP	RJ
Level 2	SP1 - SP5	AP1 - AP10	RJ1 - RJ10
Level 3	Checklists Templates Diagrams	Checklists Templates Diagrams	Checklists Templates Diagrams

Another way to use WSM is to select and adapt questions and topics while creating a questionnaire for a project that requires ideas and comments from individuals in different locations. The Level Two questions and Level Three checklists and templates provide a possible starting point for creating the questionnaire. The advantage of using these questions is their emphasis on the business goal of work system improvement.

Conclusion - the Whole, the Parts, and the Specifics

This chapter provided a summary of the work system method. WSM is designed to help you and your colleagues think and communicate about any system in your organization. Approaching systems effectively requires that you maintain a big picture view of the whole while also recognizing the parts and digging into some of the specifics that must be handled in order for the system to perform well.

Level One is nothing more than a reminder to define the system and the problem, do some kind of analysis, and produce a justified recommendation. The SP questions in Level Two help you see the whole and establish the big picture. Identification of the parts using a work system snapshot (SP5) leads to the AP questions in Level Two, which ask about problems related to the nine work system elements and the work system as a whole. These AP questions also motivate the search for improvement possibilities. The RJ questions in Level Two ask for specifics about the recommendation and its justification from different viewpoints. The Level Three checklists, templates, and diagrams help you look at the work system in more depth and help in identifying issues that a broad overview would never catch. Whether or not you need a Level Three understanding of a particular work system, anyone responsible for system effectiveness needs that level of understanding.

Well designed IT support for the work system requires delving further into details of work practices, information, and technology. WSM can help in these situations because it addresses a troublesome gap between general, but often unfocused requests for system improvements, and the technical work that IT professionals are uniquely qualified to perform. The frequent result of this gap is production of software that addresses part of the problem but ignores issues that ultimately turn a potential success into a disappointment or a failure. Because this book is directed at business professionals, it only hints at the types of documentation, analysis, and design details that IT professionals must understand. Fortunately, many books and web sites cover that material quite well.[19]

The next three chapters look at each of the three WSM steps in more detail. Chapter 4 says more about deciding a work system's scope, defining its elements, and producing effective work system snapshots. Chapter 5 focuses on five groups of concepts for looking at a work system with greater depth and clarity. Chapter 6 provides additional explanation of the various questions about the recommendation and its justification.

Chapter 4: Defining a Work System

- Deciding the Work System's Scope
- Clarifications about Work System Elements
- Effective Work System Snapshots
- Examples of Work System Snapshots

The previous chapter introduced the work system method and explained how this method helps in organizing thinking about a work system. Awareness of an analysis approach is important, but it is also important to have useful ideas that can be applied in the analysis. This chapter and the chapters that follow fill in many of those ideas.

Identifying the right work system to analyze is the first order of business in the work system method because the work system's scope is not a given. In general, the work system should take whatever shape the WSM user believes is most effective for the analysis task at hand. Several people discussing the same work system often find that their initial views of the work system differ in scope and in the specifics of the work system elements. The Work System Framework provides a general model, but each individual's initial view of the specifics in any particular situation is based on personal experience, interests, beliefs, and imagination.

The clarifications in this chapter address distinctions that arise when people try to use WSM to gain insight about a situation. The first section mentions key issues for deciding a work system's scope. Next come distinctions that may be important in defining the work system's elements. The last section discusses characteristics of effective work system snapshots, one page summaries produced before launching the Analysis and Possibilities step in WSM.

Deciding the Work System's Scope

The surface simplicity of the concept of work system sometimes obscures the need to be careful when deciding a work system's scope. Work systems do not exist in nature. Work system is a mental construct used to understand real world situations. The scope of the work system is a decision, not a given. The initial decision about a work system's scope often changes as the analysis unfolds.

The smallest work system in which the problems or opportunities occur. When using WSM to analyze problems or opportunities, the work system should be the smallest work system in which the problems and opportunities occur. If the work system is too small, some of those problems and opportunities will be beyond the scope of the analysis and will not be addressed. If the work system is too large, unnecessary topics will expand the time and effort required for the analysis.

For example, assume that a firm has identified problems related to an ineffective hiring system. Managers who need to hire employees complain that hiring takes too long, and that many qualified applicants have already taken other positions by the time they receive offers. For purposes of the analysis, the work system of hiring people might extend from the approval of a new job opening to the point when the new employee starts work. Alternatively, it might focus on the activities that transpire from the time when the human resources department receives the application until the time when the hiring manager receives a list of qualified candidates. The decision about the scope of the work system would depend on the initial view of the problem. Perhaps the problem is mostly about slow processing of resumes within the human resources department; perhaps it is related to a broader scope of activities, including negotiations about job openings, creation of job postings, and interviews. As the analysis proceeds, it may become apparent that the problem is broader or narrower than was initially imagined. In either case, the person doing the analysis should always be aware that the scope of the system is a decision that can change as the analysis process generates a better understanding of the situation.

Attention to the work system's size and location. Even a basic description of a work system should indicate something about its size. For example, although the steps in the sales process may sound similar, a sales work system that involves three sales people in a single local office is quite different from one involving 50 sales people working from 12 offices on three continents.

Relationship to other work systems. Trying to understand a work system in isolation from other work systems is usually insufficient.[20] Most work systems receive and use things provided by other work systems; most produce products and services that are received and used by other work systems; most can be viewed as subsystems of other work systems.

Relationships between systems are often expressed as "inputs" received from other systems and "outputs" directed toward other systems. Computer programs always have inputs and outputs, and can be modeled in the form of "input -- processing -- output." The physical and biological sciences often focus on inputs and outputs across a system boundary. Accountants and economists perform various types of input/output analysis. The spirit of the work system approach is based on a less mechanical, more humanistic viewpoint than any of these types of analysis. Instead of including slots for inputs and outputs in the Work System Framework, inputs and outputs are handled based on general guidelines for summarizing a work system's scope and relationship to other work systems:

- As noted above, the work system being analyzed should be defined as the smallest work system that has the problems or opportunities that launched the analysis.
- The work system's outputs are the products and services it produces for its customers. Those customers are often participants in other work systems that use the products and services.
- The work system's inputs are information and other resources it receives from other work systems and from other sources. Some of the inputs a work system receives are not important enough to list on a one-page summary. Those that are important enough to include should be mentioned explicitly in the activities listed under work practices. For example, a sales system might start with a prospect list produced by a different work system. In that case, the first activity listed under work practices would be something like "salesperson contacts prospect on prospect list."

Dividing a work system into smaller work systems. It is often possible to divide a work system into several smaller work systems. The desirability of doing so depends on the purpose of the analysis and on the degree of overlap between the two smaller work systems. For instance, assume that a manufacturing firm's customers complain about late deliveries of custom-built products. The work system's scope could start with taking orders, and could extend through manufacturing, packaging, and delivery. Alternatively, it might be more meaningful to analyze four separate work systems: taking orders, manufacturing, packaging, and delivery. If it seems likely that much of the problem involves coordination between the separate subsystems, it might be worthwhile to analyze the larger work system and consider the subsystems as single steps in a business process within that work system. If it seems likely that the problem exists within each of the smaller work systems and that coordination is not an important issue, it makes more sense to consider the smaller work systems individually.

Defining a work system in terms of work, not technology. Defining a work system in terms of the technology it uses is often counterproductive. Doing this focuses attention on the technology rather than on the work that is being done and the results that are being produced for customers. In particular, it is important to avoid organizing the analysis around the features and benefits touted by a software vendor. Doing so frames the analysis in terms of the software vendor's selling points. From a business viewpoint it makes more sense to focus on business issues related to the work system's work practices, products and services, and customers.

From a business viewpoint, the work defines the system, not the technology that is used to do the work.

An organization as a group of work systems. It is possible to view an entire organization as a single work system. In most situations, however, is better to view an organization as a combination of many smaller work systems. Viewing an entire firm as a single work system tends to produce a bloated analysis that covers too many groups of people performing too many different types of roles and activities. For example, a firm that builds and sells automobiles contains hundreds or even thousands of work systems. Viewing it as a single work system would include so many different roles and activities that it would be difficult to see the relationships between work practices, participants, information, technologies, products and services, and customers.

On the other hand, it is often worthwhile to view a small company's entire value chain as a single work system. For example, a small company that produces real estate appraisals for lenders wants to improve customer satisfaction. For that purpose, the work system could easily include the firm's primary value chain activities, such as taking orders, assigning appraisals to appraisers, performing appraisal work, producing final appraisal documents, and conveying appraisals to customers. Treating the firm's primary activities as a single work system makes sense in this case because its activities are tightly linked and are performed through a small number of work roles.

Information Systems that Support Work Systems

Chapter 2 noted that an information system is a work system whose work practices are devoted to processing information, which includes capturing, transmitting, storing, retrieving, manipulating, and displaying information. Some information systems exist to produce information products and services for internal or external customers. Others exist to support work systems that produce physical products and services for customers.

Increasing reliance on computerized information systems has led to greater overlap between work systems and the information systems that support them. Information systems and the work systems they support were often rather separate decades ago when most business computing stored information on punched cards[21] and tapes, and recorded transactions in periodic batches rather than in real time.

Today, many information systems are inextricably connected to the work systems they support. For example, the package delivery firms FedEx and UPS have extensive information systems that track each package as it moves from pick-up point to delivery point. These firms exist to deliver physical packages, and their information systems support those activities. Remove the information system and their primary work systems can't operate efficiently. Ignore the work system and the information system is meaningless. In extreme cases such as highly automated manufacturing, the information system and work system overlap so much that the manufacturing is largely controlled by the information system. Turn off the information system and this type of manufacturing grinds to a halt.

The overlap between work systems and information systems is apparent in the activities performed by work system participants. For example, the activities required to track and move FedEx and UPS packages combine data collection activities that are part of their information systems and physical activities that move the packages. The data collection activities can be treated as part of a separate information system, but they also can be treated as part of the work system of tracking and moving packages.

In most situations, it is helpful to start the analysis by focusing on the work system and assuming that the information system is an integral part of the work system. Analyzing the work system without separating out the information system provides a balanced starting point for understanding how the work system operates and how it might be improved through changes that might or might not involve IT.

Figure 4.1 illustrates some of the different forms of overlap between work systems (WS) and related information systems (IS). A set of sales-related examples illustrate each type of overlap:

- (A) A comparatively small IS provides information for a WS but is not part of it. Example: A web based IS develops a prospect list by extracting information from web sites. The WS for direct sales uses the prospect list.
- (B) A comparatively small IS is a dedicated component of a WS. Example: Salespeople travel to customer sites. During interviews and negotiations with customers they use a model to compare alternatives for the customer.
- (C) The WS is primarily devoted to processing information and the IS and WS are almost identical. Example: An ecommerce web site that customers use in self-service mode to identify possible purchases, find product descriptions, and enters orders.
- (D) One IS overlaps with several separate WSs. Example: The sales force uses an IS for sales call tracking. The IS contains models that generate projections of sales based on the stage in the sales process and historical sales cycle data. Those models are used by the sales force, but are also used by the finance department for estimating sales revenues in future quarters.
- (E) A large IS supports a number of different WSs and might be larger than any of them. Example: A firm purchases a CRM (customer relationship management) software package that provides integrated database capabilities for tracking customers, sales cycles, customer orders, and customer services. Separate WSs in each of those areas use different parts of the CRM software in their operations.

Information system success. Given the overlap between information systems and work systems they support, it is difficult to evaluate an information system's success in isolation. The entire system effort may be deemed unsuccessful if the work system disappoints or fails for reasons unrelated to the information system. For example, consider a well designed planning information system that worked effectively for years. A new CEO comes on the scene with a new agenda and changes corporate planning from a bottom-up work system to a top-down work system. The previously successful information system processes the same information, but has no impact on decisions. From a business viewpoint, evaluating this information system separately from the work system is not useful.

Clarifications about Work System Elements

The elements of the Work System Framework are broad categories that are sometimes interpreted differently by different people thinking about the same system. The brief definitions of the elements in Chapter 2 are a starting point for understanding the work system approach, but experience has shown that clarifications about how to interpret each work system element help make the analysis more insightful and useful. The following clarifications start from previous definitions and add distinctions emphasizing choices and interpretations that are necessary to use the work system method effectively.

Figure 4.1. Different types of overlap between work systems and related information systems

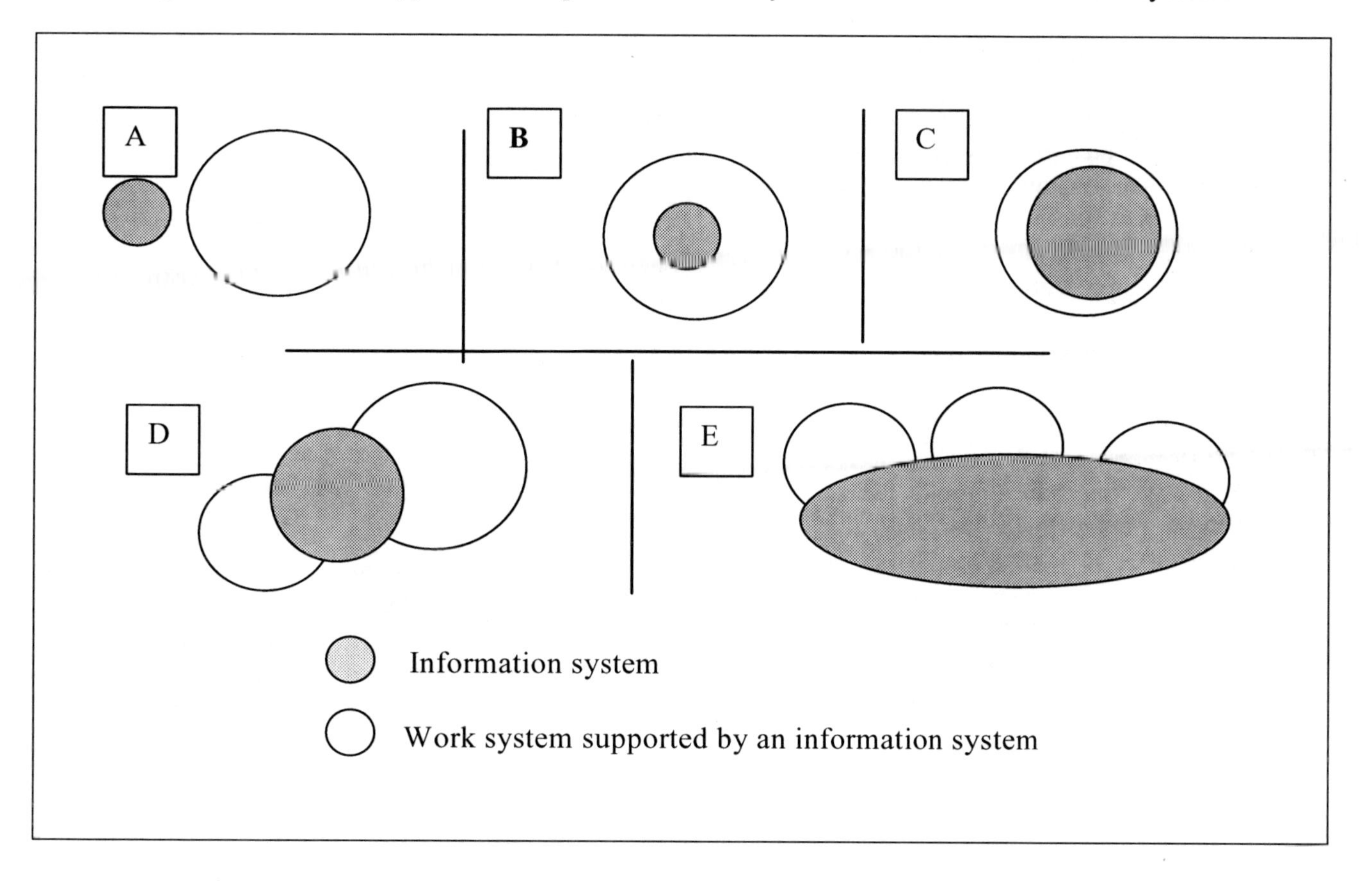

Customers

Customers are the people who receive, use, or benefit directly from products and services that a work system produces. In most cases, a work system's customers can experience or perceive the quality of products and services. Accordingly, they are in the best position to evaluate products and services.

Effective analysis of a work system avoids going overboard in identifying customers. The following distinctions and special cases clarify the way the term *customer* should be used in WSM.

Stakeholders in general. A work system's stakeholders are people affected by its products and services and by its operation. Stakeholders include:

- customers, who receive and use products and services that the work system produces
- participants, whose livelihood depends partly on work done within the work system
- managers, who are responsible for the efficiency of work and the quality of the work system's products and services
- anyone else who is affected directly or indirectly by the work system's products and services.

When analyzing a work system, it is counter-productive to treat all imaginable stakeholders as customers. Doing that diverts attention from the customers whose direct concerns and satisfaction related to the work system's products and services should be an important part of the analysis.

Internal and external customers. External customers receive and use the products and/or services that a firm produces. Internal customers are employees or contractors who receive and use a work system's products and/or services while working inside the organization. Many work systems have both internal customers and external customers. For example, the customers of a manufacturer's production system include the external customers who will receive and use the product and the (internal) packaging department that will receive finished goods from production, package those goods, and forward them to a shipper.

Some observers think that the whole idea of internal customers doesn't make sense because a company exists to produce things for its external customers. Although that viewpoint might make sense in a speech about the need for greater customer focus and less focus on internal politics, it is not helpful when analyzing specific systems in organizations. All systems in organizations exist to produce something for someone, and some of those systems are truly directed inward, such as payroll or hiring systems. In a few rare cases those systems might affect the business's external customers in a noticeable way, but trying to include external customers in the analysis of inward-facing systems usually adds little and diverts attention from more pertinent topics.

Downstream work systems. If a work system produces products and services that are used by another work system, the second work system's participants can be considered the first work system's customers. The packaging department mentioned above is an internal customer. In turn, a third party delivery service might be considered the customer of the packaging department. These internal customer relationships might be important if variability of production rates causes coordination problems, delays, and overtime in the downstream work systems. On the other hand, if a work system encompasses four sequential steps, it is not useful to list steps 2, 3, and 4 as customers because they are all inside the work system.

Customers as participants. Self-service work systems are designed to treat customers as work system participants, thereby increasing convenience for customers and/or reducing production costs within the work system. For example, an ATM user is both a customer and participant in a work system of making bank deposits and withdrawals. Similarly, the user of an ecommerce web site is a customer and participant in a work system for purchasing items online.

Participants as customers. Most work system participants are paid for the work they do, and therefore have an economic stake in the quality of the products and services the work system produces. However, most should not be considered the work system's customers because they do not use or obtain direct benefits from those products and services.

On the other hand, there are some cases in which participants should be viewed as customers. For example, both the loan applicants and the loan officer in a loan approval system should be viewed as customers because the approval or denial decision produced by the work system will have direct impacts on both of them. The loan applicant participates by providing information, and is a customer of the approval or denial. The loan officer participates by negotiating with the applicant and helping the applicant produce the loan application, but should also be considered a customer because loan approvals affect the loan officer's bonus.

Customers are participants in self-service systems.

A complete analysis of the loan approval system should treat the loan officer as a customer because the prospect of a higher bonus may motivate the loan officer to sugarcoat the applicant's situation in order to influence the approval decision. In a rather different example of participant as customer, the programmers in a software development group might be viewed as customers of their own software development work system if they will have to maintain that software over time. The potential users of that software are the most direct customers, but the programmers will feel direct effects of the quality of their own work when they fix or improve the software in the future. In contrast, programmers who produce software that they will not have to maintain usually should not be considered customers for that software.

Regulators and tax authorities as customers. Consider a work system that determines how much federal and state tax a firm should pay. Parts of that work system involve mechanical compilation of information, but other parts involve accounting decisions about how to treat specific transactions or assets. The work system's product is a set of financial statements that are submitted to tax authorities that will evaluate the statements for accuracy and completeness. In this case the regulators and tax authorities should be treated as customers because they will use and evaluate the product even though they do not benefit from it directly. In this case the company's top management should also be considered customers because their legal and economic stake in the results is quite direct.

Top management and stockholders. It is pointless to treat top management and stockholders as customers of every work system in a firm even though they can be viewed as stakeholders. A central challenge in analyzing work systems is to focus on topics and issues that are genuinely useful for understanding those systems, and to downplay topics that provide little or no insight. With the exception of work systems whose products they see and use, top managers and stockholders are usually quite distant from the situations that are being analyzed. For example, consider a work system of cleaning equipment in a factory. Top managers and stockholders may benefit very indirectly from equipment cleanliness produced by that work system, but with rare exceptions they will have no personal experience of equipment cleanliness and they will contribute nothing to the evaluation or improvement of that system. Listing them as customers of most work systems rarely adds new understanding or contributes to the analysis or recommendations.

Products and Services

Work systems exist in order to produce products and services for their customers. The term *products and services* is used instead of *outputs* because output sounds mechanical and is often associated with computers and software, which are not the central focus even though they are essential for most work systems. In addition, the term output seems inappropriate as a way to describe services that many work systems produce.

Combination of physical things, information, and services. Many work systems produce a combination of physical things, information, and services. Physical products derive most of their value from their physical form and operation, whereas information products derive most of their value from information they contain. Few products are purely physical things or purely information. For example, a DVD of a movie is a physical thing that stores the movie in the form of coded information. Most of its value resides in the information it stores, but some of its value comes from the convenient physical form of the DVD, which is smaller and easier to use than the videotapes in the previous generation of technology. For companies such as Netflix and Blockbuster, the complete offering includes the DVD and the services related to acquiring it. A future generation of technology may replace the DVD with some type of download or broadcast that shifts the balance even further in the direction of information.

Products of the work system, not products of the company. When thinking about a work system it is important to be specific about the particular products and services it produces and to exclude other products and services that other work systems produce. For example, a work system of selling medical insurance policies produces an analysis of each client's needs, the sale of insurance policies, and possibly documentation of the client's policy. On the other hand, it doesn't produce medical care and doesn't produce reimbursements for medical costs. Those products and services are produced by other work systems.

Outputs, not measurements of outputs. In the Work System Framework and WSM, products and services are the outputs that a work system produces, not the measurement of those outputs or the attainment of goals. For example, consider a work system of selling medical insurance and assume its goal is the sale of 4.8 insurance policies per week per salesperson. The products and services produced by the work system include the sale of insurance policies, the client information and analyses that are produced, and the documents that are created. Attainment of the goal is not the product of the system, but rather, confirmation that a particular level of a key metric was achieved during a particular time period. Notice also that the goal itself is not part of the work system.[22] Next month, a new manager may establish a different goal without changing the work system or the products and services it produces. For clarity about what is being analyzed, it is always important to think of the products and services as what is being produced rather than management goals that may or may not be achieved.

Intermediate products and by-products. Most work systems include a number of sequential activities in which the output of one step is the input to another step. In some cases, those intermediate outputs should be considered products produced by the entire work system. For example, the insurance sales system mentioned previously exists to produce insurance sales, but the process of producing those sales generates client profiles, contract offers, and other intermediate products that may be worth treating as products of the work system. The number and quality of client profiles produced may be an indicator of whether the sales people are generating client interest or using selling tools effectively. Similarly, highly automated manufacturing usually produces not only physical products, but also records of exactly how those products were produced. In examples such as the semiconductor chips that are used in anti-lock brakes, the manufacturing history may be a very important by-product. If an auto manufacturer discovers brake failures related to defective chips, it would much rather recall only the cars whose anti-lock chips came from the lot with the flawed chips, rather than recalling all of the cars that used any of the chips.

Intangible products and services. Some valued products and services are not described well as information or physical things or services. Examples include enjoyment, peace of mind, and social products such as arrangements, agreements, and organizations. Relatively few work systems focus on producing intangibles, but it is necessary to mention the possibility, especially since important intangibles may not conform to a general rule (explained later), that each product or service should be the output of at least one activity included in the work practices.

Work Practices

Work practices include all of the activities within a work system. These activities may combine information processing, communication, decision making, coordination, thinking, and physical actions. In some work systems, work practices can be defined tightly as highly structured business processes; in other work systems, some or all of the activities may be relatively unstructured.

Work practices versus business process. The Work System Framework uses the term *work practices* instead of *business process* because *business process* implies the existence of a structured set of steps performed in a particular order. Although many work systems operate through a well-articulated business process, in many other work systems a clear business process does not exist. A famous example is the activities that occur in an airport's control room. Air traffic controllers make frequent decisions, coordinate their actions, and communicate with airplanes. These activities are basically about minute-to-minute coordination rather than a pre-defined sequence of steps.[23] Calling these activities a business process would be a stretch.

Actual work practices versus idealized work practices. As workplace researchers point out repeatedly, the work that actually occurs often deviates from the idealized work practices that were originally designed or imagined. In many situations, different work system participants perform the same business process steps differently based on differences in skills, training, and incentives. In some situations, participants bypass or modify the idealized business process because it is too difficult or time consuming, and sometimes because it seems an obstacle to getting work done. In some labor disputes workers try to pressure management by resorting to **working by the book**, scrupulously following all rules and procedures and refusing to perform the small workarounds that allow them to get their work done. The result is often gridlock. The official rules may have made sense once, but now create unnecessary documentation, coordination, sign-offs, and delays.

Multiple perspectives for discussing work practices. The work system method views business process as only one of a number of perspectives for discussing the activities within the work system. Other perspectives include communication, decision making, coordination, control, information processing, and thinking. Each perspective has its own vocabulary and primary issues.

…. ***business process***. From this perspective, work practices are viewed as a set of steps, each of which:

- has a beginning and end,
- is triggered by specific events such as the completion of prior steps
- triggers other steps within the work system or generates part of the work system's products and services.

Business processes are often represented graphically using diagrams such as flow charts, swimlane diagrams, and data flow diagrams, which will be illustrated in Chapter 10.

…. ***communication***. Regardless of how beautifully the idealized business process seems to be designed, focusing on sequences of steps often omits important aspects of communication between work system participants, suppliers, and customers. A communication perspective focuses on issues such as the types of communication that are used, the clarity of messages, the extent to which messages are understood, the amount of effort devoted to communication, and opportunities to improve communication. For example, consider work practices related to developing information systems. Regardless of how clearly the idealized business process is defined, communication difficulties between business and IT professionals often have negative impacts on project success.

…. ***decision making***. In many situations, the details of steps in a business process are less important to analyze than a few specific decisions within the process. For example, consider the process by which a prestigious university accepts students. The details of receiving and handling applications certainly matter, but the main question is about decision making: Which criteria should be used and how effectively are those criteria applied in practice? Decision making is the central issue in many other situations including hiring decisions, strategy decisions, medical decisions, and design decisions.

coordination. In other situations, coordination is a key issue. At its heart, coordination is about managing dependencies between activities.[24] Typical dependencies are related to:

- shared resources -- who gets what and when do they get it
- task assignments -- who does what, and when do they do it
- producer/consumer relationships, such as the treatment of notification, sequencing, and tracking
- standardization of communication
- scheduling and synchronization, both in a planning mode and in a recovery mode when reality diverges from the plan.

When coordination is a key issue, focusing on the business process or communication may not be as effective as asking whether the main facets of coordination in the work system can be improved.

Business process is but one of several perspectives for understanding the activities in a work system.

…. ***control***. Efficient and effective operation of work systems requires control methods that use past and current operational information to create plans and make sure plans are satisfied to the extent possible. Although control is an essential part of a work system, many work systems have inadequate methods for planning work and for using operational information to stay on plan or respond to circumstances that require new plans.

.... ***processing information***. When analyzing a work system it is often useful to look at the processing of information. At the work system level this is not about computers. Rather, it is about whether and how relevant information is captured, transmitted, stored, retrieved, manipulated, and displayed. Each of these six information processing tasks may be the source of problems or may be an opportunity to improve the work system.

.... ***performing physical activities.*** Even in highly computerized systems it is often important to include physical activities in the analysis. For example, delays may be caused by confusion or lack of information, but they may also be caused by the inefficient layout of the workplace.

.... ***thinking***. Work practices also involve thinking, which is sometimes viewed as being so far in the background that it cannot be analyzed. At minimum, it is often possible to ask whether specific types of thinking are important, and if so, whether they are being performed effectively. For example, research on methods for supporting group deliberations have developed a number of tools to support common patterns of thinking such as divergent thinking, convergent thinking, consensus building, and organization, elaboration, abstraction, and evaluation of concepts.[25]

Participants

Participants are the people who do the work in a work system. The inclusion of human participants as part of a work system often generates much more variability than might be implied by a flow chart.

Participants, not just users. IT professionals often focus on how technology is used, and therefore speak of "users" instead of work system participants. Unfortunately, focusing on the use of technology may downplay or omit essential aspects of a work system that are not related to technology use. WSM recognizes that work system participants use technology, but views them primarily as work system participants rather than technology users.

Usually not developers of software that is used. Work system participants are usually not the people who created the software that is used in the work system. For example, a work system for hiring people may use software that was produced previously, but the software producers are not considered part of the work system unless they participate directly in its typical operation.

There are a few common situations in which software developers should be considered participants in a work system. The most common situation is when the work system itself is devoted to software development. In some other cases, a software developer plays a role in a work system because the database and/or software that is used in the work system is too awkward for non-programmers to use directly. An example involves a company whose sales and marketing information is archived in a poorly documented database. In this company, a software developer is part of the work system of analyzing marketing information because the marketing specialists need help retrieving and manipulating archived information.

Participants are people who do the work, not just people who use technology.

May or may not include work system managers. Similarly, the managers of the people doing the work may or may not be considered work system participants. They should be considered work system participants if they perform some of the activities that contribute directly to generating the work system's products and services. On the other hand, if they serve only as managers of others who perform those activities, it may be unnecessary or even confusing to view them as work system participants.

Information

Information in today's work systems involves much more than computerized databases.

Computerized vs. non-computerized information. Despite the importance of computers in most work systems, much of the information in most work systems is never computerized, and some of the computerized information does not fit into standard databases. For example, computerized information in a work system for producing a company's long-range plan includes the plan itself and the supporting spreadsheets and written explanations. Important non-computerized information exists in the negotiations and commitments between various managers. Focusing solely on computerized information misses some of the most important information in the planning process.

Types of information. Information includes:

- ***pre-formatted data items***, such as the product numbers, customer addresses, and prices on sales invoices. In databases, each of these has a specific definition and format.
- ***text***, such as email messages, memos, and other documents consisting of words rather than pre-defined data items.
- ***images***, such as X-rays, design drawings, and pictures of the damage in an automobile accident
- ***sounds***, such as the conversations that occur between call center employees and customers
- ***video***, such as a recording of a videoconference or training presentation.

Most systems analysis methods for IT professionals assume that the information will be stored in a computerized database.

Databases and documents. These larger units of information may include information of any type.

- ***Database.*** An organized collection of data. Although a database may include data of any type, the term database is usually associated with structured tables of pre-defined data items related to uniquely identified entities such as customers, products, orders, or invoices.[26] Each data item, such as a customer address or invoice number, has a definition and a pre-specified format. Some may have a range of possible values (e.g., 50 two-letter state codes or a range of valid ages for employees) or a pre-defined coding scheme (e.g., a rating from 1 to 7).
- ***Document.*** Defined in the past as "a writing conveying information." Today, electronic documents are computer files that may contain any type of information. The most familiar electronic documents are word processing documents, spreadsheets, email messages, and PDFs. Documents are usually indexed by date, author, purpose, and intended recipient(s). According to a Chief Technology Officer of Xerox, "documents have nothing to do with paper anymore." [The focus is on] "finding information and structuring it inside the document, whether it's paper or electronic, and it's about connecting the document to a workflow."[27]

Spoken communication: It is sometimes easy to forget that information in the form of spoken communication is an essential part of most work systems. In some cases, attempts to increase efficiency by computerizing record keeping have led to social isolation and have inhibited spoken communication that is essential for doing work well.

Data vs. information vs. knowledge. The common distinction between data, information, and knowledge sometimes helps in visualizing whether better information would have any effect on a work system's performance. Data includes facts, images, or sounds that may or may not be pertinent or useful for a particular task. We receive so much data every minute that our conscious minds can't possibly pay attention to all of it.

Information is data whose form and content are appropriate for a particular use. Information systems often convert data into information by formatting, filtering, and summarization. The distinction between data and information is cited frequently in explaining why vast operational databases often fail to satisfy managerial information needs. Data in these databases is information for people performing day-to-day operational tasks such as processing orders, but it is not information for managers because it is too detailed to help directly with managerial work.

Knowledge is a third basic concept for understanding the role and use of information in work systems. Knowledge is a combination of instincts, ideas, rules, and procedures that guide actions and decisions. People actually need more knowledge in many situations where they say they need more information.

Figure 4.2 shows that people use knowledge about how to format, filter, and summarize data as part of the process of converting data into information useful in a situation. They interpret that information, make decisions, and take actions. The results of these decisions and actions help in accumulating knowledge for use in later decisions. Knowledge is necessary for using information effectively regardless of how brilliantly the information is gathered and combined. Unless a work system is totally structured, system participants need enough knowledge to use the available information effectively.

Knowledge may be tacit or explicit. **Tacit knowledge** is unrecorded and is understood and applied unconsciously. **Explicit knowledge** is articulated and often codified in documents or databases. Effective knowledge management is still difficult to attain even though a great deal has been written about knowledge management. Software vendors in knowledge management tend to view it as the development and use of electronic repositories for capturing, cleansing, and retrieving information. Another important area of knowledge management involves formal and informal work practices for finding employees who have specialized knowledge that someone else in the firm needs.

The distinction between knowledge and information hints at a number of areas for potential improvement related to knowledge per se:[28]

- ***Better ability to retrieve usable knowledge from databases***. Repositories of knowledge generate no value if people don't know how to use them to find knowledge required in a current situation.
- ***Better ability to find experts***. One method is broadcasting questions and requests using email or electronic bulletin boards. Another method is using employee databases to find people who are likely to have the required knowledge.

Figure 4.2. Relationship between data, information, and knowledge

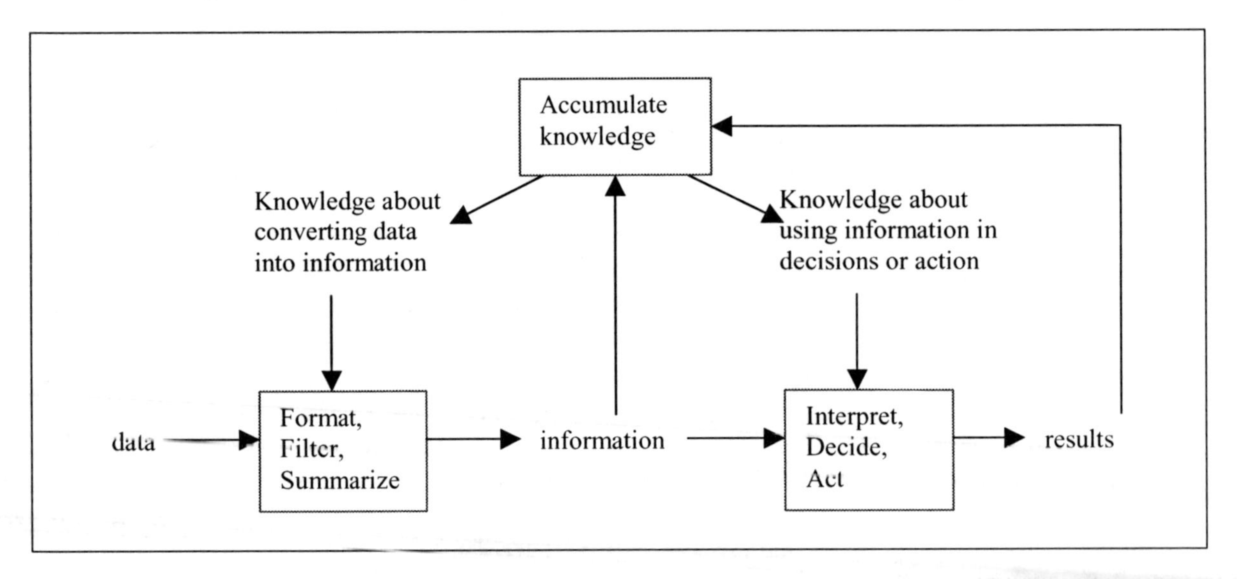

The fundamental limitation is often knowledge, not just information.

- ***Better incentives for sharing knowledge***. The mere fact that knowledge exists in a firm does not mean that the firm's employees will be inclined to share it, especially in organizational cultures that are competitive. For example, in consulting companies in which only a small percentage of associates are promoted to partner, there is little natural incentive to share information that could help someone else get promoted. To minimize this effect, some consulting firms treat evidence of knowledge sharing as a significant factor in employee reviews.[29]

- ***Retention of knowledge.*** Ideally, knowledge about company's culture, methods, procedures, history, and politics should stay with a company even when people leave. Unfortunately, downsizing and employee turnover often lead to a loss of these types of knowledge.

- ***Knowledge in work systems***. It is often said that employee knowledge is the primary corporate asset in companies that do knowledge work. The design of a work system may encourage development or codification of knowledge, or may simplify work so that less knowledge is needed.

- ***Lack of codified knowledge***. In many situations, important knowledge exists in people's heads but is never written or formalized so that it can be shared effectively. For example, this is why Xerox developed a database of insights and tips from service representatives who fix copy machines in the field. Ideas submitted by representatives are added to the database only after review by peers. Xerox says that database saved it up to $100 million per year in service costs.[30]

- ***Better knowledge about customers***. Work systems that interact with customers can produce knowledge by-products related to "knowing the customer better." Sometimes these efforts are called customer relationship management (CRM), although the specifics often are less focused on anything resembling a relationship and more focused on storage of information about customer interactions and transactions.

Technology and Infrastructure

The technologies of interest when analyzing a work system include both information technologies (IT) and non-IT technologies. Even when substantially computerized, specific tools (such as cars) and techniques (such as use of checklists) may or may not be associated with IT.

In the context of a work system, things that are considered technologies when viewed in isolation are viewed as either technology within the work system or as external technical infrastructure that the work system uses. Tools and techniques that are viewed as technologies are integral parts of a work system. Their surface details, interfaces, and affordances (such as mobility afforded by cell phones) are visible to work system participants. In contrast, technical infrastructure is used by the work system but is largely invisible to work system participants. For example, technical infrastructure often includes computer networks, database management software, and other, largely hidden technologies shared by other work systems. Thus, an ecommerce web site would be considered technology because it is visible to the shopper, whereas the Internet would be viewed as technical infrastructure that the work system uses.

Much as electricity, trash removal, and roads are taken for granted, technical infrastructure is often taken for granted except by the people who build and maintain it. The work system method mentions both technology and infrastructure because taking infrastructure for granted is often a mistake, such as when a work system cannot operate because of infrastructure problems.

In practice, the choice between treating something as technology inside the work system or as technical infrastructure should depend on what is most useful for performing a specific analysis. For example, when analyzing an order entry system, the enterprise software package SAP might be viewed as part of a firm's infrastructure and the SAP order entry software might be viewed part of the technology within the order entry work system.

Similar distinctions apply to human and information infrastructure. In each of those cases, the choice of whether to view a shared resource (people or information) as an integral part of a work system or as part of the infrastructure depends on the situation and the goals of the analysis.

Types of technologies. WSM recognizes that improving technology is often essential for improving work system performance. On the other hand, WSM does not go into depth about technology because most business professionals lack the technical background needed to understand many technology nuances that are invisible to technology users. Even though they lack that background, nothing should stop them from recognizing basic types of technology that are used in a work system, including, among others:

- ***Application software***, such as software for transaction processing, management reporting, data analysis, and specialized business functions or activities, such as designing electronic devices or selecting stocks to purchase. Concepts such as invoice or supplier that are related to specific business activities are built into application software.
- ***Non-application software***, such as operating systems, database management systems, software for data warehouses, and middleware that links applications. Non-application software contains programming and computer concepts such as query or subroutine, but does not contain concepts related to specific business activities.
- ***Computing technology,*** such as personal computers, laptops, and PDAs
- ***Communication technology***, such as email, instant messaging, voice messaging, and software supporting video conferencing
- ***Data capture technology***, such as scanners, bar code readers, and RFID readers
- ***Display technology***, such as CRT monitors, flat screen monitors, and loudspeakers
- ***Data networks*** for moving data between computerized devices
- ***Intranets*** for providing employees with Web-like interfaces to internal company information
- ***Extranets*** for providing customers or suppliers with Web-like interfaces to internal company information that they should be able to access
- ***Web sites*** for providing information that anyone can access.

Components of infrastructure. Around 55% of total IT investment is devoted to infrastructure.[31] In large companies, infrastructure services include telecommunication network services, provision and management of large scale computing (such as servers or mainframes), management of shared customer databases, and an enterprisewide intranet. (discussed further in Chapters 13 and 14)

Environment

Every work system is surrounded by many layers of environment-related issues. When analyzing a work system, it is important to identify aspects of the surrounding environment that have significant impacts on the situation. For example, organizational culture may have little bearing on a work system for generating paychecks according to highly structured rules, but may be the prime determinant of the success of a new information system for knowledge sharing. An organizational culture that includes strong expectations of cooperation and knowledge sharing tends to support initiatives that make it easier to fulfill those expectations. To the contrary, an aggressively individualistic organizational culture tends to undermine knowledge sharing initiatives that are not linked directly to personal incentives (discussed further in Chapter 14).

Strategies

Strategies can be discussed at many different levels. Ideally a work system's strategy should support the firm's strategy, but in many cases the firm's strategy is not articulated fully or is not communicated to people involved in specific work systems. Even if a firm's strategy is not articulated, it is usually possible to discuss aspects of a work system's strategy as a way of discussing big picture issues about the work system itself. For example, before plunging into details of work practices, it may be useful to discuss how structured the work should be, how many different people should be involved, how integrated the effort should be, and how complex the work system should be. Chapter 5 will present a strategy checklist that can be used in WSM. Subsequent chapters about specific work system elements will discuss strategy choices related to those elements.

Effective Work System Snapshots

A work system snapshot (question SP5) is a one-page summary designed to identify what is included within the work system that being analyzed. As the analysis proceeds, the work system snapshot remains available as an expression of the big picture within which the analysis is taking place. The work system snapshot will change if the analysis reveals that the initial scope was inappropriate, or that the work system should be viewed as several separate systems.

Figure 4.3 provides guidelines for the content of a work system snapshot. These guidelines include brief definitions of each term along with guidance about relationships between various elements of a carefully crafted work system snapshot. For example, Figure 4.3 says that products and services are produced by the activities and processes listed as work practices. This guideline is included because past users of the work system snapshot have occasionally confused the products and services produced by the work system with the products and services produced by the firm. For example, a work system snapshot attempting to describe a retail shoe store's cash management system incorrectly listed shoes among the products and services produced by the work system. Shoes were the products sold by the store, but they were not the products and services produced by the store's cash management system.

The form of the activities or processes listed under work practices is another area where past work system snapshots might have been clearer. When summarizing work practices as steps in a business process, it is important to say what is done and who does it. Thus instead of making a list of work practices such as "identify prospect, compile available information, contact prospect, and explain product," it is clearer to include both the work system participant and the action, such as:

- Phone agent identifies prospect.
- Sales assistant compiles available information
- Salesperson contacts prospect
- Salesperson explains product.

Consistency guidelines. The italicized statements in Figure 4.3 identify guidelines that make it more likely that a work system snapshot is internally consistent. These guidelines include:

- Each product or service should be the output of at least one activity or step mentioned under work practices.
- Each product or service should be received and used by at least one customer group.
- Each customer group should receive and use at least one of the products and services mentioned.
- Each participant should be involved in at least one activity or step listed under work practices.
- Each item of information should be used or produced by at least one activity or step listed under work practices.
- Each technology should be used in at least one activity or step listed under work practices.

Figure 4.3. Guidelines for the content of a work system snapshot

Note: The italicized guidelines are designed to promote internal consistency within the six parts of the work system snapshot.

Customers	Products & Services
• These are the people who make most direct use of the products & services. • Exception: In some cases it is useful to include non-customer stakeholders who have a direct and perceptible stake in the products and services produced. • Usually top managers and stockholders should not be mentioned because their stake is extremely indirect. • Customers may also be participants. Example: a self-service work system whose participants make direct use of its products & services • ***Each customer group should receive and use at least one of the products and services mentioned.***	• These are the products and services produced by activities and process steps listed as work practices. • These are not the end products of the firm unless the work system itself produces those end products. • ***Each product or service should be the output of at least one activity or step mentioned under work practices.*** • ***Each product or service should be received and used by at least one customer group.***

Work Practices (Major Activities or Processes)
• These are the activities and processes through which the work system produces its products and services. • The list is more understandable if the activities or processes are listed in sequential order. In some cases, however, the activities are not performed sequentially. • To make this summary as understandable as possible, try to summarize the activities or processes as no more than 10 steps that can be explained in more detail elsewhere. • Each step should be a complete sentence identifying a relevant role and a step or activity performed by someone in that role. Examples: "Credit analyst prepares loan write-up" or "Loan officer negotiates loan terms."

Participants	Information	Technologies
• These are the people who perform the work. • ***Each participant should be involved in at least one activity or step listed under work practices.***	• These phrases identify the information that is used or produced by the work system. • Use phrases such as "employee database" or "details of project schedules" • ***Each item should be used or produced by at least one activity or step listed under work practices***	• These are the technologies used to perform the work. • ***Each technology should be used in at least one activity or step listed under work practices.***

Adherence to the guidelines in Figure 4.3 would have improved many work system snapshots that were produced in the past. However, there are some circumstances under which it makes sense to violate a guideline. For example, when a service or an intangible product such as an agreement is being produced, the product or service may not be the output of any specific activity.

Examples of Work System Snapshots

Figures 4.4 through 4.9 are six examples of one-page work system snapshots related to specific problems or opportunities summarized in the table below. Each example starts with one or two sentences identifying problems or opportunities in the situation. The work system snapshot of the existing or proposed work system is brief, but provides enough information to help people agree about the problems and opportunities they want to address and the boundaries of the relevant work system. In all of these cases, it is likely that the preliminary discussion would lead to some modification of the work system snapshot and possibly a modification of the problem statement. As the analysis unfolds in any of these cases, the people doing the analysis might find additional issues that will lead to further revision of the work system snapshot. In some cases the original work system snapshot might be divided into separate work system snapshots summarizing different, separable parts of the situation.

Examples of work systems	*Examples of problems or opportunities for these work systems*
Company A's work system for hiring new employees starts with submitting a requisition for a new hire and ends with a hiring decision and a formal offer letter or rejection letter.	Company A is concerned that it takes so long to hire new employees that many promising recruits go to other companies.
Company B's fulfillment system for ecommerce orders starts with verification of availability of the customer's component options. It includes internal production operations, packaging of the product, and third party shipping.	Company B is concerned that its ecommerce sales are lagging behind those of its major competitor.
Company C's system for creating a regional marketing budget begins when local marketing managers propose allocations of the marketing budget across various media. These requests are analyzed, combined, reviewed, and revised.	Company C's management wants to move from an intuition-based approach to a fact-based approach that produces a clear rationale for regional marketing plans, including a way of measuring how well the marketing plan succeeded.
Company D is an insurance broker that helps corporate clients obtain group health insurance for employees. Company D's services include consolidating the corporation's health claims for the previous years, obtaining bids for the renewal from various insurance companies, and helping the client negotiate rates and terms.	Company D has received increasing complaints about inadequate customer service. It wants to understand those complaints and wants to improve its customer service system.
Company E's process of producing a new release of its commercial software starts with obtaining wish lists for new software capabilities. Management decides on primary goals for the new release. The software development organization produces and tests the software in the new release.	Company E's customers are concerned that their requests are not adequately reflected in the new releases they receive.
Company F's process for implementing new software releases for internal departments starts with identification of possibilities for improving existing work systems. After the new software release is configured and tested based on those requirements, potential users are trained and the software is implemented in the organization as part of improvements in various work systems.	Company F plans to implement a software package it purchased from Company E. Company F is concerned that its last two software implementation projects went far over their schedules and encountered a lot of resistance.

Figure 4.4. Hiring a new employee in a large company

Company A is concerned that it takes so long to hire new employees that many promising recruits go to other companies.

Customers	Products & Services
• Hiring manager • Applicants • Human Resources (HR) Department	• Hiring of a new employee • Complete information for employee database • Offer or rejection letters

Work Practices (Major Activities or Processes)
• Hiring manager submits requisition for new hire (within pre-approved budget). • Staffing coordinator in Human Resources (HR) department publicizes the position on the company intranet, on several Internet job sites, and in newspapers. • Applicants submit resumes by email. • Staffing coordinator transfers resume information into a database, and sends the best resumes to hiring managers once a week. • Hiring manager identifies applicants to interview. • Staffing assistant contacts applicants, performs additional screening, and sets up initial interviews. • Hiring manager performs initial interviews and provide feedback to HR. • Staffing coordinator or staffing assistant contacts applicants who pass the first interview and schedules additional interviews with manager, co-workers, or others. • Hiring manager makes the hiring decision. • Staffing assistant sends a formal offer letter or a rejection letter.

Participants	Information	Technologies
• Applicants • Staffing coordinator • Staffing assistant • Hiring manager • Other interviewers	• Job requisition, including job description • Resumes from all candidates • Information from interviews • Offer or rejection letters	• Email • Database software • Corporate network

Figure 4.5. Entry and fulfillment of ecommerce orders

<table>
<tr><td colspan="6">Company B is concerned that its ecommerce sales are lagging behind those of its major competitor.</td></tr>
<tr><th colspan="3">Customers</th><th colspan="3">Products & Services</th></tr>
<tr><td colspan="3">• Customer who submits the order
• Customer(s) who receive and use whatever is ordered.
• Finance Department</td><td colspan="3">• Made-to-order product received by the customer
• Information about the order and the shipment</td></tr>
<tr><th colspan="6">Work Practices (Major Activities or Processes)</th></tr>
<tr><td colspan="6">• Customer uses website to enter order.
• Computer verifies availability of component options selected by the customer.
• Computer verifies credit card information, accepts order, and transmits it to the Manufacturing Department.
• Manufacturing Department assembles and tests the made-to-order product.
• Shipping Department packages the product for shipment.
• Third party shipper ships the product to the address designated on the order.</td></tr>
<tr><th colspan="2">Participants</th><th colspan="2">Information</th><th colspan="2">Technologies</th></tr>
<tr><td colspan="2">• Customer who submits the order
• Manufacturing Department
• Shipping Department
• Third party shipper</td><td colspan="2">• Order details
• Customer credit card and verification number
• Inventory on hand, plus planned availability
• Manufacturing status of the order
• Shipping status of the order</td><td colspan="2">• Customer's PC and the Internet
• Ecommerce web site
• Corporate database system and network</td></tr>
</table>

Figure 4.6. Creating a regional marketing budget for a beverage company

<table>
<tr><td colspan="3">Company C is a major beverage company whose regional marketing plans are developed mostly based on intuition. Company C's management wants to move to more of a fact-based approach that would produce a clear rationale for the marketing plan and would have a way of measuring how well the marketing plan succeeded.</td></tr>
<tr><th>Customers</th><th colspan="2">Products & Services</th></tr>
<tr><td>• Local and regional marketing managers
• Upper management within the region
• Regional sales force</td><td colspan="2">• Regional marketing plan for the upcoming 6 months, including an approximate weekly schedule of media purchases for different media (radio, television, newspaper, billboards) and special incentives for retailers</td></tr>
<tr><th colspan="3">Work Practices (Major Activities or Processes)</th></tr>
<tr><td colspan="3">• Local and regional marketing managers receive a summary of the national marketing plan for the upcoming year.
• Local marketing managers use introspection and intuition to propose an allocation of the local marketing budget across various media.
• Marketing specialists compare requests, obtain approximate pricing from media providers, and produce a combined request after reviewing the history of units sold by product and area.
• Regional marketing manager reviews combined request and approves it or suggests revisions.
• Local marketing managers revise requests if necessary. The regional marketing manager reviews the revisions and may negotiate further.</td></tr>
<tr><th>Participants</th><th>Information</th><th>Technologies</th></tr>
<tr><td>• Local marketing managers
• Marketing specialists
• Regional marketing manager</td><td>• National marketing plan for upcoming year
• Regional budget allocation for marketing
• History of units sold by product and local area
• Requests from local marketing managers</td><td>• Spreadsheets
• Email
• Personal computers
• Corporate network and database</td></tr>
</table>

Figure 4.7. Renewing a group health insurance policy through an insurance broker

Company D is an insurance broker that has received increasing complaints about inadequate customer service.

Customers	Products & Services
• Corporate customers wanting to renew their group health policies at the best terms possible.	• Signed contract for group insurance renewal • Bids from insurance companies • Summary of the customer's claims experience for recent years.

Work Practices (Major Activities or Processes)

- Corporate customer asks insurance broker to find a group insurance renewal policy.
- Insurance broker produces consolidated summary of corporation's health insurance claims for recent years.
- Insurance broker contacts various insurance companies and requests their bids, which include policy coverage and price.
- Insurance companies submit their bids.
- Broker analyzes the bids and makes a recommendation.
- Corporate customer selects a bid to accept or negotiates further.

Participants	Information	Technologies
• Corporate customer desiring a group health policy • Insurance broker • Insurance companies	• Corporate customer's insurance claims for recent years • Summarized version of the claims experience • Bids by insurance companies • Broker's recommendation and client's decision	• Email • Word processing • Spreadsheets

Figure 4.8. Developing a new release of commercial software

Company E is a software firm. Its customers are concerned that their requests are not adequately reflected in the new releases they receive.

Customers	Products & Services
• Current customers who have purchased Company E's software • Future customers • Software development group (which must use this release as the starting point for the next release)	• New software release, including software and documentation • Original goals for the release, plus revisions of those goals

Work Practices (Major Activities or Processes)
• Marketing and Product Management Departments obtain feedback and wish-lists from customers. • Top management identifies major goals of the new release. Product Management Department converts the major goals into more detailed requirements. • Management in Software Development divides the planned changes into separate projects with priorities, and then assigns the projects to project teams. • Project teams execute the projects. • Management in Software Development monitors progress and trims the plans if necessary. • Project teams test the entire software product and convert the software to a form for release.

Participants	Information	Technologies
• Marketing Department • Product Management Department • Top management • Management in Software Development • Project teams • Customers	• Customer feedback and wish-lists • Goals of the new release • Detailed requirements • Planned scope of projects • Progress information for projects • Plan revisions	• Computer workstations • Software for programming and testing • Documentation software

Figure 4.9. Implementing a software package in a large company

Company F plans to implement a software package it purchased from Company E. Company F is concerned that its last two software implementation projects went far over their schedules and encountered a lot of resistance.

Customers	Products & Services
• Current and future users of the software • Managers of those employees • IT group that will maintain the software over time	• An improved work system that uses the software package. • Configuration of the software package to suit the situation

Work Practices (Major Activities or Processes)

- IT analysts, user representatives, and consultants analyze the current work system to identify opportunities for improvement.
- IT analysts, user representatives, and consultants identify ways in which the software package might support better work practices.
- Programmers configure the software package consistent with the company's requirements.
- Programmers test the configured software.
- Trainers train intended users of the software.
- Users convert to the new way of doing the work using the software (with help of consultants and IT analysts).

Participants	Information	Technologies
• IT analysts • User representatives • Consultants from the software vendor • Programmers • Trainers • Users	• Operation and opportunities of the current work system • Capabilities of the software • Configuration choices • Information about implementation progress and problems	• Software being implemented in the organization • Computers and networks that run the software • Terminals or PCs used by the software users.

This page is blank

Chapter 5: Identifying Issues and Possible Improvements

- Possibilities for Change
- Work System Performance Indicators
- Work System Strategies
- Work System Risk Factors and Stumbling Blocks
- Work System Principles
- Checklists of Useful Ideas

This chapter focuses on five themes that provide concepts and topics for identifying issues and possible improvements in a work system. The themes include possibilities for change, performance indicators (organized as a work system scorecard), work system strategies, risk factors and stumbling blocks, and work system principles. These themes are the basis of the checklists mentioned in the introduction to the work system method in Chapter 3. This Chapter's discussion of each theme uses the template for a work system snapshot to organize on one or two pages numerous topics associated with that theme. (See Figures 5.1 through 5.5.)

Many of the topics in Figures 5.1 through 5.4 are explained in subsequent chapters that focus on specific work system elements, such as customers, work practices, and information. In contrast, the complete discussion of work system principles appears in this Chapter. The section on work system principles identifies 24 principles (Figure 5.5) that can be used in two ways:

- to identify current issues that might need action
- to identify possible problems that might be caused by a recommendation.

This Chapter's last section reorganizes the concepts in Figures 5.1 through 5.5 in the form of ten checklists of concepts and topics that are directly related to the ten AP (analysis and possibilities) questions in WSM's Level Two.

Possibilities for Change

The purpose of analyzing a work system is to evaluate it and decide whether and how to improve it. Improving a work system is clearly not a board game like chess, but the idea of "moves" is useful in thinking about the range of possibilities for change. The moves in the system improvement game combine procedural, organizational, and technical changes that address problems and opportunities.

Figure 5.1 lists different types of changes related to each of the work system elements. Many of these possibilities will be discussed in subsequent chapters devoted to individual work system elements. As is clear from looking at Figure 5.1, work system improvements may involve some types of moves that are related to information or technology, but also may involve many other types of moves that are related to other work system elements. The range of possibilities illustrates the fallacy of assuming that the analysis of a system should focus on improving software or hardware. A much more effective initial assumption is that anything can change except whatever is explicitly identified as a constraint for purposes of the analysis.

A glance at Figure 5.1 will reveal that some of the moves overlap. For example, eliminating built-in obstacles or delays (under work practices) usually involves improving a business process by changing its steps and/or their sequence. Overlaps of this type are expected because the goal of Figure 5.1 is to make it easy to identify a large number of possible changes. With this goal, converging on the same possible change from several different starting points is much less of a problem than overlooking some of the possible changes that should have been considered.

Figure 5.1. Typical possibilities for change

Customers	Products & Services
• Add or eliminate customer groups. • Change customer expectations. • Change the nature of the customer relationship. • Change the customer experience. •	• Change information content. • Change physical content. • Change service content. • Increase or decrease customization. • Change controllability or adaptability by the customer. • Change the relationship between work system participants and customers. • Provide different intangibles. • Change by-products.

Work Practices (Major Activities or Processes)	
• Change roles and division of labor. • Improve the business process by adding, combining, or eliminating steps, changing the sequence of steps, or changing methods used within steps. • Change business rules and policies that govern work practices • Eliminate built-in obstacles and delays. • Add new functions that are not currently performed.	• Improve coordination between steps. • Improve decision making practices. • Improve communication practices. • Improve the processing of information, including capture, transmission, retrieval, storage, manipulation, and display. • Change practices related to physical things (creation, movement, storage, modification, usage, protection)

Participants	Information	Technologies
• Change the participants. • Provide training. • Provide resources needed for doing work. • Change incentives. • Change organizational structure. • Change the social relations within the work system. • Change the degree of interdependence in doing work. • Change the amount of pressure felt by participants. • Assure understanding of details of tasks and use of appropriate information and knowledge in doing work. • Assure that participants understand the meaning and significance of their work.	• Provide different information or codified knowledge. • Use different rules for coding information. • Codify currently uncodified information. • Eliminate some information. • Organize information so it can be used more effectively. • Improve information quality • Make it easier to manipulate information. • Make it easier to display information effectively. • Protect information more effectively. • Provide access to knowledgeable people.	• Upgrade software and/or hardware to a newer version. • Incorporate a new type of technology. • Reconfigure existing software and/or hardware. • Make technology easier to use. • Improve maintenance of software and/or hardware. • Improve uptime of software and/or hardware. • Reduce the cost of ownership of technology.

Infrastructure	• Make better use of human infrastructure. • Make better use of information infrastructure. • Make better use of technical infrastructure.
Environment	• Change fit with organizational policies and procedures (related to confidentiality, privacy, working conditions, worker's rights, use of company resources, etc.). • Change fit with organizational culture. • Respond to expectations and support from executives. • Change fit with organizational politics. • Respond to competitive pressures. • Improve conformance to regulatory requirements and industry standards.
Strategies	• Improve alignment with the organization's strategy. • Change the work system's overall strategy. • Improve strategies related to specific work system elements (See strategies checklist).
Work System as a Whole	• Reduce imbalances between elements. • Improve problematic relationships with other work systems. • Conform to work system principles.

Work System Performance Indicators

The balanced scorecard[32] is a commonly used management tool. The underlying idea is that a firm's performance should be evaluated based on factors other than just financial performance. At the corporate level, a balanced scorecard often contains performance indicators related to four perspectives: finance, customers, internal business processes, and learning and growth. According to the logic of the balanced scorecard, management should identify objectives, measures of performance, targets, and initiatives in each area.

Figure 5.2 provides a starting point for a balanced scorecard for a work system rather than an entire firm. At this level, application of a balanced scorecard approach starts with identifying relevant areas of performance related to work system elements. In any particular situation at least several areas of performance for at least several elements probably will be relevant.

Performance indicators and metrics. Common terminology related to performance measurement is frequently inconsistent. Terms such as measure, measure of performance, metric, and performance indicator are not defined carefully and sometimes are used interchangeably. WSM uses the following conventions:

- ***Performance indicators*** are general areas of performance, such as speed and consistency.
- ***Metrics*** are calculated or estimated numbers that summarize specific aspects of performance during a particular instant or time interval.

For example, the performance indicators speed and consistency might be important in a particular manufacturing work system. Two distinct metrics for speed might be (1) the weekly average time from start to completion for manufacturing an important product and (2) the monthly average time to recover when an important machine goes out of calibration. Similarly, two metrics for consistency might be (1) weekly average variation in tolerances at a key step and (2) monthly percentage of rework for an important step. Metrics for qualitative performance indicators such as customer satisfaction are often based on estimates rather than physical measurements. For example, two metrics for the satisfaction of the factory's customers might be (1) the average customer rating of the product and (2) the average customer rating of the factory's responsiveness, both on a scale from 1 to 7.

Ideally, WSM users should evaluate whether the work system is achieving targets for all significant metrics. They should also estimate the extent of improvement that is likely to occur after the recommendations are implemented. In many situations they cannot do a thorough job in either area. First, many analysis efforts quickly find that important metrics have not been tracked. In other words, whether or not targets have been established, no one knows whether those targets are being met. Under those circumstances, estimates of the likely performance impact of recommended changes are no more than guesses. Anyone using WSM under those circumstances is left with a quandary about how to describe current performance and how to estimate the impacts of recommended changes. In many cases it is impractical to accept lengthy delays to set up performance tracking. If action is required regardless of whether desired information is available, the WSM user needs to proceed cautiously based on estimates that may not be supported by facts. Unlike quality management methods that require tracking of metrics over time, WSM is designed to allow the analysis to proceed whether or not complete information is available. Obviously, the analysis is much more solid and convincing if performance information is available and is used effectively.

Figure 5.2. Performance indicators that might be included in a work system scorecard

Customers	Products & Services
• Customer satisfaction	• Cost to the customer
• Customer retention	• Quality perceived by the customer
	• Responsiveness to customer needs
	• Reliability
	• Conformance to standards or regulations
	• Satisfaction with intangibles

Work Practices (Major Activities or Processes)	
For business processes and work practices in general.	For communication:
• Activity rate	• Clarity of message
• Output rate	• Absorption of message
• Consistency	• Completeness of understanding
• Speed	• Signal to noise ratio
• Efficiency	
• Error rate	For decision making:
• Rework rate	• Quality of decisions
• Value added	• Degree of consensus attained
• Uptime	• Range of viewpoints considered
• Vulnerability	• Satisfaction of different legitimate interests
	• Justifiability of decisions

Participants	Information		Technologies
• Individual or group output rate	• Accuracy	• Relevance	• Functional capabilities
• Individual or group error rate	• Precision	• Timeliness	• Ease of use
• Training time to achieve proficiency	• Age	• Completeness	• Uptime
• Job satisfaction	• Believability	• Appropriateness	• Reliability
• Turnover rate	• Traceability	• Conciseness	• Compatibility with related technologies
• Amount of management attention required	• Ease of access	• Ease of understanding	• Maintainability
	• Access time	• Vulnerability to inappropriate access or use	• Price/performance
	• Controllability of selection and presentation		• Training time to achieve proficiency
			• Setup and maintenance time

Work System Strategies

Strategies express big picture choices about how resources are deployed to meet goals. A work system's strategies are conscious rationales under which it operates. Thinking about work system strategies focuses on why a work system operates one way or another, not just the details of how it happens to operate. For example, each of the following strategies, or a version of its opposite, might be appropriate for a particular work system:

- Automate work to the extent possible. (Opposite: Do everything manually.)
- Structure work and minimize application of judgment to the extent possible. (Opposite: Rely on judgment and avoid structuring work.)
- Automate information processing, but assure that system participants can use judgment in making decisions. (Alternatives: process information manually; enforce decision rules.)
- Use an assembly line approach to separate the work into discrete steps that can be performed by different people. (Opposite: Use a case manager approach to consolidate the work into steps that can be done by a single individual.[33])
- Do everything as cheaply as possible. (Opposite: Do everything possible to help work system participants do their work effectively.)
- Use segregation of duties to reduce the likelihood that people will be able to collude in fraudulent activities. (Opposite: Assume fraud won't happen and minimize the number of people who are involved.)
- Minimize costs by outsourcing parts of the work. (Opposite: Do all of the work in-house.)
- Maximize control by monitoring work closely. (Opposite: Maximize employee autonomy and assume they will coordinate voluntarily.)
- Use templates to reduce costs while allowing some degree of customization. (Alternatives: Do everything from scratch; no customization.)
- Substitute telecommunications for transportation to reduce costs and effort. (Opposite: Do as much work as possible in a single location.)

Strategies vs. goals. These examples illustrate that strategies differ from goals. Strategies are stated in terms of how things will be done. Goals are about what will be achieved. For example, a manager may decide that a work system's goal is production of 75 units per hour. The work system's strategies should guide the deployment of people and other resources to achieve its goals.

Many work systems do not have a clear strategy. For example, today's work practices may have evolved gradually from practices set up ten years ago by a group of people who are no longer with the organization. Under those circumstances, work system participants may be able to describe work practices but unable to explain why those work practices should or should not continue.

The idea of work system strategies can help WSM users visualize alternatives that may not be obvious. Experience with WSM has shown that it is comparatively easy to recommend small, incremental changes in work systems, such as eliminating an unnecessary step or computerizing information that is currently stored on paper. For most people it is much more difficult to imagine and describe changes in a work system's rationale.

Figure 5.3 uses the template of the work system snapshot to organize various strategy decisions that are relevant to many work systems. Many of the strategies in Figure 5.3 that are listed under specific work system elements will be discussed in subsequent chapters about those specific elements. The strategies that apply to a work system as a whole will be discussed next because those topics fit here better than elsewhere in the book.

Figure 5.3. Work system strategy decisions

<table>
<tr><th colspan="3">Customers</th><th colspan="3">Products & Services</th></tr>
<tr><td colspan="3">• Customer segmentation
• Treatment of customer priority
• Nature of the customer experience
• Style of interaction with the customer</td><td colspan="3">• Mix of product and service
• Product/service variability
• Mix of information and physical things
• Mix of commodity and customization
• Controllability and adaptability by customer
• Treatment of by-products</td></tr>
<tr><th colspan="6">Work Practices (Major Activities or Processes)</th></tr>
<tr><td colspan="3">• Amount of structure
• Range of involvement
• Level of integration
• Complexity
• Variety of work
• Amount of automation</td><td colspan="3">• Rhythm
• Time pressure
• Amount of interruption
• Form of feedback and control
• Error-proneness
• Formality of exception handling</td></tr>
<tr><th colspan="2">Participants</th><th colspan="2">Information</th><th colspan="2">Technologies</th></tr>
<tr><td colspan="2">• Reliance on personal knowledge and skills
• Personal autonomy
• Personal challenge
• Personal growth</td><td colspan="2">• Quality assurance
• Quality awareness
• Ease of use
• Security</td><td colspan="2">• Functionality
• Ease of use
• Technical support
• Maintenance</td></tr>
<tr><th>Infrastructure</th><td colspan="5">• Reliance on human infrastructure
• Reliance on information infrastructure
• Reliance on technical infrastructure</td></tr>
<tr><th>Environment</th><td colspan="5">• Alignment with culture
• Alignment with policies and procedures</td></tr>
<tr><th>Strategies</th><td colspan="5">• Fit with the organization's strategy
• Fit with the strategy of related work systems</td></tr>
<tr><th>Work System as a Whole</th><td colspan="5">• Centralization/ decentralization
• Capacity
• Leanness
• Scalability
• Resilience
• Agility
• Transparency</td></tr>
</table>

Strategies for a Work System as a Whole

Strategies that apply at the level of an entire work system involve a rationale that encompasses work practices and other work system elements. Six strategy variables that apply to a work system as a whole will be mentioned briefly.

Centralization/ decentralization. Organizations can be centralized or decentralized, and the same distinction applies for work systems. As a strategy issue, centralization/ decentralization is reflected in how and where work is controlled and in the distribution of responsibility and authority. Much of the current attention to virtual teams and networked organizations is part of a trend toward using IT-enabled communication capabilities to support decentralized work systems.

Capacity. A work system's capacity is the amount of work it can do in a given amount of time. Capacity is related to a combination of work practices and human and technical resources. For example, it is often possible to increase or decrease production rates by making changes related to people without changing work practices. Conversely, changing work practices may increase capacity without changing headcount or workload. Strategy issues related to a work system's capacity often involve the amount of slack designed into the work system. Almost no one touts the advantages of having excess capacity because excess capacity uses capital and human resources inefficiently and often hides sloppy planning and execution.

Leanness. Many organizations try to use "lean" approaches, meaning that they try to operate near full capacity and with minimal reserves of human or technical resources. Lean approaches are often associated with **just-in-time** (JIT). Work systems using a JIT approach avoid performing any work step unless the next downstream step needs its output in order to satisfy customer demand. Thus, work systems using a JIT approach avoid doing work to keep people and machines busy or to build inventory buffers. In addition to reducing capital costs of work-in-process, minimizing inventory buffers between steps helps in identifying production problems, which become apparent quickly because they cannot hide behind excess inventory. Another advantage of minimizing the amount of the idle inventory sitting in warehouses or on factory floors is that non-existent inventory cannot become obsolete and cannot be broken, lost, or stolen.

Lean strategies can lead to high efficiency, but they are vulnerable in a number of ways. They may be less able to respond quickly to unanticipated problems or sudden increases in demand. Lean strategies are difficult to maintain in work systems that experience peak loads, such restaurant kitchens, which have busy periods and slack periods throughout the day. When taken to an extreme, lean strategies increase vulnerability to accidents and surprises. For example, in 1997 a fire destroyed the Kariya Number 1 plant of Aisin Seiki, a large Japanese auto parts manufacturer. That plant supplied 99% of Toyota's requirements for an essential brake component. Toyota, a world leader in lean manufacturing, had less than a day's inventory on hand. It faced the possibility of a shutdown that could have lasted for months. Toyota's other suppliers volunteered to help. Aisin Seiki and Toyota engineers helped set up temporary production lines in 62 locations. Within two weeks, the entire supply chain was back in full production.[34]

Scalability. A mismatch between customer demand and work system capacity calls for adjusting capacity or adjusting demand. Scalability is the ability to increase (or decrease) capacity without major disruption or excessive costs. The most scalable work systems are those that easily incorporate incremental resources to provide more capacity, or that easily shrink to lower capacities. Unfortunately, work systems that depend on knowledgeable participants are relatively difficult to scale up because of the amount of time needed to bring new participants up to speed. These systems are also difficult to scale down because of employee concerns and loss of knowledge.

Resilience. The Toyota example is about resilience, the ability to adjust to unexpected events or problems. A work system's resilience is related to the combined characteristics of its work practices, technologies, and participants. Work systems that rely heavily on highly structured business processes controlled by IT may lack resilience if there is no practical way to work around problems related to IT. For example, reservation systems that rely on IT are dead in the water when the computers go down. In contrast, work systems that rely more on human effort and judgment may be more resilient in some situations even if they are less consistent and efficient.

Agility. Scalability is related to a work system's size and the amount it produces. In contrast, agility is about is a work system's ability to respond quickly and effectively to changes in customer needs or environmental conditions. Sources of agility include a talented, experienced staff and work practices that are relatively flexible and can change when necessary. Computerized systems may have more agility or less agility than noncomputerized systems designed for similar purposes. For example, a highly computerized production system might be more able to respond quickly to changes in customer requirements if those changes can be expressed using parameters already programmed into the automated parts of the system. On the other hand, a largely manual system might adapt more easily to a change in customer requirements related to product and service features that cannot be programmed effectively.

Transparency. Transparency is the extent to which it is clear how a work system operates, and especially how it makes important decisions. Lack of transparency is a common complaint about management decision making and organizational politics. Too little transparency often leaves external stakeholders and some work system participants wondering why the work system produced a particular result or decision. The right level of transparency is often elusive, however, because too much transparency allows people outside of the work system to obtain information that should be kept confidential.

Work System Risk Factors and Stumbling Blocks

Risk factors are recognized factors whose presence in a situation increases the risk that a work system will perform poorly or will fail totally. For example, lack of participant experience and participant dissatisfaction are risk factors because they often affect performance negatively. Similarly, some work systems contain **stumbling blocks**, features of the work system's design or of its environment, whose presence tends to interfere with efficient execution of work. Examples of built-in stumbling blocks include unnecessary inspections or signoffs that absorb time and cause delays without adding value.

The presence of risk factors and stumbling blocks is a warning sign that should lead to corrective action if possible. If a risk factor or stumbling block that has a significant effect cannot be eliminated, at minimum its effect should be mitigated if possible. For example, if there is no way to avoid having inexperienced people do the work, it may be possible to introduce tighter inspections and close mentoring. Similarly, if there is no way to eliminate inspections that are required for purposes outside of the work system, perhaps there is a way to do the inspections more efficiently.

Figure 5.4 lists risk factors and stumbling blocks that were compiled based on a review of many published articles about system risks.[35] The Figure does not distinguish between risk factors and stumbling blocks because they overlap in many cases and because their identification serves the same purpose.

Figure 5.4. Work system stumbling blocks and risk factors

Customers

- Unrealistic expectations
- Unmet customer needs or concerns
- Customer segments with contradictory requirements or needs
- Unsatisfying customer experience
- Lack of customers or customer interest
- Lack of customer feedback

Products & Services

- Unfamiliar products or service
- Products & services difficult to use or adapt Product mismatched to customer requirements or needs
- High cost of ownership
- Complex product, stringent requirements
- Incompatibility with other aspects of the customer's work environment

Work Practices (Major Activities or Processes)

- Inadequate resources
- Inadequate quality controls
- Uncertainty about how work should be done
- Excessive variability in work practices
- Over-structured work practices
- Excessive interruptions
- Excessive complexity
- Inadequate security
- Omission of important functions
- Built-in delays
- Unnecessary hand-offs or authorizations
- Steps that don't add value
- Unnecessary constraints
- Low value variations
- Inadequate scheduling of work
- Large fluctuations in workload
- Inadequate or excess capacity

Participants

- Inadequate skills, knowledge, or experience
- Inadequate motivation and commitment
- Excessive job pressures
- Inadequate teamwork
- Inadequate role definitions
- High turnover
- Disgruntled participants
- Multiple, inconsistent incentives
- Disagreement about goals and priorities
- Responsibility without authority
- Lack of accountability
- Unnecessary management layers
- Informal organization of work mismatched to formal organization
- Departmental rivalries and politics

Information

- Use of incorrect, untimely, or otherwise inadequate information
- Difficulty accessing relevant information
- Inadequate sharing of information
- Inadequate information security
- Poorly articulated knowledge about work practices
- Use of obsolete information
- Inconsistent coding of information
- Manual re-entry of previously computerized information
- Multiple versions of the same information
- Misuse of information developed for a different purpose

Technologies

- Technology complex or difficult to understand
- Technology difficult to use
- Technology difficult to maintain
- Technology unproven
- Incompatibilities with other relevant technologies
- Use of inadequate technology
- Undocumented technology
- Inadequately maintained technology

Infrastructure

- Inadequate infrastructure
- Poorly maintained infrastructure

Environment

- Lack of top management interest or support
- Mismatch with organizational culture
- Unstable environment

Strategies

- Poorly articulated corporate strategy
- Misalignment with the organization's strategy

Work System as a Whole

- Inadequate management
- Inadequate security
- Inadequate measurement of success

Work System Principles

The idea of defining work system principles and incorporating them within the work system method came from observing difficulties encountered by users of earlier versions of WSM. The work system elements provided a useful outline for identifying and describing a work system, but many teams had difficulty searching for improvements other than relatively obvious changes such as recording data that wasn't being recorded or sharing data that wasn't being shared. They seemed to need guidelines for evaluating both the current system's operation and the likely impacts of any proposed improvements. Providing a set of work system principles seemed a plausible way to support their analyses, but it wasn't clear what those principles should be.

Applicable to Any Work System

Work system principles are general statements about desired work system characteristics or results that should apply to almost any work system, not just specific tasks such as hiring employees or building software. These principles take one of two forms:

- Systems that are operating well *should* exhibit a particular characteristic.

or

- Systems that are operating well *should* accomplish a particular goal.

Work system principles can help in evaluating the current status and possible modifications of any particular work system. For the sake of brevity, they are all stated as imperatives such as "please the customer" or "do the work efficiently."

A combination of trial and error and a literature search led to a list of 24 principles. The first step was an attempt to propose a single principle for each of the 9 elements of the Work System Framework. A literature search led to additional principles that are based on ideas from sociotechnical theory,[36] general systems theory, total quality management, and general management.

All of the work system principles shown in Figure 5.5 apply to almost any work system, and therefore apply to special types of work systems, such as information systems, projects, supply chains, and ecommerce web sites. It is ironic that many authors have proposed principles for special cases (such as principles for managing projects, doing research, or selling consumer goods), but relatively few general principles for systems in organizations are discussed or used. For example, typical systems analysis and design textbooks used in business schools provide extensive guidelines for doing systems analysis for developing information systems, but say almost nothing about the principles that should apply to the work systems that are being built or supported.

Work system principles should not be confused with the common term "success factor." Success factors are factors that are statistically correlated with success, whereas principles are generalizations that apply to almost all systems of a particular type. For example, the principles "please their customers" and "perform work efficiently" apply to almost every system in an organization. In contrast, the success factors "top management support" and "prior experience with the technology" increase the likelihood of success but may be absent from successful systems, which may be invisible to top management and may use unfamiliar technology.

None of the principles in Figure 5.5 explicitly mentions topics such as speed, profitability, and competitiveness because those topics might be important in some cases but unimportant elsewhere. For example, although speed is often important, speed in delivery can be counterproductive if the recipient is not ready to receive a delivery when it arrives. Similarly, because many essential systems such as accounting systems are usually not a source of competitive advantage, a general principle about being competitively significant is inappropriate. Ideally, work system principles should be culture independent, and should apply in any national or organizational culture.[37] This criterion fits best with principles that involve characteristics of work practices and less well for principles involving people and their personal well-being.

24 Work System Principles

The 24 principles in Figure 5.5 apply to almost any work system. These principles can be used at different points when analyzing a work system. Early in the analysis they can be used to identify issues that might not be apparent and might be overlooked in the analysis. Later in the analysis, the principles can be used to sanity-check the recommendation by asking whether the recommended changes will lead to improvements or at least will do no harm in relation to each of the principles.

The 24 principles point to important issues but cannot be used mindlessly because combinations of two or more principles often present goal conflicts that require compromises. For example, principle #1 (please the customer), #4 (perform the work efficiently), and #10 (serve the participants) often push in opposite directions. Assume the system involves a service situation such as teaching or providing medical care. Typical customers would like to receive services when, where, and how they want those services, regardless of the convenience or efficiency of the providers. Unfortunately, pleasing the customer (#1) conflicts with providing those services in an efficient way (#4) that serves the participants (#10) by not placing them under undue stress. Many other combinations of principles call for similar tradeoffs.

The internal contradictions between some work system principles do not make them invalid. At work and at home our everyday lives are filled with compromises between valid, but contradictory principles (e.g., being ambitious but not trampling others, using resources efficiently but not being a miser, disciplining children but not being a drill sergeant, and so on). Although there is no formula for making tradeoffs in most of these situations, the principles can help in making well-considered decisions and not ignoring relevant factors that might have been overlooked.

Figure 5.5. Work system principles

Customers / Products & Services	• #1: Please the customers. • #2: Balance priorities of different customers.
Work Practices (Major Activities or Processes)	• #3: Match process flexibility with product variability • #4: Perform the work efficiently. • #5: Encourage appropriate use of judgment. • #6: Control problems at their source. • #7: Monitor the quality and timing of both inputs and outputs. • #8: Boundaries between steps should facilitate control. • #9: Match the work practices with the participants.
Participants	• #10: Serve the participants. • #11: Align participant incentives with system goals. • #12: Operate with clear roles and responsibilities.
Information	• #13: Provide information where it will affect action. • #14: Protect information from inappropriate use. •
Technologies	• #15. Use cost/effective technology. • #16: Minimize effort consumed by technology.
Infrastructure	• #17: Take full advantage of infrastructure.
Environment	• #18: Minimize unnecessary conflict with the external environment
Strategies	• #19: Support the firm's strategy
Work System as a Whole	• #20: Maintain compatibility and coordination with other work systems. • #21: Incorporate goals, measurement, evaluation, and feedback. • #22: Minimize unnecessary risks. • #23: Maintain balance between work system elements. • #24: Maintain the ability to adapt, change, and grow.

The current version of the work system method uses 24 principles.[38] It is not clear whether increasing or decreasing that number would help in generating better results. A much smaller number of principles would provide less guidance about common issues and problems, but too many principles could be overwhelming. Each of the current work system principles will be discussed briefly.

#1: Please the customers. Work systems exist to produce products and services for their customers. A basic tenet of quality management is that customers evaluate the product and that work system effectiveness is linked to customer satisfaction. Relevant performance indicators include cost to the customer, quality perceived by the customer, reliability, responsiveness, and conformance to standards and regulations.

Although few successful business people would argue with the general idea of pleasing the customer, a major part of today's business buzz seems to imply that the importance of the customer was discovered recently. Otherwise we wouldn't hear so much about organizing around the customer, customer-facing processes, customer relationship management, and providing a great customer experience. Obviously the importance of pleasing the customer is not new. The purpose of stating this principle and all of the other 23 principles is to put them together in a form that makes them easy to use and hard to forget.

#2: Balance priorities of different customers. Many work systems have multiple customers with different goals and needs related to the products and services the work system produces. Ideally, whatever resources are available for the work system should be deployed in a way that reflects the relative priority of different groups of customers. That may lead to different versions of the same products and services, much like the airlines' distinction between first class, business class, and coach class. It may also lead to producing fundamentally different products for different customers, such as services for primary customers and performance information for managers who are secondary customers of a work system.

#3: Match process flexibility with product variability. Work systems designed for repetitive production of a single, unvarying product or service need much less flexibility than work systems designed for production of products and services that have many variations. For example, a production line that produces cans of diet soda needs much less flexibility than a job shop that manufactures different custom-built products for different customers. The idea of matching process flexibility with product variability is expressed by the Work System Framework's double-headed arrow between work practices and products and services. That arrow says that the work practices should match the products and services, just as three other double-headed arrows say that work practices should match the participants, information, and technology.

#4: Perform the work efficiently. Effectiveness is about pleasing customers (principle #1). In contrast, efficiency is about using resources well in the internal operation of the work system. Although there are exceptions in some situations, performance indicators that are positively related to efficiency include activity rate, output rate, consistency, speed, and uptime. Performance indicators that are negatively related to efficiency include error rate, rework rate, and vulnerability.

#5: Encourage appropriate use of judgment. This is a restatement of the sociotechnical principle of "minimal critical specification," i.e., that no more should be specified in the design of work practices than is absolutely essential. In work system terms, this is reflected in the appropriate amount of structure in work practices. If work practices are structured too tightly, work system participants will not be able to use their judgment. If they are not structured enough, participants will be more likely to apply inconsistent judgments to questions and issues that have known answers and approaches.

Notice the term *appropriate use of judgment.* This principle does not say that any work system participant should be able to exercise judgment about anything. In many situations, work system participants should be required to perform the work in a prescribed manner using prescribed rules; in many other situations, the rules should be treated as guidelines that can be overridden based on judgment; in yet other situations, no general guidelines exist for important decisions even though rules may be enforced for clerical activities such as data collection.

#6: Control problems at their source. Some steps in many business processes receive inputs from other work systems. In many situations, those inputs contain errors, noise, incompleteness, and timeliness problems that should have been corrected elsewhere. Likewise, errors sometimes occur during activities within the work system. The principle of controlling problems at their source says that problems should be identified and corrected immediately instead of letting them affect other steps. Sometimes called the "sociotechnical criterion," this principle is consistent with Deming's view[39] that people should monitor the quality of their own work and should be responsible for it, rather than making after-the-fact inspectors responsible for quality.

#7: Monitor the quality and timing of both inputs and outputs. In practice, quality and timing issues might be observed in work system inputs, during specific activities within the work system, or in the outputs produced by those activities. For example, an overnight delay caused by untimely completion of a previous step may cause the same problems regardless of whether the delayed completion occurs within the work system or in a separate work system that provides inputs.

#8: Boundaries between business process steps should facilitate control. The work system method assumes a work system's work practices may include business processes consisting of individual steps. According this principle, any redefinition or reorganization of those steps should set the boundaries between steps in a way that makes it easy to check that the step is producing the right results and using resources efficiently.

#9: Match the work practices with the participants. Work practices well matched to some participants might be poorly matched to others with different interests and capabilities. This is a reason why the same business process may be successful in one site and unsuccessful in another. Even within a specific site and even when work practices are well defined, different individuals may perform at different levels of competence and ambition, and may do some of the work differently. For example, great programmers produce far better programs than mediocre programmers produce. Similarly, different managers may pay attention to different issues and may use different types of information when performing the same management role. When work system participants in the same roles have significantly different capabilities and interests, the design of the system may have to accommodate those differences.

#10: Serve the participants. This includes providing healthy work conditions and resources needed to do the work effectively and efficiently. Healthy work conditions include meaningful work, appropriate levels of challenge and autonomy, and possibilities for personal growth. Serving the participants is consistent with the sociotechnical principle of providing a high quality of work life.

#11: Align participant incentives with system goals. Participants in many systems have incentives that are inconsistent with system goals, for example, when management says that quality is the top priority but pays people based on their rate of production. Alignment of participant incentives with system goals reflects the sociotechnical principle of "support congruence," whereby systems of social support should reinforce desired behaviors.

#12: Maintain clear roles and responsibilities. Viewing a situation as a work system assumes at least some regularity about how work is done, who does it, and under what circumstances. The scope of the regularity may include different methods for different situations within the system, but it always assumes that the situation can be described as a system with identifiable elements such as work practices and participants.

Clear roles and responsibilities are part of the regularity within a work system. When roles and responsibilities are less clear, work system participants are less sure about who should do which work within the system. This uncertainty leads to continual negotiation and re-negotiation of who should do what. In unusual, extremely novel situations, the negotiations may be necessary, but in most situations clear roles and responsibilities lead to greater efficiency and effectiveness. Clarity of roles and responsibilities does not imply that each participant must play the same roles every time, however. It is certainly possible for roles to rotate depending on the situation.

#13: Provide information where it will affect action. This is the sociotechnical principle of "information flow." Participants in many systems have access to information that is never used; participants in other systems lack access to information they need. In both cases, better system performance might result from system changes that facilitate creation of value from information.

#14: Protect information from inappropriate use. As system-related information is increasingly computerized, protection of information has become more important because of heightened vulnerability to unauthorized access and modification, misuse, and theft.

#15. Use appropriate technology. Inappropriate technologies may be poorly tailored to the situation, inadequate for doing the job, or too expensive in capital costs and effort. The frequent use of inappropriate technologies implies that this should be included as a separate work system principle even though "performing the work efficiently" (principle #4) usually requires use of appropriate technology with appropriate interfaces and other features.

For example, this principle can be used to think about a customer tracking system that uses a spreadsheet to store critical information. Use of the spreadsheet may seem simple and straightforward, but the high error rates in using spreadsheets[40] imply that this may be an error-prone choice. The risk is even higher if the spreadsheet was created by a non-programmer with little skill in debugging.

#16: Minimize effort consumed by technology. Unfortunately, even seemingly appropriate technologies consume effort in learning about the technology, performing set-ups and technology tweaks, recovering from crashes and mistakes in using the technology, and generally just "futzing around" with the technology.[41] Additional effort consumed by technology often implies less effort devoted to the work system's value added work.

#17: Take full advantage of infrastructure. Fuller use of shared human, information, and technical resources may lead to better work system performance. For example, it may be possible to offload effort and improve productivity by using slack resources that are readily available in the infrastructure.

#18: Minimize unnecessary conflict with the external environment. Systems containing inherent conflicts with the environment typically operate with more stress and excess effort than systems that fit the organizational, cultural, competitive, technical, and regulatory environment.

#19: Support the firm's strategy. Consistent with the many articles that have been written about business/IT alignment, the form and operation of work systems should fit with the firm's strategy and should not contradict it unless there is a conscious reason for doing so in a particular situation. For example, a firm that positions itself as a top of the line retailer should have customer service and product returns systems that are consistent with its top of the line image.

#20: Maintain compatibility and coordination with related work systems. Every system receives inputs from other systems and produces products and services that are used by other systems. Relationships between work systems operate much more smoothly and efficiently when the producer system's product is compatible with the customer system's standards and procedures.

The analysis of how to improve a particular work system should include its impact on other work systems because the current system or a future revision might actually export problems to other work systems. Examples of exporting problems include:

- sending erroneous, substandard, or otherwise inadequate products to other work systems
- sending outputs at the wrong time
- communicating ineffectively with other work systems

From a corporate viewpoint it makes little sense to improve one system in a way that unexpectedly degrades performance of another system. However, managers under pressure to improve their own performance sometimes make decisions that cause this type of sub-optimization.

#21: Control the system using goals, measurement, evaluation, and feedback. A basic tenet of quality control is that feedback loops should help work system participants identify and evaluate gaps between goals and measured results. This type of feedback helps the work system stay on course. The related sociotechnical principle is that feedback systems should be as complex as the problems that need to be controlled. In practice, this means that a single goal with an associated metric is often insufficient for controlling a system. For example, trying to control a factory using the single metric of meeting monthly production quotas ignores other aspects of performance such as quality, productivity, and employee satisfaction.

#22: Minimize unnecessary risks. Most work systems have meaningful risks related to at least several work system elements. Risk cannot be eliminated, especially when people perform some of the work. However, unnecessary risk should be avoided by identifying important risks in a current or proposed system and deciding what can or should be done to minimize those risks.

#23: Maintain balance between work system elements. The double-headed arrows in the Work System Framework in Figure 2.1 indicate that the various elements of a work system should be in balance. Imbalances usually interfere with some aspect of performance. Typical imbalances include:

- work practices not consist with desired quality of products and services
- participants inappropriate for work practices
- work practices inappropriate for participants
- information inadequate for desired work practices
- available information not used by work practices
- technology inadequate for or too expensive for the work practices.

#24: Maintain the ability to adapt, change, and grow. Because a system's environment will probably change over time, the work system should have the capability of adapting, changing, and growing. In some cases the use of computerized information systems supports adaptability. In other cases, the computerized capabilities act like electronic concrete. They prevent change by making it excessively difficult or expensive to convert to different software that supports different work practices that use different information.

Checklists of Useful Ideas

Figures 5.1 through 5.5 list a large number of ideas that are useful when analyzing a work system. Many of those ideas are explained, rather than just listed, in Chapters 9 through 14. Even without reading those chapters the ideas can be used as checklists of topics that are often relevant for answering the AP (analysis and possibilities) questions in WSM's Level Two.

The use of these ideas within WSM can be organized in two ways.

Drill down. The AP questions in Level Two refer to a single work system element. Each of the tables on the following pages refers to one of the AP questions, and lists relevant ideas that were included in Figures 5.1 through 5.5. In effect, each table is a reminder of ideas that are often relevant to one of the AP questions. There is no obligation to use any of these ideas in answering any particular question, but the use of appropriate ideas from these tables leads toward clearer, more specific answers.

Scan across. It is also possible to convert Figures 5.1 through 5.5 into checklists that take a particular theme across all of the work system elements. For example, such a checklist could serve as a reminder of typical performance indicators or work system strategy decisions across a work system. The Appendix presents five checklists filled out for the loan approval example in Chapter 8. In each of those cases, one column of the checklist identifies the topics and other columns provide space for relevant details such as specific metrics and their numerical values.

AP1: Who are the work system's customers and what are their concerns related to the work system?

Principles	Diagrams and Methods
#1: Please the customers. #2: Balance priorities of different customers.	• Kano analysis (delighters, satisfiers, dissatisfiers) • House of Quality (from quality function deployment)[42]
Performance Indicators	**Strategy Decisions**
• Customer satisfaction • Customer retention	• Customer segmentation • Equality of treatment for customers • Nature of the customer experience • Style of interaction with customers
Stumbling Blocks and Risks	**Possibilities for Change**
• Unrealistic expectations • Unmet customer needs or concerns • Disagreement about customer requirements or expectations • Customer segments with contradictory requirements • Unsatisfying customer experience • Lack of customer feedback	• Add or eliminate customer groups • Change customer expectations • Change the nature of the customer relationship • Change the customer experience

AP2: How good are the products and services produced by the work system?

Principles	**Diagrams and Methods**
#3: Match process flexibility with product variability	• Product/service positioning diagram (See Figure 9.1)
Performance Indicators • Cost to the customer • Quality perceived by the customer • Responsiveness to customer needs • Reliability • Conformance to standards or regulations • Satisfaction with intangibles	**Strategy Decisions** • Product features • Service features • Information features • Product/service variability • Relative emphasis on product or service • Relative emphasis on physical things or information • Commodity vs. customized • Controllability by customer • Adaptability by customer • Treatment of by-products
Stumbling Blocks and Risks • Difficulty using or adapting the work system's products and services • Unfamiliar products or services • High cost of ownership • Complex product, stringent requirements • Incompatibility with significant aspects of the customer's environment	**Possibilities for Change** • Change information content • Change physical content • Change service content • Increase or decrease customization • Change controllability and adaptability by the customer • Change the relationship between work system participants and customers • Provide different intangibles • Change by-products

AP3: How good are the work practices inside the work system?

Principles

#4: Perform the work efficiently.
#5: Encourage appropriate use of judgment.
#6: Control problems at their source.
#7: Monitor the quality of both inputs and outputs.
#8: Boundaries between process steps should facilitate control.
#9: Match the work practices with the participants

Diagrams and Methods

- Flow chart (See Figure 10.1)
- Swimlane diagram (See Figure 10.2)
- Data flow diagram (See Figure 10.3)

Performance Indicators

For business process or activities:

- Activity rate
- Output rate
- Consistency
- Speed
- Efficiency
- Error Rate
- Rework Rate
- Value Added
- Uptime
- Vulnerability

For communication:

- Clarity of messages
- Absorption of messages
- Completeness of understanding
- Efficiency: Value compared to amount of information

For decision making:

- Quality of decisions
- Degree of consensus attained
- Range of viewpoints considered
- Satisfaction of different legitimate interests
- Justifiability of decisions

Strategy Decisions

- Degree of structure
- Range of involvement
- Degree of integration
- Complexity
- Variety of work
- Degree of automation
- Rhythm – frequency
- Rhythm – regularity
- Time pressure
- Amount of interruption
- Degree of attention to planning and control
- Error-proneness
- Formality in exception handling

Stumbling Blocks and Risks

- Inadequate quality control
- Uncertainty about work methods
- Excessive variability in work practices
- Frequent changes in work practices
- Over-structured work practices
- Excessive interruptions
- Excessive complexity
- Inadequate security
- Inadequate methods for planning the work
- Omission of important functions
- Built-in delays
- Unnecessary hand-offs or authorizations
- Steps that don't add value
- Unnecessary constraints
- Unclear or misunderstood business rules and policies
- Low value variations in methods or tools
- Large fluctuations in workload

Possibilities for Change

- Change roles and division of labor
- Improve business process by adding, combining, or eliminating steps, changing the sequence of steps, or changing methods used within steps
- Change business rules and policies that govern work practices
- Eliminate built-in obstacles and delays
- Add new functions not currently performed
- Improve coordination between steps
- Improve decision making practices
- Improve communication practices
- Improve the processing of information, (capture, transmission, retrieval, storage, manipulation, and display)
- Change practices related to physical things (creation, movement, storage, modification, usage, protection, etc.)

AP4: How well are the roles, knowledge, and interests of work system participants matched to the work system's design and goals?

Principles	Diagrams and Methods
#10: Serve the participants. #11: Align participant incentives with system goals. #12: Operate with clear roles and responsibilities	• Organization chart • Social network diagram[43]
Performance Indicators	**Strategy Decisions**
• Individual or group output rate • Individual or group error rate • Training time to achieve proficiency • Job satisfaction • Turnover rate • Amount of management attention required	• Management attention required • Reliance on personal knowledge and skills • Personal autonomy • Personal challenge
Stumbling Blocks and Risks	**Possibilities for Change**
• Inadequate skills, knowledge, or experience • Inadequate understanding of reasons for using current methods • Multiple, inconsistent incentives • Unclear goals and priorities • Responsibility without authority • Inadequate role definitions • Lack of accountability • Inadequate management or leadership • Unnecessary layers of management • Inconsistency between the organization chart and actual work patterns. • Poor morale • Disgruntled individuals • Lack of motivation and engagement • Ineffective teamwork • Turnover of participants • Inattention • Excessive job pressures • Failure to follow procedures • Departmental rivalries and politics	• Change the participants • Provide training on details of work • Assure that participants understand the meaning and significance of their work • Provide resources needed for doing work • Change incentives • Change organizational structure • Change the social relations within the work system • Change the degree of interdependence in doing work • Change the amount of pressure felt by participants

AP5: How might better information or knowledge help?

Principles

#13: Provide information where it will affect action.
#14: Protect information from inappropriate use.

Diagrams and Methods

- Entity-relationship diagram (See Figure 12.1)
- Data Dictionary
- Data flow diagram (See Figure 10.3)

Performance Indicators

- Accuracy
- Precision
- Age
- Believability
- Traceability
- Ease of access
- Access time
- Controllability of selection and presentation
- Relevance
- Timeliness
- Completeness
- Appropriateness
- Conciseness
- Ease of understanding
- Vulnerability to inappropriate access or use

Strategy Decisions

- Quality assurance for information
- Awareness of information quality
- Ease of use
- Information security

Stumbling Blocks and Risks

- Use of obsolete or inaccurate information
- Difficulty accessing information
- Misuse of information developed for a different purpose
- Misinterpretation of information
- Multiple versions of the same information
- Inconsistent coding of information
- Re-entry of previously computerized information
- Inadequate control of information access and modification
- Unauthorized access
- Poorly articulated knowledge about work practices

Possibilities for Change

- Provide different information or knowledge
- Use different rules for coding information
- Codify currently uncodified information
- Eliminate some information
- Organize information so it can be used more effectively
- Improve information quality
- Make it easier to manipulate information
- Make it easier to display information effectively
- Protect information more effectively
- Assure understanding of details of tasks and use of appropriate information and knowledge in doing work
- Provide access to knowledgeable people

AP6: How might better technology help?

Principles

#15: Use cost/effective technology
#16: Minimize effort consumed by technology.

Diagrams and Methods

- Network architecture (many variations)

Performance Indicators

- Functional capabilities
- Ease of use
- Uptime
- Reliability
- Compatibility with complementary technologies
- Maintainability
- Price/performance
- Training time to achieve proficiency
- Time absorbed by setup and maintenance

Strategy Decisions

- Functionality
- Ease of use
- Technical support for technology usage
- Maintenance

Stumbling Blocks and Risks

- Use of inadequate technology
- Undocumented technology
- Inadequately maintained technology
- Technology incompatibilities
- Technology complex or difficult to understand
- Non- user friendly technology
- Equipment and software downtime
- New or unproven technology
- Mismatch with needs of work practices
- Unauthorized usage
- Software bugs
- Unauthorized changes to software

Possibilities for Change

- Upgrade software and/or hardware to a newer version
- Incorporate a new type of technology
- Reconfigure existing software and/or hardware
- Make technology easier to use
- Improve maintenance of software and/or hardware
- Improve uptime of software and/or hardware
- Reduce the cost of ownership of technology

AP7: How well does the work system fit the surrounding environment?

Principles	Diagrams and Methods
#18: Minimize unnecessary conflict with the external environment	Various diagrams and methods can be used to analyze the competitive environment that affects the system.
Performance Indicators	**Strategy Decisions**
• Fit with environment	• Alignment with culture • Alignment with policies and procedures
Stumbling Blocks and Risks	**Possibilities for Change**
• Lack of management support and attention • Poor fit with organizational policies and procedures • Poor fit with organizational culture • Negative impacts of recent organizational initiatives • Poor fit with organizational and competitive pressures • Noncomformance to regulations and industry standards • High level of turmoil and distractions.	• Change the work system's fit with organizational policies and procedures (related to confidentiality, privacy, working conditions, worker's rights, use of company resources, etc.) • Change the work system's fit with organizational culture • Respond to expectations and support from executives • Change the work system's fit with organizational politics • Respond to competitive pressures • Improve conformance to regulatory requirements and industry standards

AP8: How well does the work system use the available infrastructure?

Principles	Diagrams and Methods
#17: Take full advantage of infrastructure.	• Diagram of network architecture or information architecture (many variations)
Performance Indicators	**Strategy Decisions**
• Adequacy of human infrastructure • Adequacy of information infrastructure • Adequacy of technical infrastructure	• Degree of reliance on human infrastructure • Degree of reliance on information infrastructure • Degree of reliance on technical infrastructure
Stumbling Blocks and Risks	**Possibilities for Change**
• Poor fit with or use of human infrastructure • Poor fit with or use of information infrastructure • Poor fit with or use of technical infrastructure	• Make better use of human infrastructure • Make better use of information infrastructure • Make better use of technical infrastructure

AP9: How appropriate is the work system's strategy?

Principles

#19: Support the firm's strategy

Diagrams and Methods

Performance Indicators

- Fit with organization's strategy
- Fit with reality faced by work system

Strategy Decisions

- Designing fit with organization's strategy
- Designing fit with strategy of related work systems

Stumbling Blocks and Risks

- Misalignment with the organization's strategy
- Poorly articulated corporate strategy

Possibilities for Change

- Improve alignment with the organization's strategy
- Change the work system's overall strategy
- Improve strategies related to specific work system elements

AP10: How well does the work system operate as a whole and in relation to other work systems?

Principles

#20: Maintain compatibility and coordination with other work systems.
#21: Incorporate goals, measurement, evaluation, and feedback
#22: Minimize unnecessary risks.
#23: Maintain balance between work system elements.
#24: Maintain the ability to adapt, change, and grow

Diagrams and Methods

General-purpose techniques that may be useful in analyzing an entire work system or any part of it include, among many others, histograms, scatter plots, run charts, control charts, fishbone diagrams, and Pareto charts.

Performance Indicators

- Quality of inputs from other work systems
- Timeliness of inputs from other work systems
- Effectiveness of communication with other work systems
- Effectiveness of coordination with other work systems

Strategy Decisions

- Degree of centralization
- Capacity
- Resilience
- Scalability
- Agility
- Transparency

Stumbling Blocks and Risks

- Inadequate resources or capacity
- Inadequate management
- Inadequate security
- Inadequate measurement of success

Possibilities for Change

(Change individual elements appropriately.)

This page is blank

Chapter 6: Justifying a Recommendation

- Desired Changes to Work System Elements
- Alternatives not Selected
- Comparison with an Ideal System
- Likely Success Addressing the Original Problems
- Negative Impacts of Recommended Changes
- Fit with Work System Principles
- Implementation of the Recommendation
- Stakeholder Analysis
- Costs, Benefits, and Risks
- Questionable Assumptions
- How Much Justification Is Needed?

The overview of the work system method in Chapter 4 identified ten questions (RJ1 through RJ10) for assuring that the recommendation and justification are coherent and compelling enough to explain to others. In effect, these questions express typical objections that often arise in planning meetings, such as:

- Did you consider all of the changes that need to take place, not just changes in software and hardware?
- Did you consider alternatives?
- Is this just a stopgap, or is it really a solution?
- Did you consider new problems that might be caused?
- Do you have a tentative plan that seems feasible?
- Did you identify the questionable assumptions that might be showstoppers?

Questions RJ1 through RJ10 serve several purposes. First, they identify issues that should be considered before pursuing a recommendation. Second, they make sure that the recommendation is justified from economic, organizational, and practical perspectives. Third, they can help in recognizing topics that need a deeper discussion and/or additional expertise. Although none of these questions is intrinsically technical, several require input from IT professionals. For example, it is rarely feasible to produce a cost-benefit justification without help from someone who understands the technical issues and technical resources required. This chapter discusses some of the main issues related to each of the ten questions about a recommendation and its justification. Each of the Chapter's subheadings identifies the main topic of one of those questions.

Desired Changes to Work System Elements

Because a work system is a system, changes in one part of a work system usually result in changes elsewhere. Therefore, the first step in identifying the recommended changes is to summarize the proposed changes for each element of the work system. It isn't enough to say that new software will be used or that the participants will receive training.

To clarify the nature of the intended changes and to make sure the tentative project plan is realistic, it is useful to distinguish three types of changes:

- work system changes that are primarily unrelated to information system changes, such as changing incentive structures or providing training on work practices not related to the information system

- work system changes that are directly related to information system changes, such as changing work practices that rely on new information system capabilities

- information system changes that have little bearing on how the work system typically operates, such as making hardware or software changes that are not visible to work system participants, but that help in maintaining the hardware or software.

Separating the three types of changes usually reveals that a change in technology will not accomplish the desired change in work system performance by itself. Establishing realistic expectations related to changes that do not involve IT is essential. Well over half of the costs of most IT-related changes in work systems are devoted to things other than hardware or software. A leading researcher on IT economics found that nine dollars of **complementary assets** including human capital and organizational capital are required for every dollar of IT hardware capital that a company owns.[44]

A complete recommendation considers possible changes in each element of the work system, not just technology.

Alternatives not Selected

Too many recommendations are made without serious consideration of alternatives. In such cases the assumed options are to do nothing or to adopt a particular proposal. Identifying and considering genuine alternatives other than doing nothing decreases the likelihood that a better alternative was not even considered.

In some situations, the decision is easy because one alternative is better than all other alternatives on all relevant criteria. But in many other situations, several alternatives are better on some criteria, but not all. In those cases, it is useful to identify the factors and tradeoffs that led to the selection of the preferred alternative. Examples of typical factors and tradeoffs include:

- current out-of-pocket costs versus costs of long term delays
- advantages and disadvantages of doing one large project versus several small projects
- costs and other impacts of retraining current employees versus hiring new employees
- relative emphasis on privacy versus monitoring and control
- costs and benefits of real time feedback versus delayed feedback.

Comparison with an Ideal System

Recommendations may be based on so many assumed constraints and limitations that highly beneficial options are never considered seriously. It may be unclear whether the recommendation only attempts to stop the bleeding or whether it represents a good long term approach to the problem. It may make sense to revise the recommendation or to do nothing until a more complete project can be undertaken.

Ackoff's concept of idealized redesign provides a way to talk about an ideal system. "An ideally redesigned system is one with which the designers would now replace the existing system if they were free to replace it with any system they wanted." For the idealized design to be meaningful, it must be technologically feasible with current technology and must be operationally viable in the current environment. However, implementation issues should not be considered at this stage because they might constrain creativity.[45] The attempt to describe an ideal system might lead to a realization that the recommendation is no more than a stopgap measure. Also, it might clarify that the recommendation is only a first step in a beneficial direction, but that additional steps should follow when time and other resources are available.

Likely Success Addressing the Original Problems

In some situations, clear and pressing needs that were recognized by project initiators gradually become diffused as the analysis proceeds and additional factors emerge. For example, the analysis may find that one of the initial goals is infeasible technically, that another goal is being handled by another project, and that three additional problems that were not part of the original mandate will have to be resolved in order to make progress on the original problem. Sometimes the scope of the project is reduced in order to meet a completion date. Under such circumstances it is especially important to ask how well the recommended changes address the original problems and opportunities.[46] If the original vision is not going to be realized, the project's customers and funders may prefer to cancel the project rather than attempting to achieve only part of the original goal.

Negative Impacts of Recommended Changes

Changing one part of a system often causes problems elsewhere in the system. In some situations, the new problems may be worse than the original problems that motivated the change. For example, an attempt to solve an efficiency problem may result in new work practices that lead to employee turnover.[47] Speaking about technological change at a more general level, the social critic Jacques Ellul said:

> All technical progress exacts a price; while it adds something on the one hand, it subtracts something on the other.
>
> All technical progress raises more problems than it solves, tempts us to see the consequent problems as technical in nature, and prods us to seek technical solutions to them.
>
> The negative effects of technological innovation are inseparable from the positive. It is naive to say that technology is neutral, that it may be used for good or bad ends; the good and bad effects are, in fact, simultaneous and inseparable.
>
> All technological innovations have unforseeable effects.[48]

Whether or not Ellul sounds too pessimistic, his basic point about changes generating new problems is worth heeding when evaluating a recommendation to improve a work system. An example from Zuboff's *In the Age of the Smart Machine* illustrates the point. An insurance company installed a new information system for processing dental claims, thereby reducing the knowledge demands of the job in order to speed processing. According to a manager, "a lot of quality issues are now built into the machine. It requires less thought, judgment, and manual interventions." The new work system was designed to maximize productivity by minimizing tasks that involved interpersonal coordination, such as collecting mail or answering the telephone. Although productivity of claims processing increased by 105% over two years, a manager "felt somewhat appalled at the nature of the job he had helped to create." He said, "It's reached a point where the benefits analysts can't move their fingers any faster. There is nowhere to go anymore if you don't want to sit in front of a terminal." A shift occurred in how benefits analysts related to their work. According to one analyst, "I don't know half the things I used to. I feel that I have lost it – the computer knows more. I am pushing buttons. I'm not on top of things as I used to be." Another analyst said, "You don't have to think that much because the system is doing the thinking for you. You don't have to be concerned with what is on that claim. People here have begun to feel like monkeys."[49] This example concerned impacts on work system participants. Other types of negative impacts might be related to costs, stresses elsewhere in the organization, or undesirable impacts on customers.

Fit with Work System Principles

The recommended work system should reduce any significant nonconformance to work system principles. In addition, based on the general goal of doing no harm, it should conform to each principle in Figure 5.5 at least as well as the existing work system. At minimum, if the recommendations will create new problems, the benefits of the recommendations should outweigh those problems.

The Appendix shows how a work system principles checklist can be applied to an example during the analysis and possibilities (AP) step. The same checklist applies equally during the recommendation and justification (RJ) step. During AP, it is used to identify problems that should be addressed; during RJ, it is used to evaluate whether the recommendation solves those problems and whether it creates new problems. Merely going through the work system principles checklist during the RJ step is no guarantee that the recommendation will succeed. However, looking for conflicts between the principles and the recommendation is worthwhile because significant conflicts could indicate trouble ahead unless the recommendation is changed or other corrective actions are taken.

Implementation of the Recommendation

There is little value in producing recommendations that are infeasible organizationally, economically, or technically. Therefore any serious recommendation should include a tentative implementation plan. In practice, this means that ownership and management of the project and work system should be clear. In addition, a tentative project plan should seem reasonable in terms of timing, expenses, and availability of people and other resources.

Ownership of the project and work system. Implementation of any planned change usually requires a combination of will and authority. Someone has to care about the recommended change and someone has to have the authority to make sure it happens. At minimum, this requires:

- ***A work system owner***, who has the authority make changes in the work system, assuming that the recommended software or hardware changes are feasible
- ***A project manager,*** who will make sure that the steps in the project are carried out effectively
- ***A funder***, an organization or individual who will provide funding and other resources needed to perform the project
- ***Project participants***, people who will do the work on the project.

Project scope and tentative project plan. The work system life cycle model (WSLC) that will be presented in Chapter 7 provides a useful summary of steps that might be included in a simplified project plan. For each step, the tentative plan should identify tentative resource requirements (key people, number of person days, and other resources) and start and end dates. As will be explained, use of the WSLC model instead of a software project model emphasizes the goal of improving a work system, not just producing and installing new or improved software.

Stakeholder Analysis

To increase the likelihood of success, it is important to make sure that the views of different stakeholders are heard and considered seriously. In the many cases when this is not done, the ignored stakeholders are less likely to support the change and may resist it or undermine it. Typical examples of important stakeholders or stakeholder groups include:

- ***Top management***, but only if the project is big enough that they will approve the funding or will need to provide symbolic or personal support to encourage successful implementation
- ***Business unit management***, because they own the work system and therefore are responsible for the success of work system changes
- ***Supervisors***, because they are directly involved in making sure the work system operates properly
- ***Work system participants***, because they do the work in the work system and use hardware and software directly
- ***Unions***, in case the work system changes are at all related to significant union issues
- ***Managers and key participants in other affected work systems***, because their work systems may provide inputs or use outputs from the work system that is being changed; also because they may be affected indirectly through changes in responsibilities, power, or other issues that matter to them
- ***The IT group***, because it will have to do the technical work related to computer hardware and software, and will have to maintain the hardware and software over time. If contractors are involved rather than company employees, the IT group may have to manage the contractors or maintain whatever IT capabilities they produce.

Costs, Benefits, and Risks

Economic justification includes a wide range of topics, starting with identification and timing of costs, benefits, and risks:

- ***Direct costs*** related to accomplishing the change from the existing work system to the new work system. These costs are for salaries, software expenses, other expenses directly related to the system change project.
- ***Indirect and hidden costs*** that might be missed because they are already included in other budgets. An example is the salaries of managers and work system participants who participate in the project. Their salaries are included in existing budgets, but a substantial amount of their work may be devoted to the project. In many projects, the indirect and hidden costs may exceed the direct costs.
- ***Tangible benefits*** that can be measured directly. These benefits include cost savings due to headcount reduction, increased efficiency, travel reduction, and other ways in which the direct cost of operating the new work system will be less than the direct cost of operating the old work system.
- ***Intangible benefits*** that may be very important but cannot be expressed easily in dollar terms. These include better decision making, better coordination, better information about customers or internal operations, and greater job satisfaction. In many projects, intangible benefits may be more important than the tangible benefits that are easier to measure.
- ***Risks*** of many types may affect the likelihood of attaining the originally estimated costs and benefits. The example in the Appendix includes a checklist for the work system risks and stumbling blocks identified in Figure 5.4. It also includes an additional checklist for risks that apply specifically to the initiation, development, and implementation phases in the work system life cycle (explained in Chapter 7).

One of the reasons for analyzing a project's costs, benefits, and risks is to provide a basis for comparing and selecting among proposed projects, especially when resources are insufficient for doing all of the projects that are proposed.

- ***Financial comparisons*** are used to help in selecting among potentially beneficial projects whose total costs exceed the available resources. These comparisons start by expressing each project's monthly or quarterly costs and benefits in monetary terms. Producing dollar estimates is easier for direct costs and tangible benefits, but even estimates for those categories may contain large errors due to unexpected problems or delays. The cost and benefit stream for each project is summarized using financial measures such as net present value or internal rate of return.[50] These financial measures can also be used in post-implementation audits that try to extract lessons for the future by assessing the accuracy of the estimated costs and benefits.

- ***Payback period*** is the estimated length of time until the project's cumulative net benefits catch up with its cumulative net costs. Payback period is important for deciding whether the project is too risky to undertake and whether it is better to reduce the risk of a large failure by subdividing the project into several smaller projects that are easier to manage and will produce useful results sooner.
- ***Business and strategic priority*** is important to consider because some seemingly beneficial projects may not make a significant contribution to the organization's competitive advantage. For example, a strategically significant customer service project needed to match capabilities of a key competitor might seem to have a lower rate of return than an internal accounting project that will reduce clerical costs. In this case, the project with a lower rate of return should have higher priority because it matters more to business success.
- ***Real options***. Some companies use decision trees and other tools to consider the potential value of future possibilities (sometimes called real options) that will emerge when the project is completed. For example, implementation of a new customer database may solve a current problem and may also be the first step toward new customer service capabilities that might be developed later. Doing the project might lead to future possibilities that seem impossible without doing the project, but are not yet analyzed or budgeted.

When using cost/ benefit analysis it is important to recognize that most of the numbers are estimated costs and benefits related to things that should happen in the future. Those estimates may be reasonably reliable if similar projects have been done frequently in the setting. On the other hand, they can be far off if the project is novel, involves a lot of research, or has many unknown details. Given that many of the costs and benefits are probably inaccurate, one might wonder whether it is worth the trouble of producing the estimates and rolling them up into financial calculations such as net present value or internal rate of return. The basic answer is that a genuine effort of specifying the plan and estimating the numbers often leads to greater understanding of costs, benefits, risks, and other factors that might have been overlooked by the people making the proposal. Furthermore, the review of the plan and the associated estimates of costs and benefits can help in discussing the relative merits of different projects.

Even with inaccuracies, being explicit about costs, benefits and risks provides a useful discipline that makes the project selection process more rational, at least on the surface. Lurking below the surface is the recognition that the people who proposed the project may have a vested interest in it and may have underestimated costs, overestimated benefits, and ignored risks. The need to understand cost, benefit, and risk estimates is one of the reasons for the checklists and templates in the Appendix.

Questionable Assumptions

Every recommendation is built on a set of judgments, assumptions, and facts. A final step in sanity-checking a recommendation is to ask whether a change in any of those assumptions, judgments, or facts might have an impact on the recommendation. Common examples of such assumptions include:

- enthusiasm and genuine support of top management
- genuine involvement of work system participants and their managers
- availability of key individuals whose knowledge is needed to specify the detailed requirements or to produce software
- availability of database, browser, or other software that will be used
- availability of computers or communication networks that will be used
- knowledge of the technical staff

- knowledge of IT users
- likelihood of being able to insulate the project from other pressing issues in the organization

It is important to identify assumptions such as these because a review of the recommendation may reveal that important assumptions are incorrect, and that the recommendation may be invalid as a result.

How Much Justification Is Needed?

As was mentioned in Chapter 1, the success rate for systems in organizations and system-related projects is disappointing. Great things have been accomplished, but many opportunities have been missed and many unnecessary failures have occurred. Inadequate analysis and communication in the early stages of projects are cited repeatedly as one of the main reasons for these disappointments.

The ten RJ questions serve a sanity-checking role. Before an organization devotes significant resources to a project, considering these questions can help make sure that obvious issues have been covered:

- Does an understandable recommendation exist, and does it explain how the work system will change?
- To what extent do the recommended changes solve the original problem or create other problems?
- How feasible is the project from organizational, economic, and technical perspectives?
- Are there any easily recognized weak points that should be discussed?

This chapter discussed the main issues behind each of the ten RJ questions. The questions themselves simply try to make sure that a project team evaluates the recommendation from different viewpoints that are pertinent to most system-related projects. Many smaller projects require only simple answers; some of the issues may not be important for some projects. Large, complex projects require a much longer list of questions, sub-questions, and sub-sub-questions. Regardless of whether the project is small or large, the amount of time needed to consider these issues is miniscule compared to the time and effort that could be wasted if a project proceeds from an inappropriate recommendation.

This page is blank

Chapter 7: The Work System Life Cycle

- Phases of the WSLC
- Operation and Maintenance
- Initiation
- Development
- Implementation
- Importance of Change Management
- Collaborating throughout the WSLC

The Work System Framework is useful for summarizing a work system's operation and scope, analyzing a work system, and identifying potentially beneficial changes. A different model, the Work System Life Cycle model (WSLC), is useful for understanding how those changes occur, and how work systems evolve over time. In effect the Work System Framework and WSLC are like common denominators for understanding systems in organizations. One focuses on how systems operate; the other, on how systems change.

As shown in Figure 7.1, the WSLC says that a work system evolves through multiple iterations of four phases, operation and maintenance, initiation, development, and implementation. The names of the phases were chosen to describe both computerized and non-computerized systems, and to apply regardless of whether application software is acquired, built from scratch, or not used at all. The WSLC encompasses both planned and unplanned change. Planned change occurs through a full iteration encompassing the four phases of the WSLC, i.e., starting with an operation and maintenance phase, flowing through initiation, development, and implementation, and arriving at a new operation and maintenance phase. Unplanned change can occur within any phase through adaptation, improvisation, and experimentation.

The pictorial representation of the work system life cycle model places the four phases at the vertices of a rectangle. Forward and backward arrows between each successive pair of phases indicate the planned sequence of phases and allow the possibility of returning to a previous phase if necessary. To encompass both planned and unplanned change, each phase has an inwardly curved "adaptation loop" to denote unanticipated opportunities and unanticipated adaptations.

The WSLC's iterative nature and its inclusion of planned and unplanned change illustrates that most of the "life cycle" models used by IT professionals focus on a software development project rather than a system's life cycle. Although often called **system development life cycle** models, those models represent the end of the life cycle as the point when the software goes into maintenance mode or when the software changes have been accepted.

As an example of the iterative nature of a work system's life cycle, consider the sales system in a software start-up. Its initial sales system consists of little more than the visionary CEO selling directly to the initial customers. At some point the CEO can't do it alone, several salespeople are hired and trained, and marketing materials are produced that can be used by someone other than the CEO. As the firm grows, the sales system becomes regionalized and an initial version of sales tracking software is used. Later, the firm changes its sales system again to accommodate needs to track and control a larger sales force and predict sales several quarters in advance. A subsequent iteration might include the acquisition and configuration of commercial customer relationship management (CRM) software.

The first iteration of the sales work system's life cycle starts with an initiation phase. Each subsequent iteration involves deciding that the current sales system is insufficient; initiating a project that may or may not involve significant changes in software; developing resources such as procedures, training materials, and software that are needed to support the new version of the work system; and finally, implementing the new work system in the organization. Based on that logic, the transition to the version of the sales system that uses CRM is not about implementing CRM software. Rather, it is about implementing an improved work system that is enabled by CRM software and other technology.

Consistent with a business viewpoint, the WSLC model focuses on the work system but recognizes that information systems are often essential for the work systems they support. Because the WSLC describes a work system life cycle rather than an IT project, it portrays unplanned changes and adaptations as part of business reality, not as undesirable deviations from a project plan. It recognizes that business requirements change frequently and that unanticipated adaptations may emerge during any phase even though project milestones within each phase should be identified and scheduled in advance. Accordingly, the graphical representation of each phase includes an "adaptation loop" as a reminder that unplanned changes and innovations may modify or extend systems in unanticipated directions. Including these adaptation loops instead of pretending they don't exist encourages continuing negotiation around the natural conflict between the IT group's need to define bounded, measurable projects and the business's need to improve work practices and adapt to changing business problems and opportunities.

Phases of the WSLC

The work system life cycle explains how the initial version of a work system is created and how the transition occurs between two successive versions of the same work system. The phases of the work system life cycle model are named so that they apply to any work system, regardless of whether IT is involved.

- **Operation and maintenance** is the ongoing operation of the work system after it has been implemented, plus small adjustments, corrections of flaws, and enhancements.

- **Initiation** is the process of defining the need for significant change in a work system and describing in general terms how the work system changes will meet the need.

- **Development** is the process of defining and creating or obtaining the tools, documentation, procedures, facilities, and any other physical and informational resources needed before the desired changes can be implemented successfully in the organization.

- **Implementation** is the process of making a new or modified system operational in the organization, including planning for the rollout, training work system participants, and converting from the old way of doing things to the new way.

Table 7.1 summarizes issues that are addressed in each of these phases. At the end of this chapter, Figures 7.2, 7.3, 7.4, and 7.5 combine these with many other issues that are discussed throughout the chapter.

Figure 7.1. The work system life cycle (WSLC) model[51]

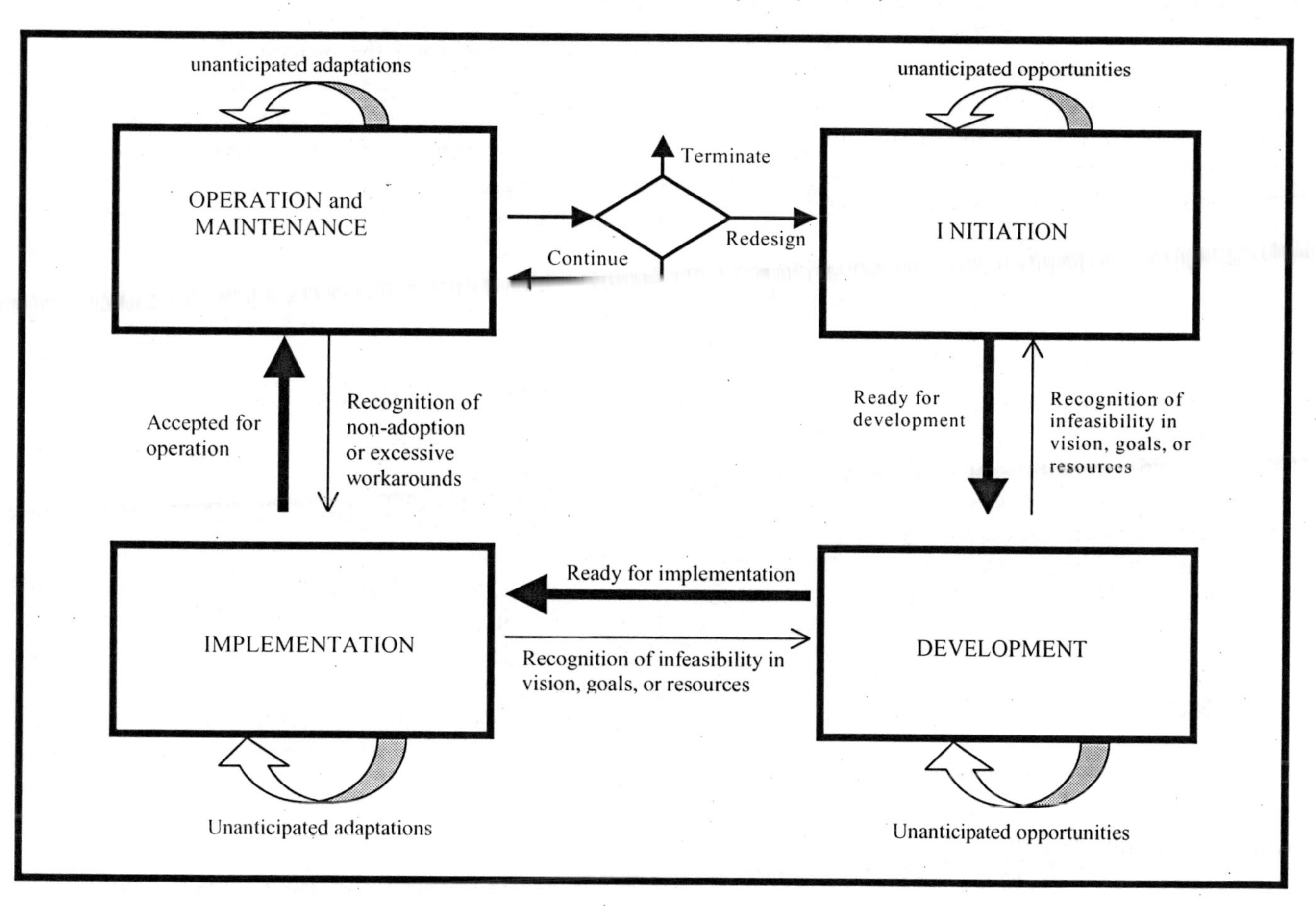

Table 7.1. Issues addressed by WSLC phases

Phase of the WSLC	*Typical issues addressed in this phase*
Initiation	• Can we agree on the purposes and goals of the proposed changes? • Are the proposed changes feasible economically, technically, and organizationally? • Are the changes unnecessarily elaborate and expensive?
Development	• Can we assure that the work system changes genuinely solve the problem? • Do the information system changes conform to the organization's expectations and standards for technical quality?
Implementation	• Can we convert effectively and painlessly from the old work system to the new work system? • Can we resolve personal and political issues related to changes in work patterns and power relationships?
Operation and maintenance	• Can we attain acceptable work system performance? • Can we maintain and improve performance with incremental changes, adjustments, and experimentation?

Operation and Maintenance

The discussion of the Work System Framework throughout most of this book is actually about work systems in operation, i.e., the operational aspects of the operation and maintenance phase. Chapter 3 explained that the work system method starts by identifying an existing (operational) work system and problems or opportunities related to that work system. The goal of the analysis is to determine how to improve that work system.

The maintenance aspects of the operation and maintenance phase involve monitoring the work system and making small adjustments and corrections to keep it operating effectively. This phase continues until the work system is terminated or until major changes are required. At that point, management allocates resources to initiate a project.

Initiation

The initiation phase is the process of clarifying the reasons for changing the work system, identifying the people and processes that will be affected, describing in general terms what the changes will entail, and allocating the time and other resources necessary to accomplish the change. This phase may occur in response to obvious problems, such as unavailable or incorrect information. It may be part of a planning process searching for innovations even if current systems pose no overt problems. When the work system involves software, errors and omissions in the initiation phase may result in software that seems to work on the computer but needs expensive retrofitting after initial attempts at implementation in the organization. Unless the initial investigation shows the project should be dropped, this phase concludes with a verbal or written agreement about the proposed work system's general function and scope, plus a shared understanding that it is economically justified and technically and organizationally feasible. This agreement might be general and informal, or might be quite specific in identifying budgets, time lines, and measurable objectives.

Key issues during the initiation phase include attaining agreement on the purpose and goals of the proposed change and assuring that the likely benefits far exceed the likely costs in terms of time and resources. The larger the project, the more desirable it is to document specific expectations along with a plan for accomplishing genuine results (as opposed to just performing specific activities at specific times). Regardless of how formal the agreement is, the details of the desired changes will be worked out in the development phase.

The initiation phase defines a new project and assures that it makes sense from different perspectives.

Business responsibilities. The main questions in the initiation phase involve summarizing the need for changes, describing the nature and scope of the desired changes, and summarizing the budget, resources, and desired approach for accomplishing those changes. Even when some of the changes will involve computerized parts of the system, business professionals need to be involved in the initiation process because they have the clearest view of business issues and the direct responsibility for organizational and economic success.

- They should be cautious about proposed changes that seem great in the abstract but may be difficult and expensive to implement due to the amount of organizational change required or due to political and cultural obstacles. Business professionals cannot expect the technical staff to anticipate and manage such issues.
- They need to watch out for techno-hype and implied promises that technology itself will solve operational, organizational, or political problems.
- They should remember that too many technology investments are based on management whim rather than on careful consideration of how the technology will be used to help generate value for internal or external customers.
- Consistent with their responsibility for work system performance, they should also make sure

that IT professionals participate actively in this phase to provide their expertise on what IT can and cannot do with differing levels of effort and expense.

Feasibility study. Work system changes often include changes in an information system and changes in other parts of the work system. In these situations, it is usually worthwhile to write a formal **functional specification** that outlines the goal in general terms and explains how changes in the work system and in the information system will lead to better work system performance. The feasibility study in the initiation phase is often more complicated if the issues about work system and information system changes are intermingled. The estimates of costs and benefits produced for these feasibility studies are often inaccurate because the development phase will clarify many details. Sometimes the development phase becomes a research project that finds problems never anticipated during the original feasibility study. For example, the mere fact that an information system is involved may broaden the scope of the inquiry if changes in the information system touch upon IT infrastructure that affects other work systems.

Development

The development phase is the process of specifying and creating or obtaining the tools, documentation, procedures, facilities, and other physical and informational resources needed before the desired changes can be implemented in the organization. This phase includes deciding how the work system will operate and specifying which parts of the work will and will not be computerized.

Most development projects involve changes related to software. Those changes may be accomplished through various combinations of programming new software, acquiring commercial software from software vendors, and developing prototypes that help in refining requirements. If new software is to be programmed, the development phase may include producing detailed specifications of how the software will operate and what the software users will see. If commercial software is to be used, that software must be configured for the current situation. In some cases additional programming will be required. Alternatively, if an iterative prototyping approach is used, representatives of work system participants will evaluate a series of prototypes to define the required functionality. In any of these cases, the IT staff will create and/or acquire the software, and will configure the software so that it operates properly and satisfies the requirements. Any new hardware will be acquired and installed, and the entire technical system of hardware and software will be tested.

Key issues in this phase revolve around creating or obtaining all required resources in a cost-effective manner and, if necessary, demonstrating that tools and procedures actually meet the requirements. Whether the work system will absorb or reject the desired changes is determined by the next phase.

The development phase for a work system involves much more than building software.

Completion of development does not mean "the system works." Rather, it only means that the tools, documentation, and procedural specifications have been produced, and that computerized parts of the work system operate consistent with those specifications. Whether or not the computerized parts of the work system operate adequately will be determined by how the entire work system operates in the organization.

Business and IT responsibilities. Inclusion of changes in the information system requires several sets of activities in the development phase. One set of activities is aimed at designing and codifying the work practice changes that will be implemented in the organization. Business professionals are largely responsible for these activities even though systems analysts who work in the IT group may be heavily involved. Another set of activities is aimed at improving the hardware and software in the information system. These changes are designed to support the desired work practices. Therefore the redesign of work practices must be completed before the information system changes can be specified in detail (unless successive prototypes are used).

Business professionals should be responsible for assuring that the IT professionals understand how work practices will change. They also should define or approve the user's view of the hardware, software, and user interface. Based on these specifications, the IT professionals will do the technical work of acquiring and installing any new hardware; acquiring, creating, or modifying software; and testing and documenting the hardware and software. If at all possible, business professionals should be involved in testing hardware and software features that users will see.

Four development models for information systems. The development phase for the information system can be handled in a number of very different ways that were mentioned in the previous paragraphs. It is useful to say a bit more about four idealized models of the development phase for information systems. Entire books have been written on each of these.

- ***Structured software development.*** The IT staff designs, programs, and tests the information system modifications based on detailed specifications created within the development phase. The work of the IT staff in the development phase is controlled by producing a sequence of deliverables that must be reviewed and signed off. The first deliverable is an external specification that explains exactly what screens and reports will look like to the user. This deliverable is produced through collaboration between business representatives and the IT staff. Next is an internal specification that provides a technical blueprint for the information system's internal structure. These deliverables provide detailed guidance for producing software and documentation. The software and documentation are produced and debugged. The final step is to test the system of hardware and software to make sure that it satisfies the requirements.
- ***Prototyping.*** Working in close cooperation with representatives of work system participants, programmers use a trial and error method to figure out how the information system should operate. They do this by creating, evaluating, and improving successive versions of a working model of the information system. When the prototyping effort has yielded the required insights, the prototype may be adopted for use, or the process may switch to structured software development based on the knowledge gained by prototyping. **Agile development** methods and various forms of **extreme programming**[52] are designed to obtain rapid feedback about a succession of prototypes or other partial versions of the information system.
- ***Commercial application package.*** Purchasing an application package from an outside vendor is the most common approach for acquiring software. This type of software is built to support a specific set of business processes, such as billing customers, accepting customer orders, or keeping track of inventory. Commercial application software has significant advantages and disadvantages. Purchasing vendor software saves a huge amount of time and may provide higher quality software than could be produced by the firm's IT group. On the other hand, the commercial software was developed to support work systems at other firms, and the business process models built into that software may not be a good fit for the needs that launched this project. It may be necessary to change work systems to fit the software rather than producing software to fit work system needs.
- ***End-user development.*** It may be possible to bypass programmers if an information system is to be very simple, and if end-users are willing and able to develop it using spreadsheets and simple database software. This approach is effective only when the situation is simple enough that the information system can be designed, programmed, and tested without the help of an IT professional. Even in these situations, the risk of mistakes is relatively high because non-IT professionals know little or nothing about how to debug and maintain software.

A key goal in the development phase is assuring that information system features genuinely solve problems that managers and work system participants want solved. This is sometimes difficult because work system participants may be unable to describe exactly how a new or modified information system should help them. They also may not see that a new information system could help them in some ways but might become a hindrance in other ways.

Another key goal is to perform the technical work in a way that makes it easier to modify the information system as new needs arise in the future.

Freezing requirements. IT professionals often face the challenge of trying to keep projects under control even though business needs change. To retain their own sanity and to make information system projects manageable, they often call for freezing software requirements so that they can do their work without aiming at a shifting target. Freezing requirements makes it possible to manage the project, but leaves the possibility that the information system that is developed will not meet the business needs that exist when it is implemented. In extreme cases, meeting the schedule becomes more important to developers than providing what their customers need. A common compromise is to attempt small, easily managed projects whose short-term deliverables can be produced quickly. This approach reduces the risk of wasting tremendous effort in a large project that builds the wrong information system.

All too often, business requirements change long after a project is underway.

Business professionals whose main exposure to computers involves spreadsheets and word processing often wonder why it takes IT professionals so long to produce what may seem like minor changes in an information system. They sometimes fail to realize that personal work creating spreadsheet calculations is quite different from development or maintenance of an information system designed for use over an extended period by an organization. Their personal work on spreadsheets involves calculations they control completely and may not need to explain to anyone else. A week or month later it may not matter if they cannot remember the precise logic of those calculations. In contrast, IT professionals building or maintaining software need to make the results of their work understandable to users and IT staff members months or years in the future.

As was mentioned earlier, completion of the development phase for the information system does not mean that a work system project is a success. Rather, it only means that the computerized parts of the work system operate on a computer. Whether or not the work system changes are successful will be determined later based on the implementation in the organization.

Implementation

The implementation phase is the process of making the desired changes operational in the organization. Implementation activities include planning the rollout, training work system participants, converting to the new work methods, and following-up to ensure the entire work system operates as it should. Ideally, the bulk of the work in this phase should occur after development is complete, implying that all tools and procedures are ready and that all software has been tested and operates correctly on the computer. The implementation phase ends when the updated work system operates effectively in the organization.

Conversion to the new system. Detailed planning for the conversion from the old way of doing things to the new is an important step in the implementation phase. The actual conversion to the new work system occurs after work system participants are trained. Conversion to the new system usually raises issues about how to move to different work practices with minimum disruption and how to deal with political questions and changes in power relationships. If a work system's development phase created or modified an information system, some parts of the conversion involve the changeover to the new or modified information system. Other parts of the conversion may involve changes in practices that are unrelated to the information system. When the conversion affects information and methods used for transaction processing, there may be some risk that the new work system will encounter unforeseen problems that jeopardize or prevent its successful operation. To minimize the risk of disrupting business operations, it is often necessary to perform the transaction work twice, once using the old work system and once using the new work system. With this approach, the old system will remain available as a fallback if the new system fails. Unfortunately, operating two systems in parallel often places substantial stress on the organization, which may already be overworked.

Importance of Change Management

Across the four phases, the success of planned changes is determined partially by the details of the changes and partially by perceptions of the development and implementation processes. It is possible for a work system change to fail organizationally even though a software change succeeded technically.

Overcoming inertia. The likelihood of success drops if the implementation effort cannot overcome the inertia of current work practices or if the implementation itself causes resistance. Involvement and commitment by participants and their managers based on benefits they will receive is often important for overcoming social inertia. Low levels of involvement and commitment make it more likely that the system will never reach its full potential or will fail altogether. If commitment is low, software that has been implemented somewhat successfully may be used for a while and then gradually abandoned, eventually making it seem that the project never happened. Unless a business problem is both evident and painful, overcoming inertia may absorb more time and energy than the computer-related parts of system development. A passage from Machiavelli's *The Prince*, written in 1513, is often cited in conjunction with system implementation: "It must be remembered that there is nothing more difficult to plan, more doubtful of success, nor more dangerous to manage than the creation of a new system. For the initiator has the enmity of the old institution and merely lukewarm defenders in those who would gain by the new ones."[53]

Resistance to change. Even with Herculean efforts to make the implementation successful, many systems encounter significant resistance from participants and outside stakeholders. This resistance may be motivated by a desire to help the organization, such as when someone believes a system change is a step backwards. Resistance may also have selfish or vindictive motives, such as when someone believes the new system will undermine personal ambitions or improve the prospects of personal rivals.

Resistance to change may take many different forms. Overt forms range from public debate about the merits of the system to outright sabotage through submission of incorrect information or other forms of conscious misuse. Between the extremes of public debate and sabotage are many less overt forms of resistance. A person resisting a system through *benign neglect* says nothing against it but takes no positive action to improve its chances of success. Resistance through *resource diversion* involves saying nothing against the system but diverting to other projects the resources it needs. Resistance through *inappropriate staffing* involves assigning people to the project who lack the background and authority to do a good job. Resistance through *problem expansion* involves trying to delay and confuse the project effort by claiming that other departments need to be included in the analysis because the system addresses problems related to their work.[54]

Many books and articles have been written about change management because it is a significant and complex topic. Although different authors use different terms, discussions of change management typically include the following components:

Vision. A planned change that will have significant consequences requires a clear vision that can be explained convincingly. There is an important distinction between a vision and a **slogan**. A charismatic leader might be able to deliver slogans such as "work smarter" or "just do it," but a vision with lasting impact needs to have enough depth that people can visualize how it will affect them, their colleagues, and the larger organization.

Planning. Change does not occur by wishful thinking. Major changes usually require careful planning related to roles and resources. A variety of roles are sometimes lumped under the general heading of **change agent**, a person who promotes change in an organization. A more detailed look at change agents often identifies a variety of roles, including executive sponsor, project manager, trainer, and manager or supervisor of people whose work practices are to be changed. The primary resources needed for most change are the time and energy of change agents and of work system participants. There is a common tendency to underestimate how much time and energy will be required for even relatively small changes that involve a large number of people.

Project success depends on many non-technical factors such as change management and effective collaboration between business and IT professionals.

Leadership. Executive sponsors and line managers who explain the vision and plan perform the most visible leadership roles, but leadership in successful change is often shared by change agents and work system participants themselves. The leadership from work system participants occurs in discussions with colleagues about the advantages and disadvantages of the proposed changes, and in honest feedback about important issues that might otherwise be repressed or ignored.

Communication. In change projects, communication is such a major success factor that "communication, communication, communication" is almost a mantra for change agents. Effective communication covers the project's rationale, its specifics, and its likely impacts on work system participants and others in the organization. Effective communication is a two-way street. People need to know what is coming, but they also need to express objections, confusions, and other concerns. That type of feedback is essential for assuring that implementation issues are identified and managed.

Training. Many change projects require carefully executed training on the details of new procedures, new information, and new computer interfaces. Ineffective training often results in confusion and errors during the conversion from the existing work system to the new work system. These problems absorb time and energy and can raise doubts about whether the work system changes make sense and will be successful.

Measurement. As with any project, the progress of change projects should be monitored and measured. Planned activities such as training sessions should be monitored for timeliness and effectiveness. The monitoring should include qualitative indicators and key incident reporting about how well the project and the changes are being accepted by work system participants and other stakeholders.

Perseverance. Changing the way people work requires perseverance. Apparent completion of the implementation phase may not guarantee that the intended changes will survive over time. Work system participants may revert to the previous way of doing their work if the rationale and details of new processes seem questionable. Therefore change management efforts should include ongoing monitoring and feedback about whether the desired changes have actually occurred, whether there are problems, and what is being done about those problems.

Collaborating throughout the WSLC

By focusing on work system issues rather than IT issues, both the Work System Framework and the work system life cycle (WSLC) model can serve as a common basis of communication that can support collaboration between business and IT professionals. Figures 7.2 through 7.5 summarize areas in which the WSLC can be used to support these collaborations.

Keeping the big picture in view. Figure 7.2 summarizes the activities in each phase of the WSLC. Effective collaboration between business and IT requires a shared vision of how current projects are related to the on-going life cycle of particular work systems. This shared vision is the basis of informed agreements about planned activities, milestones, key deliverables, willingness to incorporate unanticipated adaptations, and criteria for progressing to the next step or reworking a previous step. Key questions raised by Figure 7.2 include:

- Do we agree about which work system is being improved and why?
- Where are we in the life cycle of this work system?
- Are we able to move forward or should we to go back to rework prior assumptions and deliverables?
- What is our stance about unanticipated adaptations and opportunities for the work system, for the information systems that support it, and for projects attempting to improve both systems?

Agreeing about roles and responsibilities. Figure 7.3 represents a typical division of responsibilities between business and IT professionals. It shows that business professionals have important roles and responsibilities in each phase of the work system life cycle. Effective collaboration across a work system's life cycle requires clarification of those roles and responsibilities, especially since the IT group typically has no authority to manage or enforce changes in work practices in other departments.

Agreeing on performance indicators. Each phase in a work system life cycle should be monitored and measured. Figure 7.4 presents typical performance indicators that are often relevant within each phase. The performance indicators for the operation and maintenance phase are typical performance indicators for work systems in operation, such as efficiency, consistency, and speed. Those same indicators also apply to the initiation, development, and implementation phases because each is a work system on its own right. Figure 7.4 is designed to emphasize that a work system and an information system that supports it may have some indicators in common and others that are different. This is why the performance indicators for initiation, development, and implementation are grouped under the headings of general indicators, indicators for work systems, and indicators for information systems.

Avoiding common pitfalls. The sheer complexity of IT-related projects brings many unavoidable difficulties, but many pitfalls and risks in the work system life cycle can be named, discussed, and minimized. Figure 7.5 names a number of common pitfalls.

The ideas presented in these Figures can be used at different levels of detail. Just the names of the phases and the forward and backward arrows between the phases clarify the nature of the work system life cycle. The backward arrows emphasize that ineffective or incomplete work in one phase usually requires rework, regardless of what the schedule might say. Recognizing the range of activities within the four main phases can help business professionals appreciate why they have responsibilities in each phase, why producing or upgrading software is only part of the battle, and why no one should declare victory until the work system changes have been implemented and the work system is meeting its operational goals. Furthermore, identification of the many business and shared roles within those phases can help business professionals understand their responsibilities in projects related to IT-reliant work systems. At a more strategic level, CIOs and other IT executives can use this model to explain the overall role and function of an IT organization in both continuous and discontinuous change. They can also use it to negotiate about coordinating and sharing responsibilities throughout the work system life cycle.

Figure 7.2. Activities in each phase of the work system life cycle

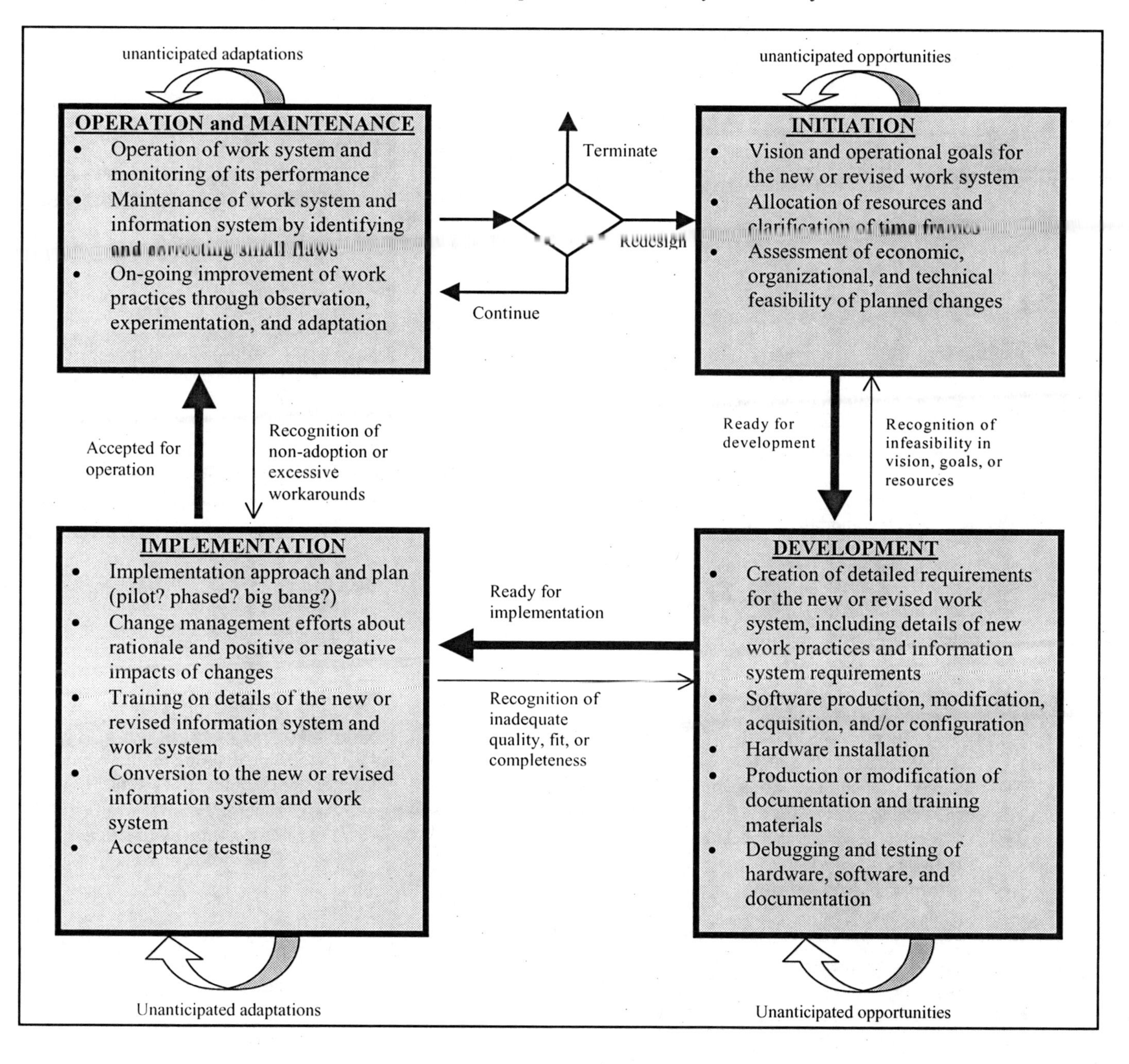

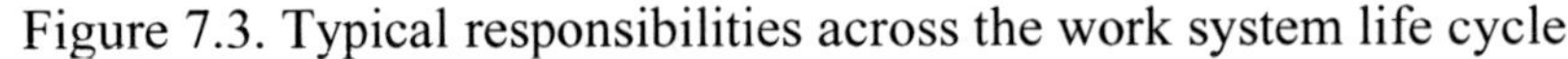

Figure 7.3. Typical responsibilities across the work system life cycle

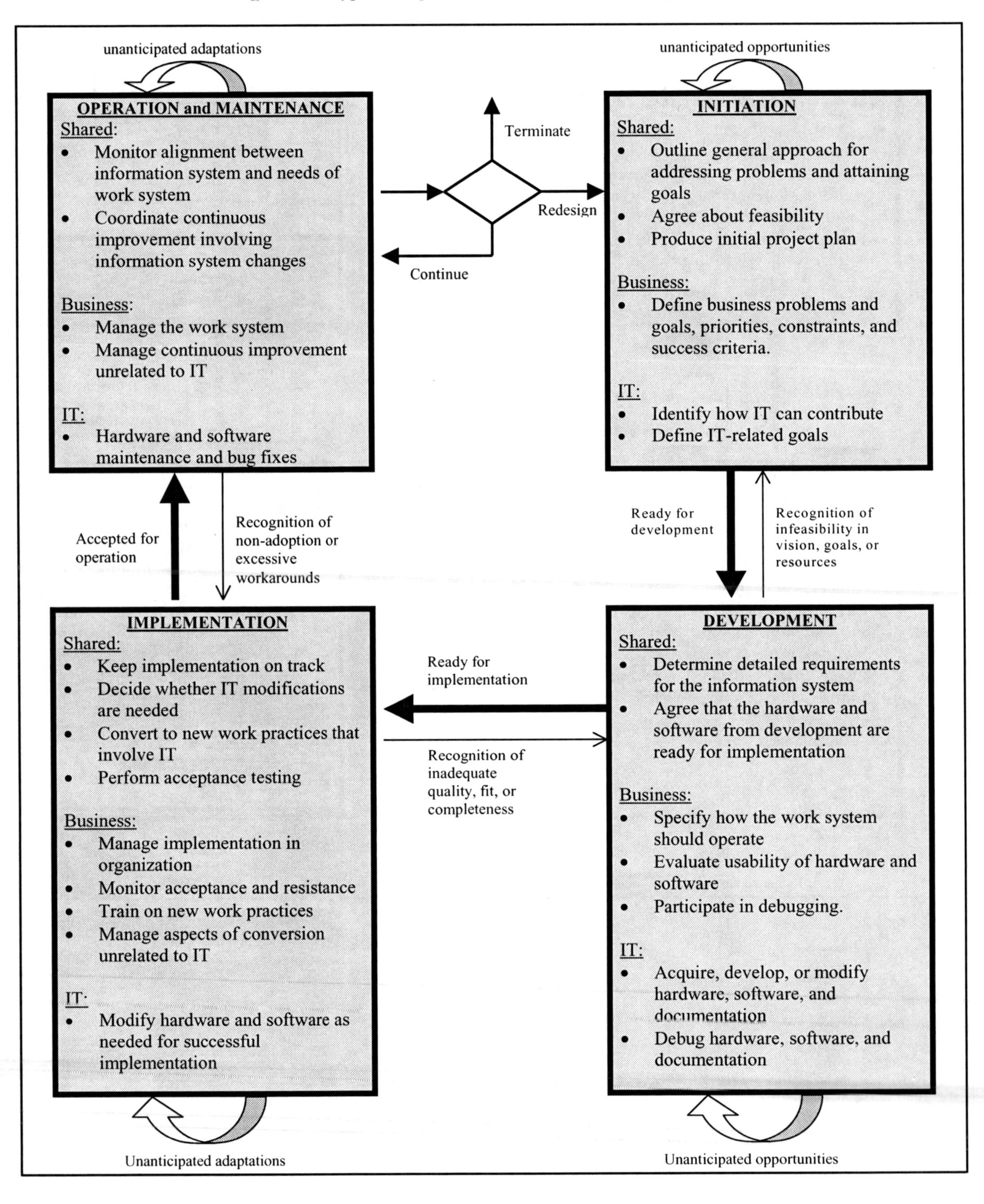

Figure 7.4. Performance indicators across the work system life cycle

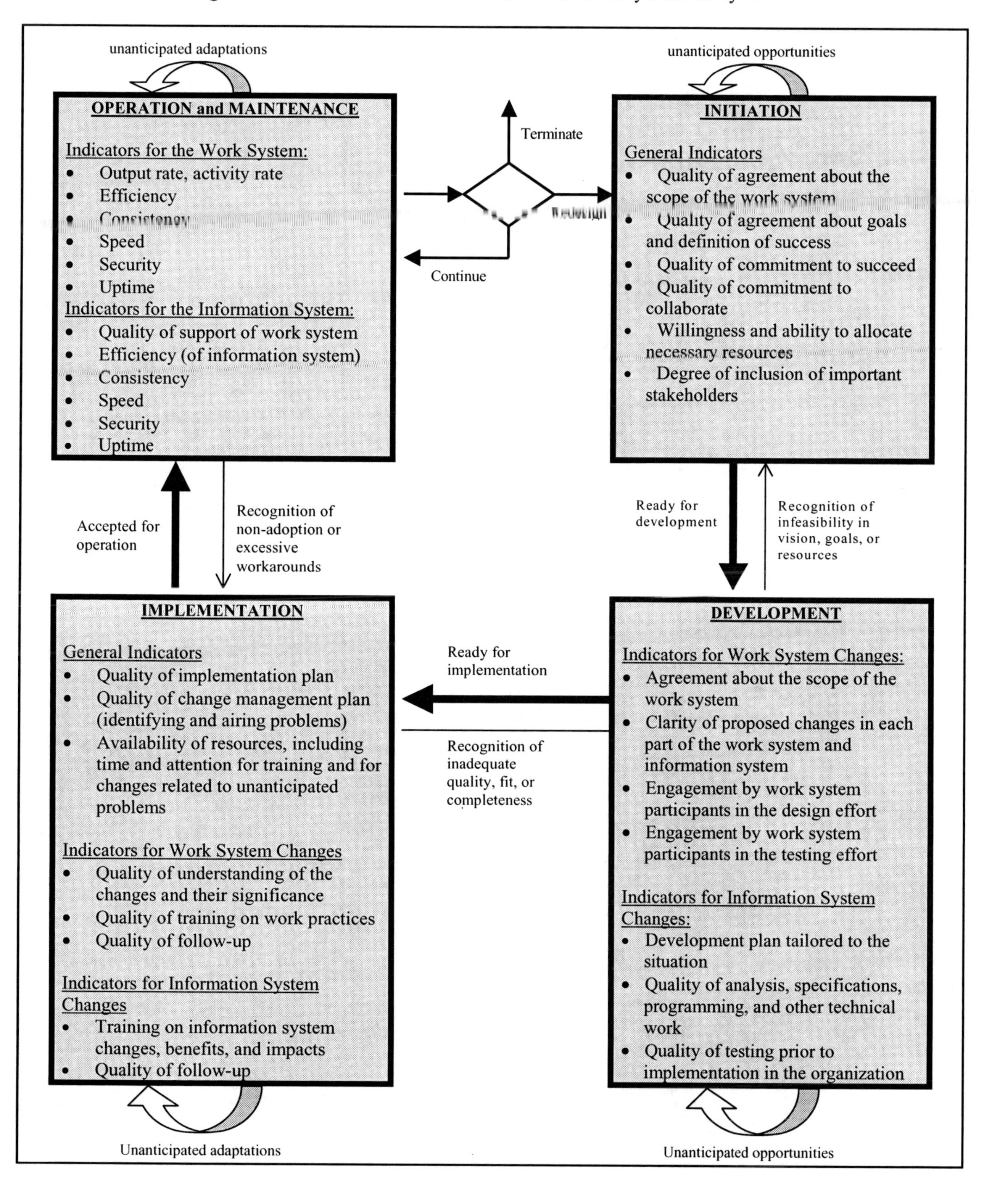

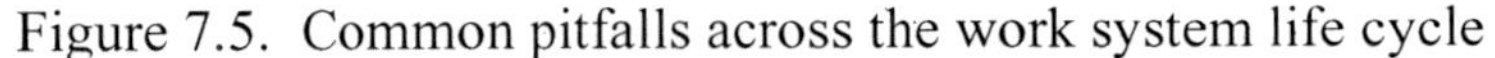

Figure 7.5. Common pitfalls across the work system life cycle

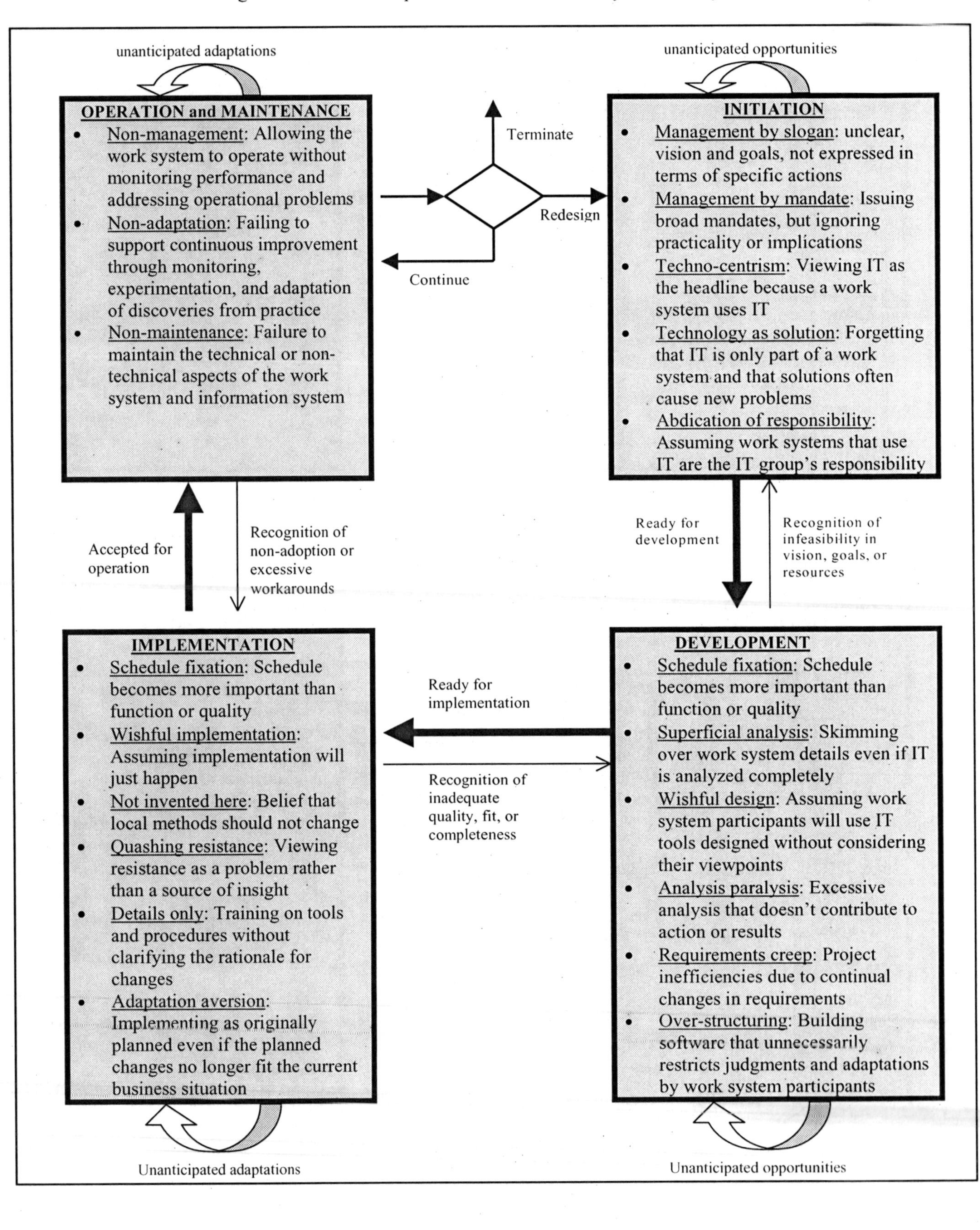

Chapter 8: Example Illustrating the Work System Method

Part I of this book consisted of Chapters 1-7, which introduced the basic ideas about work systems and the work system method. Part II goes into more depth with explanations and examples related to topics that have been summarized thus far but not explained fully.

This chapter presents an example that helps in visualizing how the work system method is applied. It also helps in recognizing the relevance of topics explained in Chapters 9-14, which are organized around work system elements. Three of those chapters are about specific elements (work practices, participants, and information) and three combine two elements (customers and products & services; technology and infrastructure; environment and strategies). Sections within those chapters provide brief discussions of components of each element, evaluation criteria, strategies, tradeoffs, and other topics. Many of those topics appear in the Level Three checklists mentioned in Chapters 3 and 5. Chapter 15 shows how the work system approach helps in identifying omissions from system- and IT-success stories and in interpreting jargon related to IT and systems. The Appendix shows how the example presented in this chapter might appear in a multi-level work system questionnaire.

The example. This chapter's example concerns a loan approval system in a large bank. It was constructed to illustrate the importance of many diverse issues that will be discussed in Chapters 9 through 14. To accomplish this goal, it combines material from a bank management textbook[55] and various aspects of case studies of repetitive decision making involving financial risk. No proprietary information is included, and all values of metrics are fabricated. Some of the issues are about IT, but others are about other aspects of work systems. The description of the situation looks at each work system element, starting with customers, products and services, and work practices.

IMPORTANT: This example was constructed to show the potential significance of a wide range of topics and issues, perhaps more than a typical real world situation would include. You may question the rationale for some aspects of the current situation, and you may or may not agree with the recommendations in the Appendix. That type of questioning is exactly what business and IT professionals need to do when they collaborate to create or improve work systems. The example illustrates the type of analysis and level of detail that would just be a starting point for producing a carefully designed and well-justified replacement for a current work system.

First-cut answers, not completed documentation. Following a brief introduction to the situation, the loan approval example is presented as a set of first cut answers to the Level Two questions identified in Chapter 3. Shown next to the answers are possible questions from a manager reviewing the incomplete analysis that has been done thus far. The goal here is not to show what a completed report would look like, but rather, to show how the initial analysis would raise new questions and would probably demonstrate the need for further clarifications.

Answers to the five SP questions for identifying the system and problem are followed by answers to the ten AP questions about the analysis and possibilities. The recommendation is not discussed here because we assume that the analysis is ongoing. The Appendix extends the example by providing tentative recommendations supported by five Level Three checklists.

Background about the situation. TransRegional Bank Group (TRBG) is a large regional bank providing a broad range of banking services to companies and individuals. TRBG's top management faces a difficult dilemma. Due to poor controls in the past, TRBG's current loan portfolio is in poor shape, with almost 50% more nonperforming loans than a typical bank of its size. (Nonperforming loans are loans whose debt payments are more than 90 days past due.) It also has a number of loans to companies that are likely to encounter serious financial difficulties if a predicted business slowdown occurs. Around 65% of TRBG's assets are in commercial loans. Substantial loan losses would reduce TRBG's assets and make it less able to repay the depositors who provided the money for those loans. The CEO and CFO fear that the FDIC (Federal Deposit Insurance Corporation), a federal agency that regulates banks, could begin to question the adequacy of TRBG's capital. The FDIC insures bank accounts up to $100,000. It inspects the quality of a bank's loan portfolio to make sure that the bank has enough high quality assets to return depositors' funds.

TRBG's top management is also under pressure to improve the bank's productivity and profitability, both of which are below the average for comparable banks. Unfortunately, improving the quality of TRBG's portfolio seems a bit contrary to increasing profitability. At least in the short term, granting more loans generates more interest payments and more loan fees. In the longer term, granting loans with higher risk increases the number of nonperforming loans and the number of loans that must be foreclosed at a loss. TRBG's main source of income is interest it receives when it loans money to commercial borrowers. Unless the risk of a loan seems high, a typical loan's interest rate is 1% more than the prime interest rate quoted by the largest money center banks to their largest and most stable clients. Those customers are too large for TRBG to handle. During 2005 the prime rate moved from 5.25% to 7%, substantially more than the low interest rates TRBG and other banks paid to depositors during 2005.

Management is aware of a variety of approaches for approving loans. On one extreme are largely interpersonal processes such as deliberations by loan committees. On the other extreme are largely automated processes based on statistical methods. Regardless of the details, any bank's method of approving a loan request should evaluate the 5 C's of credit: character, capital, capacity, conditions, and collateral.[56]

- Capacity: Does the borrower's cash flow provide enough money for both loan payments and ongoing business expenses?
- Capital: Is the borrower's capital at stake? A lender should loan funds only if the borrower would lose a substantial amount of capital if the enterprise failed.
- Collateral: Is there a secondary source of repayment in case the intended repayment method fails?
- Conditions: Are business and economic conditions favorable for the borrower and the borrower's industry?
- Character: Do the borrower and its officers have a reputation of honesty, experience, and integrity?

Evaluation of the 5 C's of credit for any particular loan is often rather subjective even though relatively standard analysis of financial and operational ratios can be used to evaluate a borrower's capacity to repay a loan. To minimize a bank's risk related to a specific loan, the loan agreement may include loan covenants, requirements that the borrower must remain within specific levels of business metrics such as current ratio, days of receivables outstanding, inventory turnover, and net worth. Other covenants might include limits related to capital outlays, additional debt, sales of existing assets, and other actions that might jeopardize the collateral.[57]

TRBG's top management commissioned studies of many areas of the bank. This study will look at TRBG's system of approving applications for commercial loans of over $100,000 to new clients. Other studies will look at increasing the profitability of existing clients and at TRBG's methods of finding new or existing clients who might be interested in obtaining a loan.

SP1: Identify the work system that is being analyzed.	
The work system is TRBG's system of creating and approving applications for commercial loans above $100,000 to new clients. The work system starts when a loan officer discusses loan needs with a prospective client. The approval process depends on the size of the loan, with the largest loans going to a loan committee or executive loan committee. The last steps in the system are communication of the approval or denial to the client, and production of loan documents. (See the work system snapshot in Figure 2.2.) New clients submit around 2100 loan applications per year, which yield around 1250 loan approvals and around 725 loans. Approximately 230 of the approved loans come from loan applications that were not approved in their original form. Note: All participants in this work system also devote substantial time to new loans and reevaluations of lines of credit for existing clients. The CEO and CFO decided that activities related to existing clients would be studied in a separate analysis.	*Does it make sense to do two separate studies if the same people are doing similar work for new clients and existing clients? Doesn't the amount of time spent on existing clients affect the time available for new clients?*
SP2: Identify the problems or opportunities related to this work system.	
TRBG's loan approval system is not generating the level of interest payments needed to meet the bank's profitability targets. Simultaneously, TRBG may need to adopt tighter loan approval methods than were used in the past because its loan portfolio has an excessive number of nonperforming or otherwise risky loans. In addition, TRBG's aggressive competitors are taking loan business that it might have taken itself.	*Is it possible to tighten credit standards and increase short-term profitability? These goals seem somewhat contradictory.*
• The current work system has produced a substandard loan portfolio for the bank. • The current system for processing and approving loan applications is too expensive because it absorbs too much time of too many employees.	
• Senior credit officers are extremely overloaded, especially after two of them quit and moved to other banks last month after complaining about overload for a year.	*How much time is too much time?*
• Credit analysts are frustrated by a combination of awkward technology, inexperience, frequent interruptions, and many starts and stops in their work as they gather information and clarifications from various sources.	
• Some of the market and financial projections provided by clients are questionable.	*What is done to verify client projections?*
• Some borrowers believe the approval process takes too long. A relatively simple loan takes a minimum of two weeks from the time of the loan officer's first conversation with the new client to the time when the client is	*What is the actual amount of time?*

notified of the approval. A more complex loan often takes a month or more, especially if an appraisal is required, if information is required from other sources, and if the application is completed just after the weekly meeting of one of the committees. • Some loan officers believe that the approval process is partly political and shows favoritism to certain loan officers. • Loan officers, senior credit officers, and loan committees believe that the loan write-ups they receive are cumbersome to use and that the current loan evaluation model provides inadequate guidance for making approval or denial decisions, for setting risk-adjusted interest rates, and for setting terms and conditions for the loan. • Some loan committee members believe they have too little time to review loan applications before loan committee meetings. • In some cases, senior credit officers and loan committees receive loan write-ups that are incomplete, wasting their time and causing delays. Management believes that some loan officers and credit analysts need more extensive training on the criteria for approving loans. • Some loan applicants complained that they were led to believe loans would be approved, only to find out that a senior credit officer or loan committee required unrealistic terms in loan covenants.	*Any benchmarks for the right amount of time?* *What do they expect? Could a better model produce a valid risk assessment?* *How often do they not have enough time?* *Who decides whether employees are adequately trained?* *What do these unrealistic loan covenants look like?*
SP3: Identify factors that contribute to the problems or opportunities. In the last five years, TRBG's credit standards were lax, resulting in a substandard loan portfolio. Approving riskier loans boosted TRBG's interest income during those years, but TRBG's current loan portfolio contains 50% more nonperforming loans than a typical bank of its size. Other existing loans have become riskier due to potentially adverse competitive conditions for certain borrowers. A significant economic downturn could lead to regulatory action by the FDIC.	
SP4: Identify constraints that limit the feasible range of recommendations. All loans must conform to the TRBG's lending procedures and must satisfy TRBG's creditworthiness standards. In the past these constraints were sometimes overridden, but TRBG's top management is committed to enforce lending procedures and creditworthiness standards in the future.	*Why were the constraints overridden? Will the new system be overridden as well?*
SP5: Summarize the work system using a work system snapshot. (See Figure 2.2)	*Does Figure 2.2 leave out anything that is important?*

AP1: Who are the work system's customers and what are their concerns related to the work system?	
• The **loan applicant** is a client firm that applies for the loan and receives an approval or denial. Approximately 70% of TRBG's loan applicants want real estate loans. Other loans fall into several groups. Manufacturers may need to finance inventories or upgrade facilities; software companies may encounter cash shortfalls between new software releases.	*Do different industries need different types of loans? If so, should the system be different for different industries or different loan types?*
• **Loan officers** receive bonuses based on the total revenue they bring to the bank, including revenue from new clients and existing clients. Loan officers are customers of the work system because they have a direct stake in the outcome of the approval decisions. Some loan officers believe that they suffered financial losses due to personal or political bias against them.	*Do loan officers misrepresent clients? Any evidence?*
• TRBG's **Risk Management Department** can be considered a customer because it is responsible for the quality of TRBG's loan portfolio. Each approval or denial decision has an impact on the loan portfolio. Four years ago, the former head of the Risk Management Department complained that lending standards were not being enforced consistently.	*When and why were the lending standards overridden?*
• The **Federal Deposit Insurance Corporation** (FDIC) is the regulatory body that oversees the bank. One of its main mandates is to insure that bank deposits are safe. It rates banks based on six categories of performance whose acronym is CAMELS: capital adequacy, asset quality, management quality, earnings, liquidity, and sensitivity to market risk.[58]	*Will FDIC issues have any direct impact on the design of the new work system?*
AP2: How good are the products and services produced by the work system?	
• The **loan application** sometimes contains errors or small misrepresentations. Only two loan applications in the last three years were erroneous enough to be viewed as fraudulent.	
• The **loan write-ups** produced by credit analysts have varied quality. Several senior credit officers complained that the calculations of the borrower's ability to make loan payments were based on incorrect assumptions in at least 20% of write-ups by inexperienced credit analysts.	*What are typical examples of incorrect assumptions? Why aren't they in a book of guidelines?*
• A typical loan write-up is 15 pages long, plus attachments that include past financial statements. Write-ups for complex loans can be twice as long. Senior credit officers and loan committees find these lengthy documents unwieldy and would prefer more direct guidance based on the calculation results of a model that reflects the loan repayment experience for firms comparable to the applicant.	
• The **loan decisions** (approval or denial of the loan applications) of the last five years maintained short-term profitability through income from loan interest and loan fees, but those decisions led to a substandard loan portfolio.	
• The **explanation of the decision** includes whether or not the loan application was approved, and if it was approved, what are the loan fees and	

interest rates. For approved applications, approximately 30% of applicants complained that the fees or interest rates were too high, or that excessively stringent loan covenants made the loan impractical to pursue. For applications that received denials or especially stringent loan covenants, approximately 25% of the applicants complained that the explanation for the denial or the stringent covenants was inadequate.

What does an adequate or inadequate explanation look like?

- The **loan documents** produced at the end of the process seem to be consistent and understandable. No significant complaints related to loan documents were noted in the last three years.

Several recent surveys of existing clients revealed a number of areas of satisfaction and dissatisfaction.

- Around 50% believed that TRBG's loan terms are highly competitive; 25% believed that the other terms and conditions for TRBG's loans are not only competitive, but also a bit loose.
- 30% complained that loan approvals take took long. 25% believed that the application itself was too long and contained some improper inquiries.
- 60% appreciated help provided by the loan officer and credit analyst in producing the loan application document and compiling supporting materials, such as financial statements and tax forms.

How can we be seen as loose if we have tight standards?

Which inquiries are improper?

A separate focus group of loan applicants identified the following priorities for what matters to them. Most important are the interest rate and the terms and conditions for loans; next comes the amount of time and effort required for the application; third is the speed of approval.

AP3: How good are the work practices inside the work system?

Almost every step in the business process in the TRBG case could be improved in some way:

- **Loan officer identifies businesses that might need a commercial loan**. (Finding clients is the subject of a separate analysis.)
- **Loan officer and client discuss the client's financing needs and discuss possible terms of the proposed loan**. All loan officers can discuss typical loan fees and interest rates, but some of them do not have enough knowledge of credit analysis to identify likely problems before bringing a potential loan to a senior credit officer or credit analyst. Initial discussions with clients are also hampered by the loan officer's inability to provide accurate guidance about the risk adjusted interest rate and about possible loan covenants that may be imposed as a condition of granting the loan.
- **Loan officer helps client compile a loan application including financial history and projections**. The loan application specifies the reason for the loan, the amount requested, and the intended source of loan payments.

Is the profitability problem basically that we don't have enough prospects?

Why can't we give clients better information the first time around?

Rough guidelines exist for compiling supporting material. Several loan officers believe that clearer guidelines including checklists and identification of typical confusions and stumbling blocks could increase the efficiency of this step. Senior credit officers complain that some loan officers come to them with prospective loans that will obviously fail the credit analysis. With senior credit officers so overloaded, perhaps the loan officers should discuss potential loan situations with credit analysts before producing an application.

Putting together clear guidelines seems a no-brainer. Are our guidelines really that bad?

- **Loan officer presents the application to senior credit officer**. The senior credit officer checks the application for likely errors and either sends it to a credit analyst for a complete analysis or asks the loan officer to resubmit an application that has different terms (different collateral, loan amount, repayment date, and so on) and therefore is less risky. Approximately 80% of the preliminary applications are forwarded directly to a credit analyst. The other 20% need corrections or additional work before being forwarded.

- **Credit analyst prepares a "loan write-up."** The loan write-up summarizes the applicant's financial history, provides projections explaining sources of funds for loan payments, and discusses market conditions and the applicant's reputation. Each loan is evaluated using standard financial ratios[59] and is ranked for riskiness based on the applicant's financial history, projections, reputation, and the quality of the collateral that is offered. (For example, a loan that is 100% covered by a deposit in a bank account would be rated 1 on a risk scale from 1 to 10. A loan on undeveloped property in a remote area might be rated 10.) In the first phase of the analysis, the credit analyst uses data from the last three years of tax returns or audited financial statements to fill in a standard spreadsheet that reveals three-year trends related to the applicant's ability to meet payments. The credit analyst orders an appraisal for all real estate loans, and often speaks to the applicant's CPAs, lawyers, or other sources of information needed to evaluate the loan request. A simple loan write-up takes around a day. A complex write-up can take a week. The senior credit officer who assigned the loan to a credit analyst checks progress several times if the credit analyst is inexperienced.

How can we validate the information from loan applicants?

How often do credit analysts make serious mistakes that lead to bad decisions?

How does this compare to credit analysts in other banks?

- **Loan officer presents the loan write-up to a senior credit officer or loan committee**. There are two loan committees and one executive loan committee. Most members of the loan committees believe that presentations from more experienced loan officers have more credibility than presentations from less experienced loan officers. Some believe that the loan officer's ability to make effective presentations sometimes has an inordinate impact on the decision. Certain past approvals that might have been denials for less persuasive loan officers are now viewed as unfortunate decisions. Members of the loan committees complain that they receive the loan write-up just one day before the meeting. If that day is busy, they don't have enough time to review the write-up and identify issues that should be discussed. The loan committees receive no direct feedback about whether they approved or denied loans inappropriately.

Is the point about persuasiveness realistic?

Would the committee members do anything different if they had information earlier?

- The two loan committees and the executive loan committee each meet once a week, on Tuesday, Wednesday, and Thursday afternoon, respectively. An average session makes 5 to 8 decisions in up to three hours. Some of those decisions are related to new clients (the topic of this analysis) and some are related to new loans or lines of credit for existing clients (the topic of a

Is there any practical way to provide feedback, such as some kind of

different analysis). Applications for smaller loans and loans with few creditworthiness issues may take only a few minutes. The largest loans and loans with complex creditworthiness issues may take over an hour. Loan committee members have a variety of views about how to reorganize the committees to increase effectiveness. Some believe that the committees should be shuffled every three months to minimize personal alliances within the committees; others believe the committees should specialize by industry to the extent possible; others believe that a special committee should be created to deal with new clients.

risk rating on loans that were approved or denied?

What are the politics here? Is there a deeper rationale behind the suggestions?

- **Senior credit officers or loan committees make the loan approval decisions.** Senior credit officers are authorized to approve or deny loans of less than $400,000; the two loan committees approve loans up to $2 million. The executive loan committee approves larger loans, plus some loans smaller than $2 million if the other loan committees are overbooked. The current guidelines for assigning loan decisions to senior credit officers or loan committees have been in place for two years. During several previous years, loan officers had discretion to grant loans up to $200,000 after analyzing a loan write-up with a credit analyst. The procedure was changed after two loan officers and a credit analyst were fired following discovery of discrepancies in some of the information in loan write-ups.

What is the real value added by the committees? Do they really accomplish anything?

- Neither the senior credit officers nor the loan committees are satisfied with the guidance provided by the current loan write-ups. Ideally they would like to have guidance in the form of verifiable, history-based probabilities that the loan will or will not become a problem. Also, they believe that the current risk-adjustments to loan interest rates are nothing more than guesses. They would like to have a more scientific basis for these adjustments.

Would the decisions be as good if 90% of them were made by a computer based on repayment data for comparable loans across the U.S.?

- New clients often apply to several different banks in the hope that the competition between the banks will result in better terms and conditions. The senior loan officers and loan committees are often confronted with the decision of whether to stay within the guidelines and deny the loan, or to stretch the guidelines and approve a loan for a firm that may become an excellent client. These decisions are especially difficult for startup companies whose brief track record may not be a good indication of their future prospects. They are also difficult when borrowers are having current market difficulties or are currently changing their business strategy.

To what extent did the attempt to beat the competition with aggressive lending get us into the current mess?

- **Loan officers may appeal a loan denial or an approval with extremely stringent loan covenants.** Depending on the size of the loan, the appeal may go to a committee of senior credit officers, or to a loan committee other than the one that made the original decision. Some loan committee members believe the appeal decisions are somewhat political, and that certain loan officers tend to get better results on appeals to a particular committee. Some loan committee members believe that certain loan officers just won't give up and that they appeal even when the loan write-up shows undue risks. Around 85 loan decisions are appealed each year.

How often does one loan committee override another? Any politics here?

- **Loan officer informs loan applicant of the decision.** The loan officer conveys the decision to the applicant. Loan officers who convey denials or approvals with extremely stringent covenants are sometimes concerned that they do not have a clear reason for the decision. The denial letter generated by the Loan Administration Department says nothing specific about the rationale for the denial.

- **Loan administration clerk produces loan documents for approved loans that the client accepts.** (No issues identified.)

AP4: How well are the roles, knowledge, and interests of work system participants matched to the work system's design and goals?

The work system in the TRBG case includes a number of participants and other stakeholders who play different roles and have different concerns:

- **Credit analysts** compile information needed to evaluate a loan application. Credit analysts with no prior experience in commercial banking usually need about six months to learn to do their job accurately and thoughtfully. Even credit analysts with one or two years of experience sometimes make significant errors. In addition to compiling information from sources including financial statements and/or tax returns, they may obtain additional information from real estate appraisers and from the applicant's CPA and lawyers.
- **Senior credit officers** typically have at least seven years of significant experience in commercial banking. They have authority to approve or deny loans of less than $400,000. Counterbalancing the loan officers' goal of maximizing loan volume, the senior credit officers' goal is to make sure that all loans meet TRBG's criteria for creditworthiness. Loan officers sometimes view senior credit officers as rivals because their goals are so different. Senior credit officers are extremely overloaded, especially after two of them quit and moved to other banks last month after complaining about overload for a year.

Is there policy guidance about these tradeoffs?

When are we going to replace the two who quit?

- **Loan committees** make the approval or denial decisions for loans of over $400,000. There are two loan committees and one executive loan committee that evaluates loans over $2,000,000. Each loan committee consists of five people including two senior loan officers, a regional or divisional CFO, and two other executives. The credit analyst who produced the loan write-up attends the discussion of each loan, sometimes by phone. All three loans committees have had some internal conflict related to the stringency of credit standards, especially since the executive members of the committees are under pressure to increase TRBG's profitability and productivity.

Do the loan committees or their members have contradictory personal agendas?

- **Loan administration clerk.** Several clerks in loan administration produce the loan documents. The templates for producing loan documents are well designed. It usually takes an office assistant several weeks to learn how to use those templates and other templates encountered in their job.

AP5: How might better information or knowledge help?

This is vague. Exactly what nonfinancial information is included? Is it the

The loan application includes income statements and balance sheets for the last three years (for firms that are audited), annual federal tax returns for three years, and information about current accounts receivable and accounts payable.

It also includes information about the company's background, management, products, market, plans, and the reason for the loan. Selected information is compiled in a spreadsheet and later in a loan write-up that averages 15 pages for a simple loan. The senior loan officers and loan committees believe they have plenty of information. The difficult question is how to make sense of that information and use information to make better decisions. They believe that the existing loan evaluation spreadsheet does not provide adequate guidance. They learned of the three companies that claim to have proprietary loan evaluation software that has been calibrated statistically using many thousands of past loans. Each of these companies claims that their software would reduce the number of incorrect approvals by 50%, and would also reduce the number of incorrect denials. Their models might be used several ways. They could be used to provide probabilities or other guidelines evaluating specific loans. The loan committees would still make the decisions based on judgment and negotiation. The models could also be used in a relatively automatic mode, with the models making tentative decisions that would be overridden only in unusual cases in which factors not included in the model might have a major impact. In this type of use, deviations from the model's proposed decision would be recorded for subsequent review. Some loan committee members believe that the information concerning current market conditions, competitive conditions, and industry growth rates should be improved in order to support better decisions for startup firms and firms in unfamiliar industries.	*same for all loans or are some groups of loans handled differently because they are bigger or are in a different industry?* *Is there any proof that these models would lead to better decisions than the current spreadsheet?* *Could we test those models against some of our bad loan situations to see whether they would have warned us to stay away?* *How would they use this information for making better decisions?*
AP6: How might better technology help? The work system uses technology in three major ways. • Credit analysts use an awkward combination of three tools to produce loan write-ups. They start by entering information from financial statements, tax returns, and other sources into a complex spreadsheet that is very awkward to use. The process of entering the information is completely manual and highly error prone. Ratios and other information calculated by the spreadsheet are entered into a loan evaluation model, which generates information related to loan risk. Information from both the spreadsheet and the loan evaluation model are copied manually into a loan write-up template in Microsoft Word. Subsequent corrections in either of the two earlier steps may not be reflected properly in a revised write-up. • Communication technology such as telephones, e-mail, and the Internet is used for contacting clients and transmitting information between loan officers, clients, and headquarters. TRBG's communication and office technology are due for an upgrade, but shortcomings in these areas do not have a major impact on this work system. • Typical office technology and databases are used to store, retrieve, and print information related to clients, the existing loan portfolio, loan applications, and loan write-ups.	*Why haven't we integrated these steps to reduce errors?* *Any problems with communication?*

<table>
<tr><td>Neither the senior credit officers nor the loan committees are satisfied with the guidance provided by the loan write-ups. Ideally they would like to have guidance in the form of verifiable probabilities that the loan would not become a problem. Also, they believe that the risk-adjustments to interest rates are nothing more than a guess. They would like to have a more scientific basis for these adjustments.

Better models could lead to better decisions by taking advantage of TRBG's existing payment history information for past loans. The loan information would be loaded into a special analysis database and then analyzed using one of the three commercially available models or using a new model that an experienced credit analyst in the bank proposes building. The resulting model based on TRBG's past experience might generate decision guidance that is consistent and less prone to oversights or judgment errors.</td><td>Is it possible to build or buy a model that would generate good probabilities?

How accurate and complete are our payment histories? How far back are they computerized?</td></tr>
<tr><td>AP7: How well does the work system fit the surrounding environment?

TRBG operates in a highly competitive environment, with many banks going after the same prospects. Most prospects recognize the competition and submit loan applications to several different banks in the hope that competition will result in better terms and conditions.

Some of TRBG's competitors have been especially aggressive in reducing fees and interest rates for new clients. Some of those banks are fortunate to have strong loan portfolios with very few problem loans. Consequently, they are more able to go after slightly risky loans that TRBG might have to deny because of its weak portfolio.

A recent letter from the FDIC indicated concerns about two issues. First, the quality of TRBG's loan portfolio is poor. (TRBG's Risk Management Department is quite concerned about this problem, and believes that inadequacies in the loan approval system may undermine TRBG's solvency.) Second, the FDIC may investigate several complaints about inequitable treatment of applications from companies owned by people in minority groups.

The bank's technology strategy is to use the cheapest possible software and to avoid spending money to provide better tools.</td><td>How much are we being undercut by our competitors? Is it possible we could spend a lot of money on a new system and still be undercut?

Is there any evidence about inequitable treatment?

We may need to change this strategy.</td></tr>
<tr><td>AP8: How well does the work system use the available infrastructure?

The work system makes some use of TRBG's technical and human infrastructure. The technical infrastructure includes communication technology, office technology, and databases that are shared across the bank. Technical infrastructure seems not to be a constraint for the work system even though TRBG's technical infrastructure is a bit outdated.</td><td></td></tr>
</table>

AP9: How appropriate is the work system's strategy?	
The work system does not appear to have a clearly articulated strategy. Current strategy choices include:	
How automated should decisions be? Currently, a standard spreadsheet used by the credit analyst generates ratios and comparisons that provide loose guidance for making decisions. However, that guidance is sometimes overridden based on judgment and competitive considerations. A more automated approach might enforce closer adherence to a model's guidance, but it might also make it less likely that judgments related to unique circumstances of particular companies would play an important role in loan decisions.	*How would the committees respond to a highly automated process?* *Any reason to believe automated decisions would make fewer errors?*
• ***What is the practical meaning of following creditworthiness criteria?*** The current work system allows competitive pressures to override creditworthiness criteria in some situations. The current loan portfolio is unacceptably risky, but TRBG is under pressure to increase profitability, and short-term profitability comes from making more loans.	*How often are the creditworthiness criteria overridden? Are there any repetitive situations in which this occurs?*
• ***What can be done two increase the work system's efficiency and speed?*** Various aspects of the current work system seem to ignore efficiency issues. Also, the lack of information several days prior to the committee meetings may make those meetings less efficient.	
• ***Are the loan committees deployed appropriately?*** The three loan committees all handle any type of loan that comes to them. Each committee meets once a week, typically for up to three hours. Given the high salaries of people on those committees, the meetings are very expensive. Assigning a particular size or type of loan to each committee might improve efficiency and decision-making. Changing the schedule of the committee meetings might also be beneficial.	*What is the rationale for the current size and schedule of the loan committees? Could we get away with 2 or 3 person committees?*
AP10: How well does the work system operate as a whole and in relation to other work systems?	
• The work system as a whole isn't producing the quantity and quality of new loans that TRBG needs. Some combination of the changes related to different aspects of the system needs to address these problems.	
• When the salaries of employees are considered, the time devoted to a loan decision costs between $800 and $3000, even with the clients paying for real estate appraisals. Reducing the amount of time devoted to making these decisions would reduce costs.	

Chapter 9: Customers and Products & Services

- The Work System's Products and Services
- Work System Customers and Their Concerns
- Performance Indicators for Products & Services

The TRBG example in the previous chapter provides three reminders about the importance of products and services when analyzing a work system. First and foremost, the analysis and recommendation should pay attention to what the work system's customers want, not just how the work system operates internally. Second, the work practices within the work system usually affect the customer's satisfaction with the products and services it produces. (Work practices will be discussed in the next chapter.) Third, the outcome of the analysis will depend on which issues are considered and on the quality and completeness of the information that is gathered about the situation. Superficial analysis tends to produce superficial recommendations that ignore important issues and may even ignore possible showstoppers such as unwillingness or inability to change.

This chapter covers the following topics related to concerns of the work system's customers and the products and services the work system produces for them:

- ***What are the work system's products and services?*** The loan approval system's most direct product is the approval or denial of a loan application. Additional products include an explanation of the decision, loan documents, and information for the FDIC. The work system also produces services related to helping the applicant produce the loan application document and compile supporting materials, such as financial statements and tax forms.

- ***Who are the work system's customers and what are their concerns?*** In the TRBG case, the loan applicants come from at least five industry groups, possibly with different concerns. The loan officers are also work system customers because their bonuses depend on loan approvals produced by the work system. The FDIC is a secondary customer because it receives information generated by the loan approval system. TRBG's Risk Management Department and top management can also be viewed as secondary customers because faulty approvals undermine TRBG's loan portfolio, and faulty denials reduce profitability.

- ***What are the performance indicators and metrics for evaluating the products and services?*** The simplest performance indicators include speed, customer satisfaction, and short-term profitability. Metrics for each of these would be relatively easy to define. More complex performance indicators (for which metrics might be difficult to define and monitor) concern the extent to which loans are approved that should have been denied, and vice versa. A long-term performance indicator is the quality of TRBG's loan portfolio.

- ***What is the work system's strategy related to customers, and how might it change?*** Different groups of customers were discussed, but nothing was said about a focused work system strategy related to customers or about potential changes in the customer base.

- ***What is the work system's strategy related to products and services, and how might it change?*** Applicants receive help putting together their loan applications, but otherwise there seems to be no strategy for this work system's products and services. However, there are a number of shortcomings that could be addressed by future changes.

To minimize repetition, the chapter is divided into three sections that focus on the first three questions. Issues related to strategies and possible changes are incorporated into those three sections.

The Work System's Products and Services

This chapter extends the brief clarifications about products and services in Chapter 4, which included the following points:

- Products and services can be viewed as a combination of physical things, services, and information.
- A work system's products and services may be directed toward internal and/or external customers.
- Products and services are the outputs of a work system, but not the measurement of those outputs. In other words, a work system may produce a particular product that a customer likes, but it does produce customer satisfaction. Customer satisfaction is a performance indicator related to the customer's evaluation of the product.
- In addition to primary products and services, work systems may also produce valuable by-products; they also produce waste that should be minimized.
- Important aspects of some products and services may be viewed as intangibles that are valued but aren't described well as information, physical things, or services. For example, a work system that produces entertainment might generate enjoyment or relaxation; similarly, a work system of providing psychotherapy might generate enhanced feelings of well-being.

Content of a work system's products and services. A straightforward way to describe products and services and to think about improvement possibilities is to view the work system's products and services as a combination of:

- information content (information produced)
- physical content (physical things produced)
- service content (services provided to customers)
- intangibles (things that are valued but aren't described well as physical things, services, or information)
- by-products (secondary products)
- waste (which might be reduced).

The products and services produced by most work systems include a combination of informational, physical, and service content. Even products that are sometimes called physical products or information products may combine other types of content. For example, an automobile is a physical product because most of its value comes from its physical features even though some of its value comes from service arrangements and warranties. Other aspects of the car's value involve information features, such as dashboard displays, manuals, and information systems that track a car's location and link to emergency services. In contrast, billing systems produce information products such as telephone bills. The information itself may be delivered electronically or as a paper document, but its primary value is in the information.

One possible direction for improving an information product is to change its information content by providing better information. Other possible directions involve changing the physical content by eliminating the paper or changing the service content by customizing the delivery time, form, or information content of the bill. In the TRBG case, the main products are information products including the decision, the explanation, the loan documents, and information for the FDIC. The documents may be in paper or electronic form, but the decision and explanation would probably be delivered verbally rather than just in the form of a document. Services in the TRBG case include help in producing the loan application document and compiling supporting materials such as financial statements and tax forms.

The incredible progress of information technology has led to many new possibilities for different mixes of physical features, information, and service features, even in information products. Information can be printed on paper, distributed on CD-ROMs or DVDs, or transmitted through the Internet or via other channels. Tickets or confirmations can take a solely electronic form or can take a physical form such as traditional tickets or documents printed on a personal computer. The value of physical products can be augmented through customization based on requirements defined with the help of an expert or entered using a keyboard without ever speaking to anyone.

Commodity vs. customized products and services. A pure commodity product or service is not customized at all, meaning that every customer receives a roughly identical instance of the same thing, such as a box of cornflakes or a ride on the 8:02 AM bus to downtown. Commodity products and services are generated to the producer's specifications, not to the customer's specifications. Off-the-shelf products that are sold using bar codes are usually commodity products because each item with the same bar code is produced to be identical.

In contrast, customized products or services are tailored to the specific needs or requests of individual customers or groups of customers. Customers may or may not prefer customized versions. For example, most customers probably would not exert additional effort to obtain a customized box of corn flakes. On the other hand, many customers prefer customized versions of products such as clothes and cars.

There are many possible degrees of customization. The simplest form is based on standard product options that the customer can select, such as the way Dell Computer allows its customers to choose among different types of processors, hard disks, and other components. Similarly, a new housing development might allow each homeowner to select one of three floor plans and one of five color schemes. More highly customized products and services allow the customer more types of options and more ways of expressing those options. The most customized products and services such as custom-made jewelry and custom-designed homes may be entirely unique. Customization often complicates production systems, as is clear from the distinction between made-to-order clothes versus off-the-rack clothes that are produced in advance in large quantities. As the degree of customization increases, an increasing amount of the value of the product resides in the customization. As customization of a product increases, customer-oriented services become a more important part of the value the customer receives.

Controllability and adaptability by customers. Customization involves the producer's efforts for a specific customer before the customer receives the product. Controllability and adaptability are about what the customer can do with the product after receiving it. For many products and services, at least part of the fit is based on forms of controllability or adaptability such as:

- Customer selects options when installing the product (for example, selecting security or display options when installing a PC).
- Customer controls the product interactively during use (for example, performing an Internet search using a PC).
- Customer adapts or programs the product for future usage (for example, programming a television recorder to record a program next Thursday).
- Product senses information and responds automatically (for example, a home thermostat senses the temperature and triggers heating or cooling when the temperature is too high or low).

These examples illustrate that customization, controllability, and adaptability are a range of approaches for increasing a product or service's fit to the requirements and desires of individual customers. Each approach is a possible direction for improving customer satisfaction.

Special characteristics of services. The distinction between products and services is surprisingly fuzzy, but is still useful for thinking about how well work systems operate and how to improve them. Consider examples and counterexamples related to common beliefs about what makes services "service-like."

- ***Customization***. Services tend to be customized to specific customers.

...Example: Surgery is customized to unique needs of specific customers.

...Counterexample: A train ride from New York City to Washington, D.C.

- ***Consumed as produced***. Many services are consumed as they are produced.

...Example: A music performance

...Counterexample: Customized legal contract that will remain in force for 50 years

- ***Co-produced:*** Many services are co-produced with the customer.

...Example: Work closely with a customer when producing custom software

...Counterexample: Housecleaning service cleans a home every week while the client is at work.

- ***Ephemeral***: Services tend to more ephemeral than products.

...Example: A medical examination

...Counterexample: Customized legal contract that will remain in force for 50 years

- ***Experience vs. goods:*** The value of services tends to be related to experiences.

...Example: A chartered one-week sailing trip

...Counterexample: Production of a personal computer based on a customer's specifications.

- ***Intangible:*** Services tend to be intangible.

...Example: Music lessons

...Counterexample: Garbage removal

- ***Relationship-based:*** Services tend to be relationship-based.

...Example: Check-up with a personal physician

...Counterexample: Treatment in an emergency room by a physician who is a stranger

The examples show that the difference between products and services often becomes fuzzier as one looks closely at specific situations. Davis and Mayer summarized the general trend toward combining product and service features in economic offerings in their book *Blur*.[60] "The difference between products and services blurs to the point that the distinction is a trap. Winners provide an offer that is both product and service simultaneously."

Product-like vs. service-like characteristics define a set of dimensions for improving customer satisfaction.

From a business viewpoint, the distinction between products and services is much less important than providing a mix of product and service features that internal or external customers want. Figure 9.1 shows a list of product/service value dimensions and shows hypothetical preferences of three committee members in an ERP project. In addition to detail-oriented issues concerning software features, vendor strength, and price, the committee might consider high-level issues about the extent to which they want product or service features. On the other side, software vendors might think about their own offerings in similar terms, and might use these value dimensions to adjust to customer preferences.

As illustrated in Figure 9.1, many suggested characteristics of service can be viewed as end points of dimensions that make something more product-like or more service-like. The continual development of new IT capabilities increases the flexibility in customer offerings and creates many new possibilities of combining product-like and service-like features.

Figure 9.1. Product-like versus service-like features

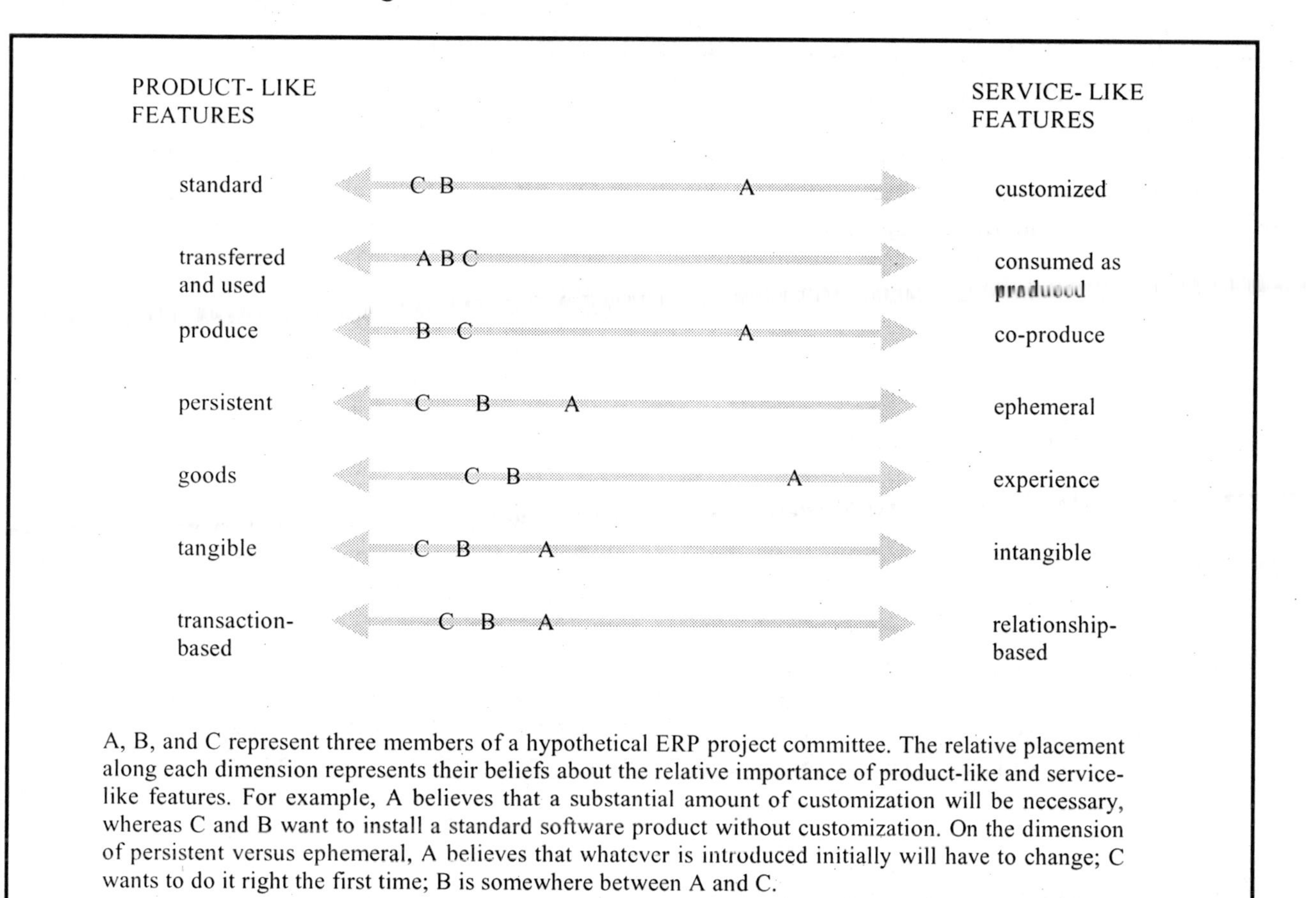

A, B, and C represent three members of a hypothetical ERP project committee. The relative placement along each dimension represents their beliefs about the relative importance of product-like and service-like features. For example, A believes that a substantial amount of customization will be necessary, whereas C and B want to install a standard software product without customization. On the dimension of persistent versus ephemeral, A believes that whatever is introduced initially will have to change; C wants to do it right the first time; B is somewhere between A and C.

The customer experience. A customer's evaluation of a work system's products and services is often linked to the quality of the customer experience, which can be described in terms of five stages:

- providing requirements for a product or service
- acquiring the product or service
- using the product or service
- maintaining a product
- disposing of a product or its remnants.[61]

Although parts of the customer experience in any particular situation may fall outside the scope of the work system, the character of the customer experience is relevant because it may affect the customer's evaluation of the work system's products and services. For example, in unpleasant experience changing a reservation can set a negative tone for a traveler's entire experience of traveling with an airline.

Form of interaction with the customer. The form of interaction between the customer and the service provider is an aspect of the customer experience that should not be taken for granted. Service through a relationship occurs when a customer has repetitive contact with a particular provider and when the customer and provider get to know each other personally and in their roles in the situation. Service through encounters occurs in isolated interactions between customers and providers who are strangers.[62] In a classical bureaucracy all customer interactions are designed as encounters because bureaucracy is supposed to be based on rules rather than relationships. In reality, however, some customers find ways to deal repeatedly with the same person in a bureaucracy, thereby creating at least a rudimentary relationship that may result in better service. The trend in medicine is away from relationship-oriented medical care provided by personal physicians and toward encounter-oriented medical care offered in a less personal environment by tightly scheduled providers with little time for building relationships.

Customer interactions can be designed as relationships or encounters.

Possible changes. Construing products and services too narrowly or assuming they must be static is a mistake when analyzing a work system because extensions in a variety of directions might improve customer satisfaction. The evaluation of a work system's products and services starts with understanding their content, looks further at the features and benefits of the content, and then moves to the customer's evaluation criteria and possibilities for improvement.

The mix of product and service features in IT-intensive products and services is especially crucial because IT is so malleable. For example, an information system that produces management reports can be positioned as a reliable source of pre-programmed reports or as a work system in which the customer (the user of the management report) can use online tools to retrieve and format information. Alternatively, it can be positioned as a service including information analysts who help clarify changing information needs and who figure out how to generate new reports that address management issues as they arise. Whether the analyst role exists at all, and whether it is placed in the IT organization or the customer organization may have a significant effect on customer satisfaction and system cost. Advances in hardware and software have made the range of choices much wider because it is now more feasible for users to manipulate the information themselves.

Work System Customers and Their Concerns

Some work systems have homogeneous customers with similar concerns and expectations for the products and services produced. Many other work systems produce different products and services for distinct groups of customers. For example, customers of a manufacturing work system may include the end customers who use the product, the shipping department that ships the product (and has to coordinate with manufacturing schedules), and managers who receive management information generated as a by-product of the manufacturing effort. While thinking about customer groups it is also worthwhile to identify secondary customers and other stakeholders who may care greatly about the work system and what it produces even though they don't use the product directly. They may include corporate executives, business unit managers, supervisors, work system participants, unions, managers of other affected work systems, and the IT group. Any one of those stakeholders may have important views about the work system and also may be able to block or subvert changes that contradict their views.

Different groups of customers may have mutually contradictory goals. Consider the work system of determining a U.S. citizen's income tax liability. Regardless of whether a paid tax preparer does the calculations and produces the tax return, the mechanical steps include gathering the relevant information, sorting the information into different income and deduction categories, calculating the totals in each category, entering the totals on the tax form, and determining whether tax payments to date are sufficient. The product of the exercise is a tax return document including a calculation of the amount of tax due. The Internal Revenue Service receives and evaluates the return, and therefore is a customer. The taxpayer is also a customer, however, because decisions about how to allocate income and expenses on the tax return affect the amount of tax that must be paid.

Customer priority. It is sometimes important to ask whether some groups of customers are more important than other groups. For example, businesses that are heavy users of electricity are far more important to large electric utilities than individuals who use a small amount of electricity in their homes. It is not surprising that the customer service representatives for large businesses provide more responsive service than individuals receive.

Asking what "the customer wants" is inadequate when a work system has multiple groups of customers

Some work systems let customers pay for priority. For example, telephone companies, electric utilities, and computer centers all use peak load pricing to encourage customers to shift usage to off-peak times. Airlines use a version of this pricing strategy when they charge full price for business travelers while giving discounts to vacation travelers who can book their flights weeks in advance. Their goal is to shift the demand pattern in a way that maximizes profit by balancing average revenue per seat sold against unsold seats.

Customer expectations. The design of a work system should take account of customer expectations. A work system that sets the expectation of impersonal, commodity service will probably receive fewer complaints than a similar work system that sets higher expectations but doesn't deliver.

Desired improvements. In addition to recognizing different goals of different groups of customers, the analysis should also recognize different concerns related to unmet expectations, wish lists for major changes, and other things that may seem unrelated to the work system itself. For example, the end customers may be concerned about the quality of the product, the shipping department may be concerned about whether it will be swamped with a month-end surge of shipments, and managers may be concerned about receiving usable management information as a by-product of the production process.

Possible changes related to customers. Analysis efforts often assume that the system's customers will not change. Even when external customers are involved, it is at least worth considering the possible advantages of adding new customers or eliminating existing customers. For example, an organization whose high level of service for all customers is too expensive may decide to supply different levels of service for different customers. Less important customers faced with a lower level of service may decide to move to another supplier, but that may be beneficial to the original supplier for whom those customers were unprofitable. In a different type of situation, a sales force tracking system designed to provide information for headquarters might be more successful if it makes the sales force more of a customer by providing information that is useful in selling.

Performance Indicators for Products & Services

Customers evaluate products and services in terms of a number of metrics that may be mutually reinforcing or mutually contradictory. A thorough analysis should identify important performance indicators, should include quantitative measures or estimates of the current values for the related metrics, and should calculate the gap between the current values and the desired values.

As a reminder that any performance indicator may have a variety of metrics, Table 9.1 identifies two metrics for each performance indicator. At least one or two of these performance indicators is relevant in most situations, although the metrics that are actually used should be tailored to the situation. Notice that some metrics can be calculated objectively based on observable events or characteristics, whereas others are subjective. The fact that a metric is subjective does not mean it is unimportant.

Table 9.1. Performance indicators and related metrics for products and services

Performance indicators	***Typical metrics related to each performance indicator***
cost to the customer	• Total cost of ownership • Total cost of acquisition, including cost of specifying requirements
quality	• Ranking of this product or service relative to alternatives in terms of fit for the desired function • Subjective perception of fit to expectations for this category of product or service
responsiveness	• Delay between contacting the producer and receiving a meaningful response • Subjective perception of the extent to which the producer tailors the product or service to the customer's needs
reliability	• Frequency of product failures or defects • Usable lifetime of the product
conformance	• Whether or not the product or service conforms to specific standards or regulations (yes or no) • Extent to which the products or services actually received conform to what was promised
intangibles	• Subjective perception of the extent to which the product or service is aesthetically pleasing • Subjective perception of the extent to which the product or service instills feelings of satisfaction and well-being

Many of the performance indicators listed in Table 9.1 apply equally to internal and external customers:

- ***Cost to the customer*** is whatever the customer must give up across the entire customer involvement cycle of obtaining, using, and maintaining a work system's product. In addition to money, this includes time, effort, and attention that might be used for other purposes.

- ***Quality*** of the product or service is the customer's perception that the product is fit for its use, that it has desired features and that these features are well configured.[63] For physical products, this is related to form, function and aesthetics; for information products, it is related to accuracy, accessibility, usefulness, and completeness; for service products, it is related to the completeness of the features and attentiveness to delivery.

- ***Responsiveness*** is the customer's perception of whether the delivery and content of products and services reflect unique needs of individual customers. Delivery responsiveness is important for any type of product. Content responsiveness tends to be more important for customized products and services because customization brings an expectation that the customer's unique requirements will be satisfied.

- ***Reliability*** of a product or service is the likelihood it will meet expectations related to delivery and operation. Delivery reliability is about whether the product or service will be available when promised. Operational reliability for a service is about whether the service is performed as expected. For a product, operational reliability is about whether it operates correctly when it is used.

- ***Conformance to standards or regulations*** may be essential, especially when non-conformance makes a product useless or illegal. For example, when the product is a tax return, non-conformance to government regulations is a serious matter. Similarly, conformance to standards determines whether pharmaceutical products are saleable.
- ***Satisfaction with intangibles*** is often difficult to measure, but sometimes has an important impact on overall customer satisfaction. An example of intangibles is the impact of a doctor's interpersonal manner and the appearance of the examining room. The substance of the medical exam is related to the doctor's knowledge and thorough efforts, but the patient's satisfaction and feeling of well-being may be related equally to the intangibles.

Interactions between Performance Indicators

Different aspects of product/service performance interact with each. For example, poor reliability often generates unwanted costs for the customer. On the other hand, higher reliability may be unrelated to higher or lower purchase prices for commercial products or transfer prices for internal services.

Performance indicators may also overlap in a variety of ways. For example, responsiveness, reliability, and conformance are sometimes viewed as partially overlapping aspects of quality. In certain situations, however, responsiveness and reliability may push in opposite directions because the product variations needed to increase responsiveness might decrease reliability by making the production process less repetitive.

Consider several examples involving adherence to standards and responsiveness to customers. In many situations, adhering to standards reduces costs and improves quality, but it may have positive or negative effects on responsiveness. Adhering to standards supports responsiveness when it results in quicker and easier production of commodity or customized products. It reduces responsiveness when the customer wants something that does not fit standards. This type of situation occurs when building codes force builders to install features that customers don't want. In many cases, the unwanted features are installed, inspected, and later removed or replaced with nonconforming features that the customer wants.

In other situations, attempts to be highly responsive to customer desires may lead to cutting corners that eventually result in lower product quality and reliability. For example, a programmer trying to produce a set of programs quickly for an insistent customer might shortcut debugging and documentation. The customer might receive the functioning software sooner, but with greater risk that it may not be reliable and may be difficult to maintain.

The dimensions of customer satisfaction interact in different ways in different situations.

The examples illustrate that there is no general formula or rule of thumb describing how an upward or downward change in one aspect of performance would typically affect another aspect of performance. The conclusion is simply that attempts to improve a work system should consider current values and likely future values of important metrics. It should not be surprising that an improvement in one area, such as cost to the customer, might come with a drop-off in another area.

This page is blank

Chapter 10: Work Practices

- Starting Points for Identifying Possible Changes
- Strategy Decisions for Work Practices
- Performance Indicators for Work Practices

The most common way to summarize the work practices in most work systems is in the form of a **business process**, a structured set of steps, each of which is triggered by an event or condition such as the completion of a previous step or the detection of a problem. Focusing on the business process emphasizes the sequence of steps, the measurement of how well steps are performed, and understanding why they are performed in one way versus another. In many situations, steps in a business process might be performed differently, merged, or eliminated.

Although a business process (or "workflow) perspective is often very useful, other perspectives provide important ideas for understanding work practices and thinking about possible improvements. For example, focusing on decision making in the TRBG case moves the business process to the background and emphasizes questions about how approval decisions are made and whether a better model or better negotiation method might improve decisions. Focusing on communication identifies possible communication gaps between loan officers and clients, and reveals that the loan officer's communication skills may have undue influence on some decisions. Focusing on coordination leads to asking whether better coordination between loan officers and credit analysts could hasten the loan approval process. Focusing on information processing leads to possible improvements in how information is captured, compiled, analyzed and presented throughout the process. These and other perspectives often overlap to some extent, but in combination they provide a number of valuable viewpoints for naming problems and identifying improvement possibilities.

This Chapter is divided into three major sections. The first identifies a number of perspectives that are introduced briefly. The second looks at big picture issues in the form of work system strategies. The third looks at performance indicators for work practices and tradeoffs related to those indicators.

Starting Points for Identifying Possible Changes

Strategy decisions for work practices will be discussed later. First, we will look at other useful starting points for thinking about possible changes in work practices:

- business process (work flow) details
- functions that are missing
- different contexts for doing the work
- built-in problems
- risks and stumbling blocks
- coordination between steps
- decision making practices
- communication practices
- processing of information
- practices related to physical things

Business process. The Work System Framework uses the term work practices instead of business process because *business process* usually treats the work as a **workflow**, a set of definable steps that are triggered by specific events or conditions, such as the completion of a previous step or the detection of a problem. A business process perspective is especially useful in looking at the details of workflows and examining possibilities such as:

- adding steps
- eliminating steps
- combining steps
- changing the sequence of steps
- triggering subsequent steps in different ways
- changing methods used within each step

For example, a business process perspective might encourage asking about whether delays between steps can be reduced or whether set-ups in a factory can be reduced though better scheduling of the work. During the reengineering boom of the mid-1990s, frequently mentioned examples such as the reengineering of IBM Credit's loan approval process[64] concerned switching from an assembly line approach to a case manager approach. In an **assembly line**, individuals perform specific activities within a predictable sequence of highly structured steps, and then hand off the work-in-process to someone else. In a **case manager approach**, handoffs are minimized because an individual uses computer software to perform most of the steps related to a specific transaction, such as handling of a loan application.

Process modeling: Methods for documenting how a process operates or should operate.

Documenting business processes. It is usually possible to summarize a business process as a list of major steps, or even steps and substeps, but that level of detail is insufficient for clarifying how most business processes actually occur. Although a list of steps is useful for clarifying the scope of the system under consideration, identifying and justifying beneficial changes in the system usually requires looking at the workflow and work practices in more detail.

Different organizations use a range of techniques for documenting how a business process operates now (the "**as-is**" process) and how it should operate in the future (the "**to-be**" process). **Process modeling** is the general category for these techniques, whose basic goal is to document processes by breaking them into tasks, clarifying the sequence of tasks, and dividing tasks into subtasks if necessary. Some techniques go further by documenting the logic by which decisions, outcomes, or conditions in one task or sub-task trigger the initiation of other tasks. Other techniques add detail about what specific information is needed or generated by a task, who executes each task, and who is responsible for the outcome. Frequently used techniques include:

- **Flowcharts** are diagrams that use standard symbols to represent the sequence of steps, decision points, and branching logic of processes. Some flowcharts also use standardized symbols to represent different types of devices and media for input/output, processing, and data storage. (See Figure 10.1)

- **Swimlane diagrams** list the various steps in a process and use a set of parallel columns as "swimlanes" for identifying who will do each step. This type of diagram is one of the ways of representing a "use case," which is a basic unit of analysis in object-oriented analysis of systems. Use cases are but one of a number of representations included in UML (Universal Modeling Language).[65] (See Figure 10.2)

- **Data flow diagrams** document the flow of data between processes or sub-processes. DFDs use just four symbols that identify processes, data flows, data stores (location where data is stored) and external entities that provide data to a process or receive data from it. (See Figure 10.3)

Figure 10.1. Flow chart for the TRBG loan approval system

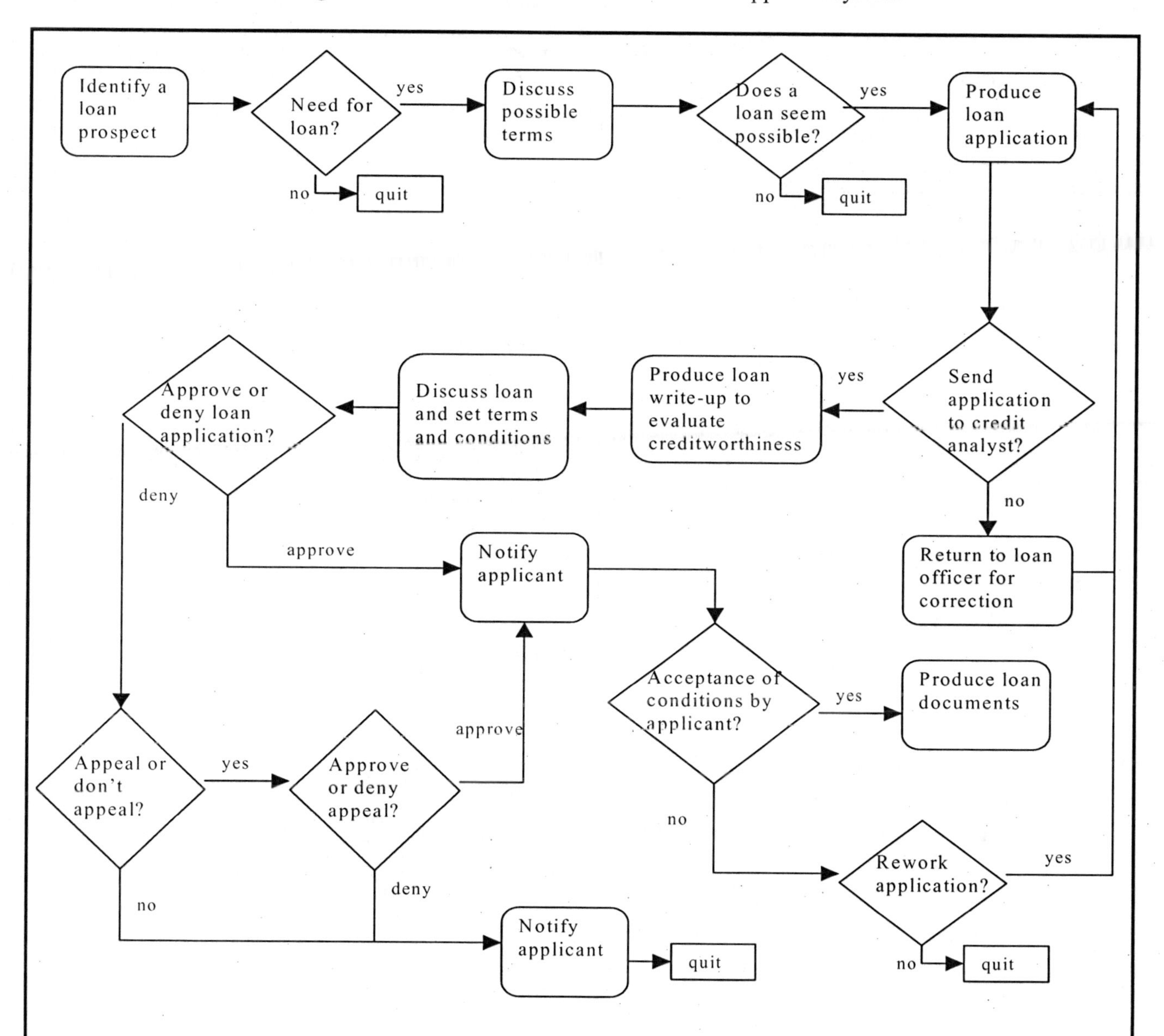

This flow chart summarizes the loan approval process in the TRBG example in Chapter 8 using a minimum number of flowchart symbols. Traditional data processing flow charts use specialized symbols related to specific data processing entities such as manual operations, punched cards, paper documents, magnetic tape, hard disks, and online displays.[66] Notice that the diamond icon for logical choices does not exist in the representations in the next two Figures.

Figure 10.2. Swimlane diagram for the TRBG loan approval system

Loan applicant	*Loan officer*	*Credit analyst*	*Senior credit officer*	*Loan Committee (or Executive Loan Committee)*	*Loan administration clerk*
Identify possible needs	Identify possible needs				
Discuss possible terms	Discuss possible terms				
Produce loan application	Help the loan applicant produce the loan application				
	If the application has obvious flaws, the loan officer needs to do rework.		Review initial application before forwarding it for credit analysis.		
Provide answers or sources of answers to questions from credit analyst	Provide answers or sources of answers to questions from credit analyst	Produce loan write-up	Check progress as inexperienced credit analysts work on write-ups		
	Present loan application to senior credit officer or loan committee	Participate in committee meeting, possibly by phone	(For loans less than $400K) Determine terms and conditions for loan	(For loans equal to or greater than $400K) Determine terms and conditions for loan	
			(For loans less than $400K) Approve or deny loan application	(For loans equal to or greater than $400K) Approve or deny loan application	
	Appeal loan denial if appropriate				
	Notify client				
Accept or reject loan terms and conditions					
					Produce loan documents

A swimlane diagram identifies the participants in each major step. Swimlane diagrams can take various forms depending on the situation.

Figure 10.3. Data flow diagram of for the TRBG loan approval system

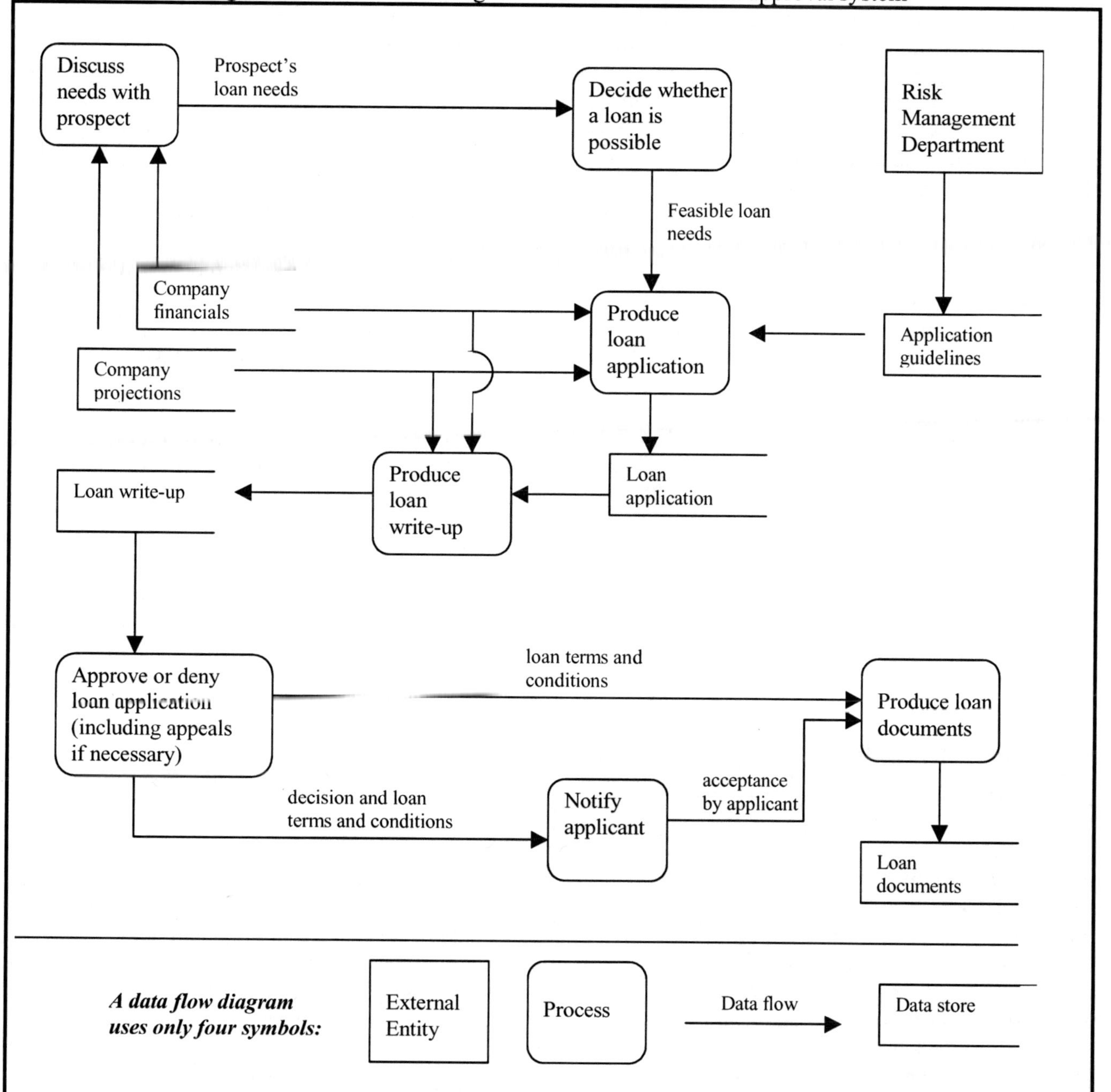

The purpose of a data flow diagram is to clarify the relationship between information and process steps. Therefore it identifies process steps that use or produce information, and identifies the primary flows of information (data flows) and the storage of information (data stores). Unlike flow charts, data flow diagrams do not attempt to represent process logic in detail.

NOTE: Aside from using different icons, this data flow diagram provides less process detail than the two previous figures. Regardless of which combination of these methods and other methods is used, business professionals should make sure that the diagrams represent reality well enough to minimize confusion about what actually happens.

A detailed discussion of documentation techniques is beyond the scope of this book, but numerous examples of each can be found using simple Web searches. It is worthwhile, however, to say something about what happens when any of these approaches is applied to a large system. Suddenly, what looked neat and straightforward on a single page becomes more awkward as it is spread across multiple pages. Regardless of which technique is used, it is necessary to have a method for decomposing a large process into subprocesses and decomposing subprocesses into smaller subprocesses. This method is called **successive decomposition**. To keep track of the various layers it is necessary to give them names such as layer 2, layer 2.3, layer 2.3.1, and layer 2.3.1.4. Just naming and numbering things is an essential discipline for keeping track of the analysis itself, especially in complex systems.

Process documentation is essential, but there are many other approaches for finding problems and identifying possible improvements.

Relevant functions that are missing from current work practices. In some cases, potentially relevant functions simply aren't performed within current work practices. For example, simple notifications of completion, shipping, and receiving may not be provided. Similarly, quality control information that is used for checking current work might not be collected and used for long-term tracking of work practices or the products and services they produce.

Different contexts. An important question for thinking about work practices is whether the work system encounters fundamentally different contexts that need to be handled differently. If so, the different contexts need to be included in the business process description in order for the analysis to be realistic. For example, the context in an emergency room treating a heart attack victim with a life-threatening problem is quite different from the context when treating someone whose problem is not life-threatening. Similarly, a local zoning commission should handle a home remodeling project quite differently from the way it handles construction of a large office building. An analysis of either work system would be incomplete without considering the different contexts that occur.

Context and environment are sometimes used as synonyms in everyday speech, but have different meanings in WSM. As used here, context refers to a temporary situation that occurs within a work system and requires activities that may not apply to other contexts. In contrast, environment refers to persistent conditions that exist outside of the work system but have impact on the work system's operation and success. Environment includes the organizational, cultural, competitive, technical, and regulatory environment within which the work system operates.

Problems built into the current work practices. In some cases, current work practices have built-in problems and obstacles that might be minimized or eliminated by changing the rules or by eliminating, re-sequencing, or combining steps. Common built-in problems and obstacles include:

- ***inattention to details***. Small details that are often unobserved can have major impacts. One of the reasons for Toyota's excellence in manufacturing is that Toyota's managers "ensure that all work is highly specified as to content, sequence, timing, and outcome." For example installation of the right front seat into a Camry is designed as a sequence of seven tasks, all of which are expected to be completed in 55 seconds as the car moves at a fixed speed through a worker's zone." If the worker installs the rear seat-bolts before the front seat-bolts, then the job is being done differently and something is wrong.[67] The importance of seemingly small details is evident in health care. For example, critically ill patients in intensive care often receive medication through catheters that are placed directly into major veins or arteries. Many of these patients develop serious bloodstream infections. Pittsburgh hospitals observed the placement and maintenance of these catheters over a period of weeks. They found dozens of small details that increased the chances of infection, such as storing gloves on the bottom of catheter kits, "causing nurses to fish through sterile material with bare hands."[68]

- ***built-in delays***. Many work systems are designed with built-in delays. For example, all new hiring by a large social service agency required signoff by a department head who traveled frequently. In many cases, hiring was delayed several weeks because the department head was out of town; some potential hires took jobs elsewhere instead of waiting.
- ***unnecessary hand-offs or authorizations***. Assume one person could perform a task. Dividing it among several people usually creates built-in delays and coordination effort, especially if each individual has many other tasks to perform. Hand-offs and authorizations may be necessary for a variety of reasons, however. For example, requiring prior authorization of certain levels of expenditures is a common approach for controlling budgets.
- ***steps that don't add value***. Systems analysts often try to distinguish between process steps that add economic value, steps that are necessary for other reasons, and steps that are unnecessary. Steps that are unnecessary waste resources and should be eliminated. Steps that don't add economic value should be questioned. Prime candidates for re-thinking include inspection steps and authorizations. For example, the editor of an internal newsletter at a General Electric plant complained at a company meeting that each new edition required seven approvals. The plant manager's response: "This is crazy. No more signatures."[69] This example is consistent with the general belief of many quality experts that people should check the quality of their own work instead of waiting for inspections and authorizations by others.
- ***unnecessary constraints***. Admission to an EMBA program was governed by a constraint that all students needed an undergraduate degree. A highly successful 40-year-old manager who applied to the program was three courses short of completing his undergraduate degree when he joined the Air Force 19 years ago. The EMBA director had to decide whether the undergraduate degree was an absolute constraint or whether it could be bypassed for this otherwise highly qualified individual.
- ***low value variations***. In the past, many large companies operated with numerous email systems and different types of personal computers. Those variations provided little or no value, but added to overhead costs because the IT staff performing maintenance work and upgrades needed to keep track of differences between technology brands that served the same purposes. Today, most large companies try to minimize low value variation by standardizing technology choices across the company.
- ***inadequate scheduling of work***. A weekly meeting is scheduled at 4:00 PM on Friday even though most attendees desperately want to be on the road before the commute traffic peaks at 5:00 PM. It is unlikely that everyone will be fully attentive.
- ***large fluctuations in workload***. Monthly output schedules for factories commonly create a problem for the shipping department. As month end nears, the factory finishes everything possible in order to meet its schedule, creating a large bubble of work for the shipping department, which may need to pay overtime to work through the bubble, but may have relatively little work to do a week later.

Built-in problems are common because many work systems were not designed carefully and have not been re-analyzed or improved.

Waste. Focusing on waste is another way to think about built-in problems. Waste is anything that uses resources but doesn't add value. The quality management guru Taiichi Ohno, a former chief engineer at Toyota who developed the Toyota production system, identified "seven wastes." Each of these wastes is often a direct result of a work system's structure.

- ***over-production***. Producing more than is needed by the next operation leads to inventory build-up, poor usage of space, and other problems.
- ***unnecessary transportation***. No value is added when work-in-process is moved.
- ***excess motion***. No value is added when work system participants move unnecessarily to get tools or materials.
- ***waiting***. No value is added during idle time when people or machines are doing nothing, or when material is not being used or transformed.
- ***inappropriate processing***. Any effort that does not add value for the customer should be minimized.
- ***unnecessary inventory***. Inventories should be minimized.
- ***defects***. Work systems should be designed to minimize defects, rework, and mistakes.

There are a number of approaches for minimizing the seven wastes. Just-in-time (JIT) systems (mentioned in Chapter 5) try to minimize over-production and unnecessary inventory. The design of workplaces and workflows should try to minimize unnecessary transportation, excess motion, and inappropriate processing. Careful scheduling can minimize waiting and reduce inventory levels. Quality control techniques such as control charts, fishbone diagrams, and Pareto charts can help in analyzing defects.

Perspectives for analyzing work practices also include:

.... coordination

.... decision making

.... communication

.... information processing

.... physical activities.

Coordination problems related to dependencies. In some cases, thinking specifically about dependencies within or across work systems is a good way to recognize potential areas of improvement. Typical aspects of coordination that might be considered include:

- ***timing of events or business process steps***. For example, delivering lumber and other materials to a construction site weeks before they are needed increases the chances of theft or damage, and may simply get in the way of people doing work. On the other hand, late delivery can prevent construction workers from doing their work even though they must be paid.
- ***sequencing of events or business process steps***. For example, it is usually beneficial to place customization steps as late as possible in a production process. Late customization reduces the probability of producing something that cannot be used if the customer cancels the order.
- ***resources required before a business process step can occur***. For example, bottlenecks sometimes occur in manufacturing when several different orders need the same expert or the same equipment simultaneously. Effective coordination of the work on those orders requires an informed decision about which one should go first and which should be delayed.
- ***responsiveness to events elsewhere***. For example, a manufacturer's sales department learns that a customer order was cancelled. Coordinating with the production department requires notification of the order cancellation and help in deciding how to redirect or reuse any part of the order that was already in production.
- ***quality of ongoing coordination efforts***. High quality coordination efforts identify and resolve coordination issues quickly and efficiently, leaving positive feelings about effective teamwork. Coordination efforts that dissolve into unpleasant squabbling have the opposite effect.

Phases of decision making. When decision making is the crux of a work system, it is worthwhile to look at how well specific phases of decision making are performed. Consider how each of the decision phases listed below appears in the TRBG case:

- ***defining the decision problem***. In the TRBG case, the decision problem is to approve or deny the loan application, or to request more information.
- ***gathering information***. In the TRBG case, the information is gathered by the loan officer and client, and submitted in the form of a loan application. The credit analyst gathers additional information such as real estate appraisals.
- ***analyzing the situation***. In the TRBG case, the loan officer does the preliminary analysis to determine whether a prospect might need a loan and might qualify for it. The credit analyst does a more detailed analysis based on information in the loan application and other information that is obtained as needed. Additional analysis may occur in a loan committee.
- ***defining alternatives***. The loan committee has three basic alternatives (yes, no, or request more information), but within yes it can set risk-adjusted interest rates and can insist upon specific loan covenants to reduce risk.
- ***selecting the preferred alternative***. The current decision process uses calculations in the loan write-up, but does not dictate how the decision will be made.
- ***implementing the decision***. A *yes* decision that is accepted by a client is implemented by generating the loan documents and executing the loan. A *no* decision or request for more information is implemented by notifying the client.

When analyzing a work system that includes important decision steps, just looking at the list of phases may reveal areas for potential improvement. In some cases, the basis for the decision is often shaky because the information is insufficient and little analysis occurs. Although the phases of decision making include identification and selection of alternatives, in many cases only one possibility is considered, and the decision may be no more than a snap judgment that ignores many relevant factors. In addition, the available information systems (including "decision support systems") often address only one or two of the phases and provide little or no help with the other phases.[70]

Issues involving communication. "They are having communication problems" is such a common refrain at home and at work that the analysis of work practices should always look for issues related to communication between work system participants. In some cases, a seemingly logical analysis of a work system founders on communication issues that combine personal preferences, organizational politics, and inconsistent incentives and goals. A widely publicized example is the difficulty attaining agreement about how the various U.S. agencies that collect security and terrorism information can and will share that information effectively.[71] An analysis of systems within and across those agencies would be incomplete if communication issues were ignored. Many of the following communication issues probably would appear:

- ***conscious misrepresentation***. This is listed first as a reminder that people providing information within systems may have many reasons to shade the truth, remain silent, or lie.[72] For example, a manager may sugarcoat a report to keep supervisors unaware of problems that might be embarrassing. Another manager trying to obtain funding for a pet project might exaggerate its potential benefits and underestimate its costs. An ecommerce customer trying to discourage unwanted advertisements might enter incorrect responses on an online form.
- ***inflexible communication patterns.*** In traditional organizations there is a tendency for communication to flow up and down the organization chart rather than going directly to individuals most likely to need the information that is being communicated. This pattern creates delays and may filter information or introduce biases. The alternative may seem chaotic, especially to managers who want to maintain control of their organizations.
- ***communication blockages.*** Rival organizations or individuals within the same firm sometimes avoid communicating with each other. In other situations, communication blockages exist by design, such as the attempt to create "firewalls" between stock analysts and investment banking

groups in large brokerages, thereby minimizing self-serving bias in stock price forecasts. In some cases these blockages are required by government regulations. For example, HIPAA, the Health Insurance Portability and Accountability Act of 1996, sets boundaries on the use of health records, thereby making those records more difficult to obtain in some legitimate situations. Similarly, FERPA, the Federal Educational Rights and Privacy Act of 1974, limits a parents' right to obtain educational records after a student is 18 years old.

- ***communication overload.*** Many people receive over 100 e-mails a day, absorbing time and making it more likely that some important messages will be overlooked. Another aspect of communication overload is the existence of so many different communication technologies and media, such as e-mail, voicemail, instant messaging, and paper documents.

- ***conflicting communication styles.*** Some people prefer to communicate through extensive written analyses; others prefer summaries; yet others tend to communicate around relationships between people.

- ***communication across distances or time zones.*** People working on distributed, international teams often need to arrive at work very early or stay very late in order to talk to colleagues in Europe or Asia.

- ***faulty communication about technical details of work.*** Many information system projects are plagued by the difficulty of communicating about technical issues with people whose backgrounds and interests are non-technical.

- ***ineffective communication about personal or organizational issues.*** The analysis of systems often focuses on procedural or technical issues and tends to downplay personal or organizational issues even when those issues are central to a system's effectiveness. For example, assume that some of the participants in a customer service system believe that one of its major problems is the inattention or incompetence of the department manager. Mentioning that problem may be a career-limiting move for them. Keeping quiet may contribute to development of system changes that will have little impact unless the manager changes in some way.

- ***impact of technologies or media used for communication.*** Technologies and media used for communication can have important impact on the richness and understandability of the messages being communicated. For example, email and phone communication filter out body language and therefore are not as rich as face-to-face communication. On the other hand, non-native speakers may find e-mail or document-based communication more effective because it helps them refine their message and minimize any impact of having an accent or not speaking as fluently as their colleagues.

- ***difficulties in communication across cultures.*** Different cultures have vastly different expectations about communication with superiors, peers, and subordinates. For example, looking someone in the eye is expected in American culture, but is quite inappropriate in other cultures if the other person is a superior. Similarly, the American tendency to get to the point quickly about the issues at hand might seem rude or inappropriate in a culture that expects greater attention to relationships between the people who are communicating.

Aspects of information processing. The work that IT performs within a work system falls under the heading of information processing.[73] It might seem surprising that everything IT does reduces to six functions:

- capturing information (e.g., using a keyboard, bar code reader, or digital camera)

- transmitting information (e.g., using a wired or wireless network to move data between locations or to have a phone conversation)

- storing information (e.g., storing information on a computer's hard disk, a memory card, or a recordable CD-ROM or DVD)

- retrieving information (e.g., retrieving information from any storage device)
- manipulating information (e.g., performing calculations or combining or reorganizing information)
- displaying information (e.g., using a computer monitor or printer)

The use of IT to automate information processing tasks is the basis of modern business. However, the fact that information processing tasks are automated does not imply that the entire work system is automated. To the contrary, even when IT is used extensively, many important activities within work systems may not involve processing information.

Aspects of physical activities. Even in highly computerized work systems, physical activities play an important role and should not be forgotten. Specific types of physical activities that might present problems or improvement opportunities include:

- creating physical things
- moving physical things
- storing physical things
- finding physical things
- modifying physical things
- using physical things
- protecting physical things
- getting rid of physical things.

Strategy Decisions for Work Practices

Although analysis efforts usually start with some consideration of big picture issues, typical analysis techniques and tools for computerized systems focus much more on work system details or technical details. This is unfortunate because even extensive tweaking of the details is unlikely to attain substantially better performance if the work system's strategy is poorly calibrated or non-existent. For example, many small details might be changed to solve an inconsistency problem, but those changes might have more coherence and force if there is agreement that the work system's strategy does not impart enough structure to the work. On the other hand, too much structure in a work practices might prevent work system participants from using their judgment in situations where judgment is needed.

Table 10.1 identifies a series of system design issues related to work system strategy. It shows that the goal for each strategy decision is to find an appropriate balance point between problems of deficiency and problems of excess. This section will look at each strategy variable in turn.

Amount of Structure

The tension between structure and flexibility in work practices is the crux of recurring controversies related to work systems that attempt to reconcile conflicting needs and interests. Examples include systems for admitting students to college, authorizing payments for medical procedures, and approving loans. For instance, banks are sometimes accused of using mortgage approval processes that discriminate against people in minority groups. Although there are many viewpoints about commercial and social goals of a mortgage approval process, from a system design viewpoint the question can be framed as an issue about how structured the decision process should be. Work practices are highly structured if:

- Information requirements are known precisely.
- Methods for processing the information are known precisely.
- The desired format of the information is known precisely.
- Decisions or steps within the work practices are clearly defined and repetitive.
- Criteria for making decisions are understood precisely.
- Success in executing the work can be measured precisely.

Table 10.1. Balancing problems of deficiency or excess in a work system's strategy

Strategy decision	***Problems of deficiency***	***Problems of excess***
Amount of structure	Too little: Work is done or decisions are made inconsistently. Work system participants have insufficient guidelines and methods for doing work.	Too much: Guidelines and methods for doing work or making decisions are so restrictive that work system participants cannot use appropriate judgment.
Range of involvement	Too little: Decisions are based on limited individual or departmental interests because too few opinions are considered. People who do the work lack important skills or knowledge that others might have brought.	Too much: Involvement of too many people slows the work and may diffuse responsibility.
Level of integration	Too little: Work system components operate too independently. Whether or not subsystem performance is good, overall performance suffers. Example: The sales staff makes promises that the production staff cannot keep.	Too much: Work steps become more difficult because more factors about other parts of the work system must be considered each time anything is done. Possible increase in the risk of catastrophic failure because mishaps in one area propagate to other parts of the system.
Complexity	Too little. Poor performance because important factors are ignored in order to simplify the work.	Too much: Poor performance because the work system is so complicated that it is difficult to understand, operate, and manage.
Variety of work	Too little: Performance and participant engagement suffer because the work is extremely repetitive and boring.	Too much: Performance suffers because too much of the work is novel. Work system participants cannot rely on previously learned routines and are continually figuring out things from scratch.
Amount of automation	Too little: Performance suffers because work that should be automated is done manually.	Too much: Performance suffers because an automated approach is unreliable or too difficult to control.
Rhythm - frequency	Too infrequent: Example, the train comes only once a day, causing inconvenience for passengers.	Too frequent. The train comes 10 times a day, and is often almost empty.
Rhythm - regularity	Too irregular. The bus leaves whenever it is full, implying that no one knows when it will arrive at its destination.	Too regular. The business leaves every hour on the hour, whether or not anyone is on board, implying that its operation may be uneconomical.
Time pressure	Too little: Tasks are finished at the whim of work system participants, resulting in low efficiency and poor schedule performance.	Too much: Time pressure is so great that work system participants feel excessive stress and become exhausted or quit.

Interruptibility	Too little: Work system participants may feel disconnected from real world issues faced by peers and customers. Example: The factory sets its schedule for two weeks and doesn't change it, regardless of whether priorities change in the sales department.	Too much: Work system participants are interrupted continually and have difficulty getting their work done. Example: The sales department seems to juggle its priorities every day. Efficiency suffers in the production department because it continually juggles its production schedule.
Prominence of feedback and control	Too little: Work is done inefficiently and production schedules are missed due to inadequate planning about who will do which work and when. Work system participants feel that work practices are disorganized.	Too much: Planning and control absorbs excessive time and effort. Work results suffer because of too much attention to planning, tracking, and providing feedback, and too little attention to performing value-added work.
Error-proneness	Too little: In a few rare cases the error rate may be so low that the work system participants become complacent and believe that nothing can go wrong.	Too much: Too many errors occur because work practices are poorly designed, poorly managed, or not well suited to the work system participants.
Formality of exception handling	Too little: Work system participants correct errors and exceptions in whatever way seems to make sense, but do not record incidents. They learn little or nothing about the frequency and impact of different types of errors.	Too much: Handling of errors and exceptions becomes ritualistic. So much time and energy goes into error handling that the work system's production suffers.

Formulas, models, and explicit constraints play a predominant role in highly structured work. In a highly structured loan approval process, the decision might be so structured that it could be automated. In a semistructured loan approval process, decision makers might use guidelines to identify loans that will obviously be approved or denied, and might use their judgment to approve or deny applications that are in the middle of the range. In an unstructured process, decision makers would make the decision based on whatever methods and intuitions seem appropriate to them. A highly structured loan approval process could be better or worse than a less structured process depending on the factors that are included in the decision and how those factors are weighed.

Well-designed work systems impose the amount of structure that is appropriate for the work being done. In some cases it is important to impose structure to maintain consistency or to be sure that the best methods are used for making a decision. In other cases, it makes more sense to provide tools and let people decide how to do the work. Imposing too much structure stifles creativity, prevents system participants from using their judgment and ingenuity to solve problems, and leaves them feeling little responsibility for the outcome. Anyone who has struggled with a large bureaucracy about a minor exception to a seemingly arbitrary rule understands that over-structuring is demoralizing for system participants and customers. On the other hand, imposing too little structure results in too much variability, often resulting in inefficiencies, delays, errors, and poor quality and reliability that the customer perceives.

The fact that different amounts of structure can be imposed says nothing about whether imposing more structure is better or worse in general. IT is now powerful enough that it can be applied in structured as well as unstructured processes. It can be used to automate decisions based on prespecified rules and criteria. It can be used to implement techniques that impose a great deal of structure by specifying exactly how work will be done and exactly what the outcome of individual work steps should be. For less structured situations, it can provide information and analysis tools that work system participants can apply in whatever way is most effective for a particular situation.

Range of Involvement

It sounds simple to say too many cooks spoil the broth, but the right range of involvement in work systems is often elusive. With too few participants, the work progresses too slowly and work system participants may feel excessive stress. With too many, work bogs down as system participants get in each other's way. Many state and local governments feel this issue acutely in complaints about the large number of different permits and approvals that are needed to open a new business or modify a building. Even small organizations may have the same problem, such as when the owner feels obliged to make every decision.

As with the amount of structure in a process, decisions about the proper range of involvement respond to a variety of concerns. The total quality movement generally pushes for reducing the range of involvement in processes by insisting that work system participants have the knowledge to do their work and the responsibility for checking it. In contrast, finance departments try to reduce the likelihood of financial fraud through the **segregation of duties**, which increases the range of involvement in financial processes. Having one person create a purchase order, another approve it, and a third approve the payment makes it less likely that people will collude, but also means that three people are involved in the work that one or two might do.

Strategy decisions for work practices address big picture issues that may be missed when focusing on details.

Having the right range of involvement is important when it is necessary to enforce organizational standards and quality expectations. This is why some large programming groups try to maintain quality and prevent unwarranted software modifications through a process called **software change control**. One person documents the desired change; another checks out a program and changes it; another verifies that the program change accomplished the desired functional change; yet another person replaces the program in the program library and documents the details of the change. A smaller range of involvement might seem more efficient, but it would not enforce quality or security standards as effectively. Individual work might proceed more quickly and might be more fun, but the software produced might contain more errors and might not be as maintainable.

Range of involvement also raises questions about supervision versus interference. For example, managers in a national clothing chain overrode product planners' decisions because "the stores weren't displaying enough denim." These managers assumed the responsibilities of work system participants. Whether or not they had identified the right issues, their interference probably undermined the original participants' feelings of engagement and responsibility.

Level of Integration

The level of integration within a work system describes the degree of mutual responsiveness between the different parts of the work system. Insufficiently integrated systems are disorganized and unproductive, but overly integrated systems are complex and hard to control. Although mutual responsiveness is the basic issue, integration occurs at five levels. The first two levels are conditions that make it easier to work together but are not inherently related to responsiveness and typically exist for reasons totally outside of any particular business process. Each of the other three levels of integration is designed into the way related steps or processes operate together.

- ***common culture***. Sharing the same organizational culture makes it easier to communicate and work together as needs arise.
- ***common standards***. Using common standards make it easier for the organization to maintain processes and technologies, whether or not they are related.

- ***information sharing***. The least obtrusive way to attain responsiveness between processes is to share information, as happens when a sales department accesses a manufacturing database to know what capacity is still available for additional orders.

- ***coordination***. This is a more obtrusive approach because it calls for negotiations or two-way information flows that provide mutual responsiveness even though the processes operate separately. In a production planning cycle, for example, Sales tells Manufacturing what it can sell; Manufacturing responds with a tentative output schedule; then they negotiate to create a mutually feasible plan before going about their individual work.

- ***collaboration***. At this fifth level of integration, major parts of the value added work are performed collaboratively. The interdependence between subprocesses may be so strong that their unique roles begin to merge. For example, many firms have moved toward product development processes involving close collaboration between Marketing, Engineering, and Manufacturing in order to bring easily manufacturable products to market sooner.

The difference between information sharing, coordination, and collaboration is especially important for understanding the potential benefits of information system investments. Information sharing is often touted as a major benefit of IT investments, but shared availability of information may or may not generate genuine coordination or collaboration, which require active commitment by participants. The design of the work system should be consistent with the level of integration the participants are ready to accept.

Lack of integration causes extra work and delays because pertinent information or knowledge from another work system is not accessible. For example, in traditional libraries the paper-based system for tracking borrowed books was not integrated with the paper-based card catalogue system. Library users often wasted time looking for books that were already checked out. In contrast, current library information systems present an integrated view of the catalogue that includes a field indicating whether the book is checked out and when it is due back. In a more complex situation that existed for decades, local telephone companies suffered from inadequate integration of their internal information systems for customer accounts, physical phone lines, and scheduled service calls. Each function had a separate database. Typical activities such as entering new telephone orders and troubleshooting customer account problems involved unreasonably long delays as service representatives logged into separate, non integrated information systems.

Excessive integration also generates problems. It forces separable subsystems to respond to each other too frequently, thereby making it difficult for each subsystem to get its work done. This is one of the reasons why many factories "freeze" their schedules a week or two at a time. Their managers believe that changing production schedules continually in response to daily events in the sales department causes too much chaos and inefficiency in production.

At the extreme, tightly integrated systems (also called **tightly coupled systems**) may be more prone to catastrophic failure than less integrated systems. Tightly coupled systems have little slack, require that things happen in a particular order, and depend on all components operating within particular ranges. When one component fails, the others may not be able to operate. The most tightly coupled systems in our society include aircraft, nuclear power plants, power grids, and automated warfare systems.[74] In contrast, **loosely coupled systems** are decentralized, have slack resources and redundancies, permit delays and substitutions, and allow things to be done in different orders. These systems may not be as efficient or powerful as tightly coupled systems, but failures tend to be localized and therefore can be isolated, diagnosed, and fixed quickly.

Complexity

The U. S. federal income tax system demonstrates the issue of complexity. A simple system might just collect a fixed or sliding percentage of personal income and leave it at that. But such a system would not explicitly address additional social, political, and economic goals, such as making it easier to own a house, making it easier to send children to day care, or recognizing depletion of oil wells. Each additional tax code feature recognizes another goal, for better or for worse, but also adds to the number of differentiated components and creates a wider range of interactions between components. The resulting complexity has become a sore point for taxpayers and a bonanza for accountants, lawyers, and people who make a living preparing tax returns. More than 60% of all individual income-tax returns in the U.S. are signed by paid preparers, up from 46% in the mid-1980s.[75]

Remove all the references to the tax system in this story and we are left with the fundamental tradeoff about complexity. Systems that are too simple don't address the variety of foreseeable situations that might occur (much like a word processor that can't number pages automatically.) Each additional function or feature shifts the balance toward complexity. Systems that are too complex are difficult to understand, control, and fix, especially when unanticipated situations emerge. As complexity increases, the ripple effects of changes in one part of the system become more difficult to trace.

Variety of Work

"Practice makes perfect" but "variety is the spice of life." The contradiction between these two proverbs creates a dilemma for work system design. Specialization and repetition are essential for learning and maintaining routines for doing work accurately and efficiently, but excessively repetitive routines result in boredom and disengagement.

Process documentation says little about design issues such as complexity, variety of work, amount of automation, rhythm, and time pressure.

The challenge for designing work systems is finding the right balance between repetitiveness and novelty. Efficiency concerns often lead toward increasing specialization, minimizing personal scope, and using rules to govern decisions. The efficiency benefits include less time spent figuring out what to do about each situation and greater accuracy and consistency in doing the work. Unfortunately, work systems designed for repetition may not even identify exceptions when they occur, and may be less able to respond to exceptions than if they had been designed to handle novel situations.

Amount of Automation

Like other characteristics of work practices, the extent of automation is a design decision. As is apparent when shopping at a store with a non-computerized inventory system, doing things manually that could be done by computers is slow and inefficient, and often makes it difficult to provide excellent service for customers. On the other hand, over-reliance on machines makes systems inflexible and may also lead to disengagement by human participants.

Although computers and other machines have enabled work practices that might have seemed unbelievable in the past, people have often relied too heavily on machines to solve problems machines alone cannot solve. A widely discussed example from the 1980s is the way General Motors spent over 40 billion dollars on highly automated factories, but reaped few rewards. Many of the robots never worked properly due to a variety of technical, social, and political circumstances. The NUMMI joint venture between GM and Toyota demonstrated the fallacy of viewing automation as a silver bullet. The NUMMI plant used comparatively little automation and had many of the same workers as a previous GM plant at the same location, but attained much higher productivity mostly by changing the expectations for both labor and management.

The general approach for the **division of labor** between people and machines is to assign tasks in a way that emphasizes strengths and deemphasizes weaknesses of each. In general, tasks assigned to computers are totally structured, can be described completely, and may require a combination of great speed, accuracy, and endurance. Tasks with relatively little information processing (such as keeping track of orders and invoices in a small business) can also be assigned to computers to assure organized and predictable execution. In contrast, people must perform tasks that require common sense, intelligence, judgment, or creativity. People handle these tasks better than computers because they are flexible and can identify and resolve previously unknown problems.

Although it is technically possible to automate many types of decision making, automating a decision is risky unless every aspect of the decision is so well understood that any mistake will be minor. The fundamental issue is that no system design team, regardless of how brilliant, can anticipate every condition a work system will face. Even if the software and human procedures in a work system operate perfectly relative to its design specifications, there is no guarantee that the work system will operate correctly when unanticipated circumstances occur. The significance of not delegating major decisions to computers is apparent from a potentially apocalyptic example. On October 5, 1960, a missile warning system indicated that the United States was under a massive attack by Soviet missiles with probability exceeding 99.9%. Fortunately, the human decision makers in the system concluded that something must be wrong with the information they were receiving. In fact, the early warning radar at Thule, Greenland, had spotted the rising moon[76], but nobody had thought about the moon when specifying how the early warning system should operate.

Rhythm

Work systems have a built-in rhythm. Ideally, a work system's rhythm should match the rhythm of customer's need for its products and services. An ecommerce web site operates almost 24 X 7; trains and other scheduled transportation systems operate on a regular schedule; other work systems operate sporadically; sometimes the rhythm and scale of operation varies by week, month, or season.

Rhythms that support internal efficiency are often inconsistent with rhythms that optimize customer satisfaction. For example, the supply chain of the Spanish clothing manufacturer Zara, has a precise rhythm that begins in its 650 retail stores. Store managers in Spain and southern Europe place orders twice weekly, by 3:00 PM Wednesday and 6:00 PM Saturday. If a store misses the Wednesday deadline, it must wait until Saturday. Order fulfillment follows a similarly strict rhythm, with trucks and airfreight mirroring the biweekly orders so that store managers know when shipments will arrive.[77]

Advances in IT enable work system rhythms that were never possible in the past. In particular, the Internet made it possible for companies to have a 24-hour presence for inquiries and transactions that can be carried out automatically. For example, online brokerages allow customers to analyze their accounts and enter orders at their convenience, even through transactions on the largest stock and commodity exchanges actually occur during specific trading hours. In a very different situation, rhythm matters for the effective operation of geographically distributed teams. Research about these teams found that scheduled telephone or videoconference meetings help these teams maintain a higher level of cohesion than they would have if they didn't meet regularly.[78]

Unfortunately, the relentless and disjointed rhythm of today's anyplace, anytime business environment often has negative effects on those who play the game wholeheartedly. Common use of cellphones, PDAs, and email means that people receive more information faster than they ever did before. The fact that the information can be sent anywhere, anytime sometimes leads to an expectation of a quick response anywhere, anytime, even from home. This is not always a welcome situation for managers and professionals already overloaded with information and work. Many business professionals complain that electronic anywhere, anytime communications has generated higher workloads, more stress, and an inability to get away from work.

Time Pressure

A work system's treatment of time pressure is basically a design decision, with different amounts of time pressure built into different types of work systems. An emergency room portrayed in a television drama is often a place of extreme time pressure, with each passing second increasing the risk of death. Most work environments have more reasonable time pressure.

In many situations, time pressure is used as a management tactic for maintaining efficiency. For example, production schedules try to satisfy customers and also create pressure to produce output on a timely basis. On the other hand, an approach for relieving unnecessary time pressure is to set reasonable expectations for responsiveness and turnaround ("sorry, our lead time is one week") so that people can do their work comfortably. It is also possible to schedule staffing so that peak loads can be handled without undue stress.

Interruptibility

Visualizing work practices as a business process with specific steps occurring in a particular sequence often ignores the fact that some people face frequent interruptions while doing their work. Interruptibility can be viewed as a design decision, with typical office work designed to be highly interruptible and surgery in an operating room designed to be interruptible only under the most extreme circumstances.

Some observers believe that the many communication technologies in today's workplace enable unnecessary interruptions. For example, the author of a book called *In Praise of Slowness* claimed that a typical office worker "is interrupted every three minutes by a phone call, e-mail, instant message or other distraction." … A [2005] study by Hewlett-Packard found that 62 percent of British adults check e-mail during meetings, after working hours and on vacation. "Half of workers felt a need to respond to e-mails immediately or within an hour."[79] People who can control their own work patterns sometimes try to limit technology-enabled interruptions by blocking off "work time" during which emails will not be sent or read.

In some cases it is possible to set limits or expectations on the frequency of interruptions by controlling particular forms of communication and access. For example, a factory that produces complex products might operate on a two-month production schedule that is reviewed every two weeks. To permit the factory to operate efficiently, each revised schedule for the upcoming two weeks is considered frozen, meaning that the sales department cannot interrupt short-term schedules by requesting changes within those two weeks. In some situations work system participants wish their own managers were more aware of the impact of interruptions. One of the complaints about micro-management is that the people doing the work find that unnecessary interruptions by managers absorb energy and impede progress.

Feedback Control

Feedback control is a very general approach for making sure that work systems accomplish their goals. The idea is simply to establish goals and to provide information that reveals whether the goals are being accomplished. Effective feedback control requires careful design. For example, assume that a work system has a weekly goal of producing 500 completed items in a typical 5-day workweek. A simplistic view of feedback control might say that the output should be 100 units per day and that the daily feedback should note any deviation from that daily goal. That view might ignore reality, however. For example, a required component might not be available, or an excess of work in process (partially completed items) near the end of the production process might make it more advisable to complete 500 units on Monday and Tuesday and complete no additional units for the rest of the week. Similarly, even if the daily goal is 100 per day, it is not necessarily obvious that the output should be 12.5 units for each hour of an 8-hour workday. These examples show that feedback control should start with planning and should take into consideration whatever factors are relevant to control the work system effectively.

Although it is a basic system concept, feedback control is often ineffective in systems in organizations.

An explicit cycle of planning, execution, and control helps work system participants know what to do, when to do it, and how to make sure work is done properly. **Planning** is the process of deciding what to do, **execution** is doing the work, and **control** is the use of information about past work to assure that future goals are attained. In the cycle of planning, execution, and control, information is produced as the work is executed. That information provides feedback on the work that is being done, provides information for management, and forms the basis of realistic assumptions for the next round of planning.

The advent of real time information systems created many new design options for the cycle of planning, execution, and control. In some situations, real time monitoring of work makes it possible to provide immediate feedback about quality. Real time comparison of execution versus the plan may provide short-term insight about what needs to be done to stay on plan. Accumulation of real time tracking information may provide valuable control information for management.

Although it may be technically possible to gather and display near real-time feedback, tight cycles of planning, execution, and control may or may not be desirable. Providing immediate feedback may be empowering or may make work system participants feel like cogs in a machine. Providing real time tracking information to management may help in achieving goals or may make work system participants feel they are being micro-managed. In both instances, the question is not so much about the nature of real time information, but rather about how real time information is incorporated into work practices of work system participants and their managers. As is usually true, the design of the work system is much more than the design of the technical capabilities.

Error-proneness

Errors are inevitable, both in the design of work systems and in the execution of the work within the systems that are designed. Execution errors can come from anywhere in a work system. The participants can make mistakes, the information can be wrong, the technology can malfunction, and the work practices can include rules or assumptions that are incorrect or are misunderstood in a particular situation.

In many cases, foreseeable errors can be identified and work systems can be adjusted to minimize their likelihood of occurring. In some cases the error minimization approach is a checkoff list, such as the checkoffs that are done before an airplane takes off. Unfortunately, errors sometimes occur even with something as simple as checkoff lists. For example, in 2005 preflight maintenance activities left a pressurization controller knob out of place in a commercial airliner. The airliner depressurized in flight and a fatal crash occurred because the crew did not catch the mistake during preflight checks.[80]

Error checking is far from perfect, even in computerized systems. For example, "a trader at Morgan Stanley ... made a clerical error in 2004 that caused a 2.8% jump in the Russell 2000 index of small-capitalization stocks just after the markets opened. [The trader] ... put in an order to buy a large basket of stocks valued at several billion dollars, rather than the intended purchase price of tens of millions."[81] In 2005, an employee at Mizuho, a subsidiary of the Mizuho Financial Group in Japan, mistakenly typed an order to sell 610,000 shares at 1 yen, around a penny each, instead of an order to sell one share at 610,000 yen ($5,057) as intended. Mizuho's information system didn't flag the error immediately. The person who entered the transaction recognized the error within a minute and a half, but the Tokyo Stock Exchange's information system blocked efforts to cancel the trade. Investors bought around 100,000 shares, eventually leading to a loss of $330 million by Mizuho.[82]

This type of incident is largely preventable by flagging unusual transaction requests and requiring an additional confirmation from the person entering the transaction. On a more mundane level, perhaps the email command "Reply to All" should require a similar confirmation to help avoid the embarrassment of sending a contentious reply to all recipients of an email instead of just the source.[83]

The avoidance of foreseeable errors involves tradeoffs between allowing errors to occur too frequently and devoting too much attention to avoiding errors. Ways to avoid foreseeable errors include designing systems that are not error-prone, identifying error conditions before they become serious problems, and making sure work system participants are fully aware of foreseeable errors and their consequences.

Making systems less error-prone is easier said than done. A 1984 report on accidents in complex systems such as nuclear plants, dams, tankers, and airplanes found that 60 to 80% of major accidents were attributed to operator error, but that many factors other than operator carelessness contributed to the problems. These factors included flawed system design, poor training, and poor quality control.[84] One of that study's main examples was the partial meltdown at the Three Mile Island nuclear plant in Pennsylvania. A commission investigation blamed the problem on operator error, but the operators were confronted with enormously complex technical systems, incomplete or contradictory information, and the necessity to make decisions quickly. The design of the system created a high likelihood of operator error.

Human error is more likely in work systems that use error-prone technology.

Unfortunately, subsystems for detecting and correcting errors are just as prone to errors as any other part of a work system. An example well known to software developers is a bug in the software that reset Therac-25 X-ray machines when the operator tried to correct an incorrectly entered setting. Several patients received fatal doses of radiation because of this bug.[85] In another example, the safety systems in a Lufthansa Airbus 320 caused a crash landing in Warsaw. The airplane received faulty wind-speed information and came in too fast, creating lift that lightened the load on the landing gear. The flight control software concluded the plane was not on the ground and prevented the jets from slowing the aircraft by reversing thrust. The resulting crash killed two people and injured 45.[86]

Although automation-related malfunctions such as these are often dramatic, inattention and carelessness by human participants is probably a far more significant factor in the failure of safety systems. On an everyday level, many Americans do not use seat belts even though the National Highway Traffic Safety Administration says that wearing a seat belt substantially reduces the risk of death for a front-seat passenger. At a professional level, the news media contain many accounts of managers and other business professionals who ignore safety guidelines related to employee safety or investments. In an extreme example, the Chernobyl meltdown occurred when operators broke many essential safety rules, including the one about not shutting down the safety system. The fact that most safety systems seem to work most of the time is reassuring, but examples such as these show that methods for avoiding errors are an important part of any work system that involves substantial resources or risks.

Exception Handling

Discussions of work systems often focus on how they are supposed to work in typical situations rather than how they should respond when an error, exception, or malfunction occurs. Since real work systems encounter problems and unanticipated situations, a system's design should include those possibilities. Attention to likely errors, exceptions, and malfunctions is especially important when computers are involved because computerized processes are often more structured and less flexible than manual processes.

An example is an order processing system whose software incorporates the assumption that incomplete orders cannot be processed and that a complete order includes the customer's shipping address. What if the customer wants to order a product that takes months to build but has not decided where it should be shipped? Should the order be accepted without the shipping address? Regardless of the software developer's original intention, some real world order takers might treat this situation as an exception and might try to trick the software by entering a "temporary" shipping address so that the software would accept the order. The fact that the shipping address was actually a workaround might be long forgotten by the time the material was ready to ship, perhaps to a fictitious location.

As with error avoidance, the key tradeoffs involve the allocation of attention and effort. Insufficiently formalized work systems treat errors and exceptions carelessly and leave them undocumented. The result is inconsistency and difficulty explaining why the work was done one way or another. On the other hand, devoting too much time to documenting and analyzing errors and exceptions diverts energy away from accomplishing the system's goals. The result is less responsiveness and lower efficiency if the exceptions occur frequently.

Performance Indicators for Work Practices

It is extremely difficult to manage work without measuring performance, and single measures are often inadequate. For example, speed and consistency are typical performance indicators for a business process. Typical metrics associated with speed and consistency include average time from start to finish and average number of deviations. Overemphasis on one metric such as speed often leads to problems in other areas such as consistency or productivity. Consequently, a typical analysis of work practices should look at the current and desired value of metrics for at least several different performance indicators.

Performance indicators for business processes. Table 10.2 lists a number of typical performance indicators for business processes along with two related metrics for each. Inclusion of two metrics for each performance indicator is a reminder that both management of work and analysis of work practices call for identifying practical metrics that capture the essence of the performance indicator and can be measured conveniently.

Each of the performance indicators and metrics in Table 10.2 is relevant to a wide range of business processes. When the analysis focuses on decision making or communication, some process-oriented performance indicators such as speed and efficiency may be useful, whereas others such as consistency or activity rate might not matter.

Table 10.2. Typical performance indicators and related metrics for business processes

Performance Indicators	***Typical Metrics***
activity rate	• number of steps performed per hour • number of units started per day
output rate	• number of completions per day • number of shipments per week
defect rate	• number of defects per 1000 units • number of defects per day
rework rate	• number of units reworked per week • percentage of labor time per week devoted to rework
consistency	• number of deviations from standard per 1000 units processed • number of significant deviations from standard per week
speed	• average time from start to finish (sometimes called lead time or cycle time) • average cost of sales divided by average cost of inventory during a period (often called inventory turns)
efficiency	• units per labor hour or machine hour • time efficiency: time devoted to value added activities divided by total lead time
value added	• value of completed product minus total cost of ingredients
uptime	• percentage of time in operation • percentage of time available for operation
vulnerability	• number of security-related incidents per month • number of known security-related weaknesses, weighted by seriousness

Table 10.3. Typical performance indicators and related metrics for decision making

Performance Indicators	*Typical Metrics*
quality of decisions	• Estimated probability that the decision will lead to unsatisfactory outcomes • Number or percentage of important subproblems that are not addressed • Difference between the expected outcome of the proposed decision and the expected outcome of a recommendation produced by a mathematical model based on simplifying assumptions (and therefore likely to produce a theoretically better answer)
satisfaction of stakeholder interests	• Average satisfaction rating among important stakeholders • Number or percentage of dissenting votes or dissenting views among important stakeholders
range of viewpoints considered	• Number of distinct viewpoints considered • Number of distinct viewpoints reflected in the decision
justifiability	• Financial benefits, measured as net present value, internal rate of return, or other calculation • Implementation feasibility, measured by payback period, (the length of time until cumulative benefits begins to exceed the cumulative costs related to the decision)

Performance indicators for decision making. Some of the performance indicators for business processes are usually pertinent to decision making. For example, consistency, speed, and efficiency are often important in repetitive decision making. Nonetheless, when decision making is the crux of a work system it is often worthwhile to look at performance indicators specifically related to decision making. Just looking the list of decision making phases may reveal that information gathering is insufficient, that little analysis takes place, or that only one alternative is considered. In addition, in many situations the available information systems support only one or two of the phases and provide little or no help with the other phases.

Table 10.3 lists a number of performance indicators and related metrics that might be used when focusing on decision making rather than business process steps. Each of these is relevant in some situations and irrelevant in others. For example, reflecting the views of most stakeholders might be important in a planning or design decision, but of no significance in a decision involving optimal flow through a factory. The three performance indicators for quality of decisions are possible to measure in some situations and difficult or impossible to measure in others. The main point of the table is that a decision making perspective often calls for performance indicators and metrics specifically related to decision making issues, rather than just typical business process issues, such as speed and efficiency.

Interactions between different aspects of performance. In many cases different aspects of performance interact with each other even though they can be tracked independently. For example, a consistently higher activity rate within a well-managed work system usually results in a higher output rate. Similarly, greater consistency typically encourages higher output rates, efficiency, and speed, while reducing defect rates and vulnerability.

In other cases, the impacts could go in opposite directions. For example, a higher activity rate may be more efficient because the same labor and equipment produce more, or may be less efficient because the work becomes disorganized or generates more defects. As with the performance indicators for products and services, there is no reliable rule of thumb for the interactions. There are many cases where an improvement in one aspect of performance leads to a decline elsewhere.

Chapter 11: Participants

- Strategy Issues Related to Participants
- Performance Indicators
- Possible Changes Related to Participants

Work system participants who do the non-automated work in work systems are an integral part of those systems. Without them, those work systems simply cannot operate. Whether system goals are achieved totally, partially, or not at all depends on the participants' skills and commitment. Even a technically spectacular system may fail if its human participants are unwilling or unable to perform their roles effectively. On the other hand, technically primitive systems often succeed when supported and understood by committed participants. Furthermore, when change is needed, the successful implementation of a new or improved work system depends on participants' ability and willingness to switch to a new way of doing things.

Just as work systems depend on their participants, they also affect participants in a variety of ways. Work system changes may make work easier or harder, more interesting or more boring. They may lead to redistribution of personal and political power in the organization. New tools and techniques may help participants develop new skills or may make their longstanding skills obsolete. New information in a work system may help participants do their work, but may also cause harm by infringing on participants' privacy or misrepresenting their performance.

Analysis of the human aspects of a work system starts with identifying participants, describing their roles and organization, understanding their incentives, and clarifying the knowledge, skills, values, and interests they bring to their work. Although issues such as lack of training are relatively easy to recognize and discuss, other issues related to participants are much more difficult, such as questions about competence, commitment, social relationships, and appropriateness of behavior.

Strategy Issues Related to Participants

Many work system strategy decisions such as work roles and organization are related to both work practices and the system participants who perform the work. These strategy decisions are discussed here along with choices related to incentives, reliance on personal knowledge and skills, and working conditions.

Roles and Responsibilities

Examination of a work system's organization chart is usually essential for understanding how a work system operates and how it is managed. The organization chart shows the structure of management, but it may not clarify roles and responsibilities within a work system. Therefore, it is always useful to identify roles and responsibilities and look for issues and conflicts, which may include:

- ***Lack of accountability due to diffuse responsibility.*** Some organizations seem to be designed to keep responsibility diffuse, such as by operating through committee decisions for which no individual is fully accountable. Decision making by committee may lead to greater consensus and commitment, but it may also create opportunities to hide. Committee structures are not necessarily inconsistent with accountability, however. For example, the United States Supreme Court makes decisions by voting, but each member's vote is reported.

- ***Responsibility without authority***. It can be extremely frustrating to have responsibility for something without having authority to make it happen. For example, many chief information officers (CIOs) found themselves responsible for the success of projects but lacked authority to enforce the intended work system changes.[87]
- ***Unnecessary layers of management***. Extra layers of management may slow decision processes without improving decisions. For example, some organizations that cut a layer of management during a downsizing found that the efficiency and speed of decision making increased.
- ***Inconsistency between informal organization and the formal organization chart***. In some situations the formal organization chart does not reflect how work is actually done. In such situations, the people doing the work may feel frustrated about keeping unnecessary managers in the loop. Simultaneously, their managers may feel insecure about their inability to influence the outcome. An example is a non-technical manager whose many responsibilities include "managing" several members of the corporate technical support staff who were assigned to the manager's division. The manager may help them with various general management issues, but is probably unable to comment meaningfully on the technical quality of their work. In this situation, the technical support staff members may try to be more efficient by doing more of their work through informal channels.
- ***Departmental rivalries and politics***. Rivalries and political conflicts between departments may stem from a variety of causes and may have a variety of effects on roles and responsibilities in an organization. Departments within the same company often compete for resources, such as funding for software development projects or permission to hire additional employees. Inadequate communication and lack of empathy between departments with different goals (e.g., the accountants versus the auditors) sometimes leads to an "us versus them" mentality in which people in other departments are belittled and even treated as competition or as the enemy. In other cases, personal rivalries between managers of different departments lead to bad feelings that are eventually absorbed by department members. Regardless of the history or reason, departmental rivalries and politics should be recognized when analyzing systems because seemingly rational recommendations may be undermined if political issues are not dealt with.

Goals and Incentives

Goals and incentives for participants are a prime determinant of how they perform their work. Goals establish and communicate expectations for what work system participants are to accomplish, both individually and collectively. Incentives are the rewards that participants receive for meeting or exceeding goals. A number of examples will be presented here because incentives play such an important role in work system success.

Alignment of goals and incentives. Most managers would agree that a work system's rationale and operational details should be aligned with participants' incentives. For example, poorly designed supply chains often exhibit misaligned incentives because different individuals and companies are trying to accomplish different individual or local goals. In some cases companies create misaligned incentives for their own employees. A Canadian bread manufacturer offered deliverymen commissions based on deliveries to stores. The deliverymen kept the shelves fully stocked even when rival bread makers offered deep discounts. As a result, the manufacturer had to throw away a lot of stale bread. An incentive scheme based on profitability would have motivated the deliverymen to observe what competitors were doing and to stock the shelves consistent with likely sales.[88]

The tension between higher output levels and higher quality is often a key issue when setting incentives. If the system's goal is to produce items of the highest quality, work system participants should be rewarded based on achieving that goal.

Too often, the stated goal or mission statement is to produce items of high quality while the work system participants themselves are rewarded based on the number of items they complete, regardless of quality.[89] For example, a California telephone company spent $170,000 on training programs for 850 customer service reps after an internal employee survey revealed the belief that getting customers off the line quickly was more important than resolving customer problems. The employees liked the way the training program acknowledged their responsibility to help customers, but they became disillusioned when the company persisted in measuring their individual performance based on time per call.[90]

Misalignment of participant incentives and work system goals is a set-up for poor performance.

At least the data collection was automatic in that case. In many other cases people are asked to collect data that doesn't help them personally and may be used against them. Their natural instinct is to undermine or sabotage the data collection, as occurs frequently in systems that collect data involving sales calls, field service calls, and other work done by people who move between customer sites. Mobile workers who report the content and results of sales or service calls often believe the information will provide them no personal benefit and will consume time they should be spending more productively. Worse yet, the information may be used against them. One field service representative explained what he called "pencil whipping" the data. Since providing a specific code for each repair step would take too long with the voice response system he was using, he simply gave a vague blanket code that applied to the entire visit. He understated the actual repair time for one visit by about half because he "got beat up last month" when repairs of this type took too long. To make sure his labor utilization rate would not be too low he then overstated the hours on installations or preventive maintenance calls. He believed people at headquarters didn't really pay much attention to these numbers anyway.[91]

Medical care is an area in which the effect of incentives has been studied many times. Assume you are an emergency room physician and you treat a patient who is drunk. "Most of the nation's emergency rooms and trauma centers don't routinely run blood-alcohol tests or 'tox screens' on patients thought to be intoxicated." An obscure law adopted in 38 states allows insurers to deny reimbursement to patients under the influence of alcohol or narcotics. "Aware that health-insurance policies can contain this exclusion, hospital staff seldom run the tests or urge counseling, for fear the results will appear in claims records and reimbursements will be denied."[92]

Assume you are a physician or a manager of clinic and you know that performance is evaluated based on data generated from medical records. What could you do to manipulate the results? A 2003 article[93] in the *British Medical Journal* reviewed "strategies to optimize data for corrupt managers and incompetent clinicians." Unethical methods for meeting targets related to waiting time include:

- Add staffing during the single week of the year when audits occur.
- Schedule appointments during vacation time for the patient; when the patient does not appear, the counter goes back to zero.
- Delay registering ambulance patients until the staff is ready to see them.
- If the hospital is full, remove the wheels of a trolley to make it a bed.

Other areas for manipulating results include:

- Upcode the diagnosis to increase the reimbursement, i.e., enter a numeric code for a related diagnosis that will generate a higher reimbursement.
- Upcode patient risk factors to make medical results seem more impressive.
- Transfer dying patients to reduce the hospital's death rate.

- Upcode the type of operation.
- Provide insurance to patients with least need.

To see the effect of upcoding, consider the results of a study of the coding of strokes in over 3000 patients in communities in Massachusetts and Rhode Island where "diagnosis related groups" (DRGs) were implemented two years apart. (The *International Classification of Diseases* (ICD) is a detailed description of different diagnoses. The use of DRGs is an American classification scheme that groups closely related ICD codes for purposes of reimbursement. DRGs are broader diagnosis categories that use similar amounts of resources.)

> "In each state, concurrently with the introduction of DRGs, the proportion of strokes classified as cerebral occlusion (ICD-9 codes 434.0 to 434.9) increased, and the proportion classified as acute but ill-defined (ICD-9 codes 436.0 to 436.9) decreased. Before DRGs, 30.0% of strokes in Rhode Island and 26.6% in Massachusetts were classified as cerebral occlusion, whereas 51.8% in Rhode Island and 51.7% in Massachusetts were classified as acute, ill defined. After DRGs were instituted, the proportions of cerebral occlusion and acute, ill-defined stroke, respectively, were 70.9% and 8.5% in Rhode Island and 74.1% and 7.7% in Massachusetts." [94]

Nothing changed in the health status of people in nearby communities, but proportions of diagnoses that fell within DRGs 434 and 436 changed dramatically when the system of reimbursing based on DRGs was introduced.

In another perverse example related to incentives, "an overwhelming majority of cardiologists in New York say that, in certain instances, they do not operate on patients who might benefit from heart surgery [angioplasties] because they are worried about hurting their rankings on physician scorecards issued by the state. ... [The survey] demonstrated the difficulty that many doctors have with the public disclosure of their performance data."[95]

At a much broader level, consider how health care providers face conflicting incentives within the American medical system. Its ideal goal is to keep people healthy, but traditional monetary incentives for health care providers are based on the number of procedures and patient visits, rather than on health-related outcomes. With these incentives, doctors earn less if the system performs better and earn more if it performs worse. The move toward HMOs and other arrangements paid on a per capita basis shifted the nature of conflicting incentives. Forcing medical providers to absorb the incremental cost of each unit of medical care gives the providers an incentive to limit medical care as much as possible. The incentives in the old system drove the providers to encourage additional visits, tests, and procedures, regardless of genuine need. The alternative system drives them to provide fewer tests and procedures, regardless of need. Both approaches suffer from internal incentive conflicts and neither brings about an ideal alignment of participant goals with customer goals.

Similar issues about alignment of goals and unanticipated consequences of systems apply in many other systems. For example, arbitrary date cutoffs in budgeting and control systems create counter-productive incentives. Ideally these systems should assure proper resource allocation and use, but departments trying to avoid future budget cuts often feel pressure to consume whatever is left in the budget before the next cycle begins. The examples of inconsistent and counterproductive incentives could go on and on. The clear challenge for work system design is to take incentives seriously and to recognize that misguided incentives may undermine a system's purposes.

Reliance on Personal Knowledge and Skills

A mismatch between participants' skill levels and work system requirements is a recipe for poor performance. Some skills are highly specialized, but even basic literacy and numeracy skills are often a major obstacle. Productive employees of many firms have found it difficult or impossible to adjust to new business processes involving automated equipment because they could not understand the instructions. Similar problems occurred with some total quality initiatives in which companies discovered employees could do everyday work but lacked basic numerical skills needed to analyze quality information. For example, in the mid-1990s the National Association of Manufacturers claimed that 30 percent of companies couldn't reorganize work activities because employees couldn't learn new jobs, and 25 percent couldn't upgrade their products because their employees couldn't learn the necessary skills.[96]

The other side of the skills dilemma involves high skill participants who demand that work systems use and develop their skills. Regardless of how good a job may seem today, the firm may operate very differently tomorrow. This type of environment motivates people to view themselves as contractors providing skills for hire. If a job provides little opportunity to utilize or extend an individual's unique skills, it may be seen as a step backward. And if specialized skills are being codified in computerized systems, even the long-term value of those skills is questionable.

Designing for participants' skill levels. One approach for dealing with low or inadequate participant skill levels is to design work systems that require minimal skill. McDonald's uses this approach to attain consistent results by reducing work to procedures requiring little or no judgment. For example, timers are used the produce consistent hamburgers by controlling work on the hamburger grill. The system for producing golden brown French fries in consistent portions monitors the boiling grease and beeps to tell the worker to remove the fries. The worker then uses a special fry scoop designed to produce 400 to 420 servings per 100-pound bag of potatoes and make the fries fall into the package attractively.

Although burger flipping can be systematized in this way, total structuring of work is impractical in most work situations because the work itself requires skill and judgment. Many work system participants would not endure the types of systems that fast food workers accept. Even some fast food workers feel they need more control over their own work. One former McDonald's employee said he quit because he felt like a robot. Timers controlled every step of his work on the hamburger grill to produce consistent burgers in 90 seconds. He said, "You don't need a face, you don't need a brain. You need to have two hands and two legs and move them as fast as you can. That's the whole system. I wouldn't go back there again for anything."[97]

The McDonald's system represents an extreme, but it helps in seeing the range of system design choices related to skills. Many transaction processing systems, such as order entry and accounts payable, are highly structured but still call on employees to exercise judgment. In contrast, systems for performing professional work, such as designing buildings or analyzing financial statements, require a high level of knowledge because the work is much less structured.

Store operation systems at Mrs. Field's Cookies provide an example related to designing systems for participants' knowledge level. Founded as a single store in 1977, Mrs. Fields Cookies grew to 600 stores within a decade. The company appealed to customers through warm cookies, friendly service, and reasonable prices. As it grew, the Mrs. Fields Cookies faced the problem of training and motivating relatively inexperienced store managers to use the standards and procedures Debbi Fields developed when she operated her first store in California. The firm's business strategy called for encouraging store personnel to spend more time pleasing customers, and less time doing paperwork. Years of experimentation and development work created a unique information system designed to minimize paperwork, to help with repetitive forecasting decisions such as how many batches of cookies to bake, and to permit headquarters to monitor day-to-day operations at each store.

Based on the belief that it is increasingly expensive to find people qualified to perform all the management functions required in retail outlets, Randy Fields founded a separate software company to sell a generalized version of the software to other businesses. Detailed training about specific management tasks is possible, but he believes that training on topics such as optimal scheduling of part time labor is ineffective. He suggests it is more efficient and effective to structure this type of repetitive decision using software that incorporates techniques developed by the firm's best experts in each area. This approach reduces the paperwork load on the store managers and helps them focus on what they do best, namely, selling cookies and managing employees. In a 2004 interview Fields noted that the managers of retail stores need skills in cost control and skills in sales management. He argued that many of the cost control skills could simply be computerized. [98]

De-skilling. Highly skilled employees sometimes worry that computerization will undermine their earning power. De-skilling is the codification of judgment and automation of work procedures in a way that devalues hard earned job skills and redefines what it means to do a job. Work systems that move in these directions replace human clerical work with computerized data processing and reduce the individual's autonomy and authority through computer-enforced consistency and control. The newly codified processes often permit a less skilled person to produce the same outputs; previously valued skills become less important.

As technological advances make traditional ways of doing that work obsolete, workers with a vested interest in traditional skills often come to believe that de-skilling is a major purpose of technological change, not just a secondary result. Their concerns recall the story of the Luddites, English weavers who rebelled in 1811 and 1812 and tried to smash the machines that were displacing them. Whether or not de-skilling is a goal or a secondary result today, concerns about de-skilling may be a key issue when computerization disrupts traditional work practices.

Whether intentionally or not, automation projects may make participants' skills less valuable.

Tasks most susceptible to de-skilling call for repetition, endurance, and speed, rather than flexibility, creativity, and judgment. Such tasks are highly structured and can be described in terms of procedures. They may involve the processing of data or may involve physical actions such as spray painting a new car or cooking. In some cases, de-skilling accompanied the partial automation of decision processes that once required years of experience. For example, managers of a major insurance company once believed it took five years for group health insurance underwriters to become proficient in determining renewal premiums based on complex statistical calculations. The mystery in training new underwriters disappeared when a new information system automated standard underwriting calculations. Although the system's purpose was to provide better customer service and reduce the stress of yearend peak loads, it also de-skilled the job. New underwriters could be productive on simple cases within months, and the knowledge of the more experienced underwriters was less valued.[99] The project was originally aimed at reducing operational problems and seasonal overloads faced by underwriters, but its eventual effect was to de-value their skills.

The design dilemma in these situations involves finding ways to exploit technological advances to the fullest without treating employees unfairly. In the best cases, system participants apply their existing skills as the basis for learning new skills that use judgment more fully, while leaving the completely programmable parts of the task to the machines. The job becomes larger rather than smaller, and the personal and professional rewards increase. In other cases, however, the challenge of the job disappears along with the old skills. The work may continue, but at the cost of lower involvement and commitment.

Working Conditions and Commitment

Whether or not an employer has the moral obligation to apply and develop employees' skills, and whether or not work system design should reflect this goal, positive steps in this direction probably contribute to employees' feelings of involvement and commitment in their work. According to a 2005 article in *Harvard Business Review*, a Gallop poll found that "in the United States just 29% of employees are energized and committed at work. ... Perhaps more distressing is that 54% are effectively neutral – they show up and do what is expected, but little more. The remaining employees, almost two out of ten, are disengaged."[100] Among others, three general conditions foster genuine involvement and commitment in work.

- a feeling of empowerment
- a good match with the participant's personal sense of pleasure, accomplishment, and ambition

- sufficient interest and variety to keep the participant engaged and attentive.

Empowerment. Employee empowerment has become a cliché that is discussed and recommended more frequently than it is defined carefully and realized in practice. Genuine empowerment requires a combination of the right information, the right tools, the right training, the right environment, and the authority needed to do the job. Lack of any of these aspects of empowerment leads work system participants to wonder whether the firm or its management is serious about supporting their work. After all, why set up a work system to operate in a particular way and then not provide the things that are needed for it to succeed?

Entire books have been written on topics related to empowerment, but we will discuss just one topic that is especially important in highly computerized work systems. Management's view of empowerment is often revealed in its treatment of the information generated within these work systems. That information can be treated as a tool that helps the participants do their work or as a tool that management can use capriciously for setting rewards and punishments.

As an example, consider the computerized customer service operations run by mutual funds and brokerages. These work systems start by automatically routing phone calls to the next available agent. They provide scripts that agents follow, display an existing customer or prospect's information on a screen, and provide forms for entering new information such as the customer's investment instructions or requests for brochures. Because the work occurs through a computer, it is easy to record the exact length of each phone call along with every interval when the agent is off the phone either completing transactions, waiting for calls, or taking a break. It is even possible to monitor whether the agents adjust their average call times during busy periods when customers wait on hold the longest. Effective use of the information generated by such systems can help improve the group's performance by showing when repetitive problems occur and tracking the effectiveness of training and scheduling. Punitive use of the information can make it seem as though Big Brother is watching all the time. Recordings of phone conversations are required for backup in case a customer disputes what was said. Those recordings can be used to help train agents for improvement or can make them feel spied upon. The same information can be empowering or disempowering depending on how management uses it.

Engagement in work. Defining meaningful jobs is an important aspect of work system design. Many people want variety in their work environments and become bored if the work becomes too routine and repetitive. To attain greater variety at work, some manufacturers moved toward production systems in which teams work together to produce major subassemblies instead of asking individuals to do one task repeatedly. These teams often have the authority to redistribute work among their members, both to work more efficiently and to keep work system participants engaged.

In contrast to those efforts, computerization of many data processing systems has made their participants feel more like components of a machine. A previously mentioned example involves a computer-based dental claims system at an insurance company. With the previous paper-oriented work system, the benefits analyst pulled information about each account from a set of paper files, checked contract limitations, completed the necessary paperwork, and returned the account information to the files. Analysts were often hired based on their prior knowledge of dental procedures, and frequently discussed cases with their supervisors and other analysts. With the new computerized work system, much of the information was on the computer, which also ran programs that assured claims were processed in a standard way. The analysts spend more time entering claim data into computers and less time using their knowledge and judgment. Claims analysts who previously knew a lot about each account started saying things like: "I don't know half the things I used to. I feel that I have lost it - - the computer knows more. I am pushing buttons. I'm not on top of things as I used to be."[101] Productivity increased 30 to 40% within a year, but at the cost of lower job satisfaction for the claims analysts.

Healthy jobs. The idea that a non-hazardous job could affect a person's health might be pooh-poohed by macho managers, but researchers have actually found relationships between psychological well being at work and physical health. People with active jobs involving initiative, discretion, and advancement have the lowest heart attack rates, even though these jobs often involve stress. People in high strain jobs at the bottom of the job ladder have the highest rate of heart attacks. Even when such risk factors as age, race, education, and smoking are accounted for, those in the bottom 10% of the job ladder are in the top 10% for illness. These workers have four to five times greater risk of heart attack than those at the top 10% of the ladder, whose jobs give them a high sense of control.[102]

People in healthy jobs use their skills in meaningful work, enjoy autonomy and social relations with others, have personal rights including some control over the demands of the job, and have enough time and energy to participate in family and community life.[103] Based on these characteristics, the least healthy types of work are those with continual pressure to perform but little personal control. Examples include assembly-line workers, clerks who process business transactions, and telephone operators. These are jobs with rigid hours and procedures, threats of layoff, little learning of new skills, and difficulty in taking a break or time off for personal needs. Stereotypical high stress jobs such as manager, electrical engineer, and architect are healthier because professionals have more control over their work.

Autonomy and power. Work systems can be designed to increase or decrease autonomy and discretion in planning, regulating, and controlling work. Providing effective tools is a way to increase autonomy if the individual knows how to use them. For example, data analysis software might permit totally independent analysis work by a manager who previously had to ask for assistance to analyze data. Likewise, professionals such as engineers and lawyers can use information systems to do work themselves that previously would have required more collaboration and negotiation with others. In contrast, many work systems are designed to limit autonomy. This is widely accepted in transaction processing and recordkeeping, where everyone involved taking orders or producing paychecks uses the same rules for processing the same data in the same format. Tracking systems and accounting systems would quickly degenerate into chaos if individuals could process transactions however they wanted to do so.

Information systems that decrease existing levels of autonomy are often experienced as threats. For example, information systems that help supervisors monitor their subordinates generate this response if the monitoring seems excessive. Introducing such systems may lead to resistance and may result in turnover of personnel, especially if autonomy is a traditional in the work setting. Consider the trucking industry, which traditionally permitted truckers to set their own schedules while requiring them to occasionally call a dispatcher from a pay phone. Major trucking companies use on-board computers that control gearshift settings, maximum idling times, and top speeds. Many trucks contain antennas that permit the central office to monitor the truck's location at all times. Some drivers take evasive action by wrapping tin foil over their satellite dishes to obtain some privacy. One long-distance driver who occasionally parks under a wide overpass to escape satellite surveillance was quoted as grumbling, "Pretty soon they'll want to put a chip in the drivers' ears and make them robots."[104] A driver for FedEx expressed similar sentiments when he complained that strict rules from new management "dictated everything from how he dresses to how he holds his truck keys as he approaches a customer's door." He says that "managers shadowed him on his route, then slapped him with a citation for leaving his truck door open as he sprinted to drop off a package on a customer's front porch. "They are absolutely controlling everything I do," he says.[105]

Competitive pressures and the drive to cut costs have led to a general trend toward more on-the-job monitoring of individuals in many jobs. Electronic surveillance is especially common in situations where computerized systems are used continually as part of work. Every keystroke in data entry jobs may be monitored and statistics taken for speed and accuracy of work, and even time spent on breaks. Supervisors may be listening in when jobs use telephones extensively. For jobs involving frequent sales transactions or anything else that can be recorded automatically, every completion of a unit of work may be recorded and made available for analysis by someone at a remote location. For people who use computer networks, every keystroke, email, and Internet access can be recorded.

System Components or Secondary Customers?

Designers of systems should strike the right balance between viewing participants as work system components or as secondary customers. The components view assumes that participants will play their assigned roles and focuses on whether the roles are well defined and whether the participants have the right skills, knowledge, and training. The secondary customers view assumes that the form, operation, and performance of the work system affect job satisfaction, skill development and personal rewards.

Describing participants as system components sounds heartless, but tensions between "participant as component" and "participant as secondary customer" raise many questions that affect system success and participant morale. Treating system participants as cogs in a machine may increase productivity, at least in the short run, but it may also have negative impacts on system participants and their loyalty to the organization. Treating them as secondary customers may make employees more comfortable, but may not generate competitive levels of productivity and responsiveness.

Responding directly to these issues is often difficult. Fierce competition in many areas of business seems to call for a no-nonsense approach of defining exactly what should be done and relentlessly measuring everything that happens. System designers trained in technical/rational methods for building computer systems may also feel little desire to slog through seemingly irrational human issues that are felt differently by different people in different organizations. It is easier and more secure to concentrate on how the work system and technology should work in theory, and to leave unexplored both opportunities for innovation and foreseeable risks of poor system performance or outright failure.

The possibility that the participants will **work by the book** is one of the most obvious reasons to be sure systems are not designed as though the participants are cogs in a machine.[106] Social scientists who study work systems typically report that human participants do not follow system rules precisely, and that if they did they could never get their work done because the "rule book" often fails to reflect what the person's job really is. For example, when Xerox sent an anthropologist on service calls in the mid-1980s as part of an effort to improve its training programs for service technicians, the anthropologist was surprised to find that repairing machines was only part of the job. A large number of calls came from customers who couldn't figure out how to run the increasingly complicated machines. Although not emphasized in training or in manuals, a major part of a technician's job involved teaching customers to use machines properly.[107] System participants who are not treated as secondary customers may be less inclined to find workarounds that permit them to finish their work and also provide other services their customers need.

Performance Indicators

There are two types of performance indicators for work system participants. The first type involves measurement of their direct contribution to work or results during a particular time interval. The second type involves measurement related to participants' characteristics and perceptions, and to other factors that affect performance.

Performance indicators related to work or results from individuals or groups are measured using the same types of metrics that are used for work practices. Typical performance indicators include activity rate, error rate, rework rate, and consistency. Typical metrics include units per hour, errors per week, and average deviation from standard. When looking at the performance of an individual or group, these metrics are calculated at the individual or group level rather than for the entire work system.

Attributing results to individuals or groups makes sense only when they have reasonably complete control over their own work. For example, it is possible to measure a truck driver's mileage and average speed when driving on a highway, but in some situations the truck driver might have little or no control over those numbers. A traffic accident might occur as the truck driver is crossing a bridge. The resulting two-hour delay would be totally unrelated to the truck driver's efforts. Similarly, a factory worker cannot produce good finished items if the components are defective.

Measuring a particular individual or group's recent speed, output, or defect rate is useful and necessary, but good or bad results in those areas are often related to issues that existed before the day, week, or month during which performance was measured. Typical indicators related to factors that drive performance include job satisfaction, morale, turnover rate, fit between skills and task requirements, and amount of management attention required. Indicators such as these cannot be calculated directly from the measurement of work practices. Instead, they might be measured by employee surveys and other data collection that is separate from transactions recorded for financial accounting or management control of operational results. Surveys of this type might ask for numerical responses between 1 and 5 or 1 and 7 on questions such as "How satisfied are you with the support you receive from your manager?" or "How interesting is your job?" In some cases, nothing is measured, but related issues are observed and discussed, such as when a manager tries to understand why a group of employees is unhappy.

Participant-related risk factors are also important to monitor because their presence tends to degrade work system performance. Examples include:

- inadequate management or leadership
- employee or contractor turnover
- poor morale
- lack of motivation
- inadequate skills
- inadequate understanding of the work
- inattention
- failure to follow procedures.

As should be clear from earlier comments about commitment and autonomy, measuring performance and factors related to performance is only a starting point. The value of performance measurement results from decisions about how to use performance information and how to manage work system participants. For example, assume that you, your boss, and your boss's boss all receive your performance summary for the previous week at 8:00 AM on Monday morning. An immediate call from the two levels up every time something goes wrong would contribute nothing to your feeling of autonomy and control. To the contrary, it would leave a feeling of being micromanaged. A delay of a day or two in non-emergency situations would allow you to analyze the situation and then inform others about what happened. Thus, the mere availability of the information does not imply that it should be used in any particular way.

Possible Changes Related to Participants

Many systems analysis and design efforts view systems as technical objects and therefore overlook or downplay possible participant-related changes such as:

- change the participants
- provide training
- provide resources needed for doing work
- change incentives
- change organizational structure
- change social relations within the work system
- change the degree of autonomy in work
- change the amount of pressure participants feel.

Changing the participants and changing the organizational structure may be quite difficult, but if the real problems are in those areas, changing the information, technology, or intended work practices may have little impact. On the other hand, seemingly simpler changes such as providing training and providing the right resources may be surprisingly difficult if the proper knowledge for doing the job is not codified or accessible. People often look to IT first when they think about improving IT-enabled systems. Changes related to participants are just as valid as other possibilities, and often are essential for improving work system performance.

Chapter 12: Information

- Data, Information, and Knowledge
- Defining Information in a Database
- Linking Information to Action
- Evaluating Information in Work Systems
- Causes of Information Risks and Stumbling Blocks
- Strategy Issues for Information
- Too Little Information or Too Much?

Discussions of information in systems usually start with exhortations about the importance and value of information. Several decades ago, the striking examples involved corporations using large databases to take advantage of previously unavailable information. More recently, the striking examples are about third world farmers who can negotiate better local prices for their produce after using cellphones to learn the prices in distant markets.

Information is essential. Everyone knows that having the right information at the right time is essential, and that information systems and IT play a role in providing that information. The work systems of four hundred years ago needed information to operate, just as today's computerized work systems need information to operate. The main difference is that IT provides capabilities that allow today's work systems to be vastly more efficient and powerful.

Today's information processing capabilities start with the ability to capture information using devices ranging from keyboards and bar code scanners through digital cameras and sound recorders. Four hundred years ago, books printed by a printing press were still a relatively new medium for transmitting information. Now we can use high-speed cable and various forms of wireless transmission. Where paper was once the main medium for information storage and retrieval, now we have disk memories and flash memories. Where information in books was static after the book was printed, today we have enormous capability to manipulate and display information.

Advances in technology for capturing, storing, and retrieving information have led to many current IT applications that were infeasible ten years ago. For example, early business information systems processed only numerical information and text because representations of images and video occupied prohibitive amounts of disk space. Advances in the last decade made it not only possible, but also inexpensive to store digital photos and movies. In a sports application, the San Francisco Giants used to store films of games on videotapes and DVDs. "The team now stores video from six different cameras at each home game, plus the broadcast feed from away games. … A player can quickly pull together the video he wants -- say, his last five at-bats against certain pitchers -- from computers in the clubhouse." A pitcher can study videos of opposing batters; a batter can study the strike zone of different umpires.[108]

Current information technology is a wonderful achievement, but it is not a panacea for information problems or issues.

Better information may not help. Many well known examples illustrate the fallacy of assuming that having better information would certainly make things better. Creating value from information means attaining better work system performance than would be possible without the information. This may not happen for a variety of reasons. Information may be collected and made available, but not genuinely incorporated into work systems. Work system participants may ignore better information or may not know how to use it to attain better results. Even if available information is relevant to decisions, people may not act on it.

- People still smoke despite the health warnings on cigarette packages.
- People still build homes on flood plains despite the past history of periodic floods.
- The night before the space shuttle Challenger blew up, engineers warned that a disaster could occur because freezing temperatures at the launch site might cause the O-rings to malfunction and endanger the shuttle. Unfortunately, their warning was overruled.[109]
- Before the terrorist attacks on 9/11, a number of reports identified potential dangers related to lax airport security, suspicious people taking flight training, and the possibility of using commercial airplanes as weapons.[110]

Thus, plentiful, well-documented information may have little effect on important decisions (as with the cigarettes and flood dangers). When the information is controversial or sketchy (as with the space shuttle and the terrorist threat), the likelihood of information affecting action is lower. In other situations, people acting on only limited information may make good decisions by using whatever is available.

The need to look deeper. Given that information is essential and that better information may not help, what can be said about information? The issues ignored by each of the following clichés show why it is important to think about information carefully:

- ***Information is a resource***... Yes, but it is a resource only if people can access it when it is needed, and if they know what to do with it. And it may not be a resource at all if people are swamped with undigested information that is not directly useful for accomplishing their goals.
- ***Information is power***.... Yes, this is true under some circumstances, but see the two previous sentences.
- ***We need information, not just data***... Yes, but it is often unclear how to convert data into information for specific purposes.
- ***We want the facts, just the facts***. ... Yes, we want the facts, but people with facts often do nothing about them. Furthermore, facts are often presented in a biased manner that highlights facts favorable to a viewpoint and obscures contradictory facts. In addition, facts are about the present or the past. For many purposes projections about the future are required.
- ***We want to use technology to share information***... Yes, but meaningful sharing occurs mainly when work systems and organizational culture support sharing of information.
- ***Garbage in, garbage out***.... Yes, but organizations often do very little to identify and correct erroneous information before it causes problems.
- ***It's a digital world***.... Not really. Information is extremely important, but we live in the physical world.

This chapter looks at information in work systems. It summarizes basic ideas about information, such as the distinction between data, information, and knowledge, the different types of information and knowledge, and the necessity of pre-defining the information in computerized databases. It discusses the importance of linking information to decisions or action. It covers major issues in evaluating information in work systems and common causes of risks and stumbling blocks related to information. It closes by discussing strategy issues related to information.

Data, Information, and Knowledge

Failure to convert data into actionable information is a common explanation of why information systems containing vast amounts of data often fail to satisfy information needs. Data includes facts, images, or sounds that may or may not be pertinent or useful for a particular task. Information is data whose form and content are appropriate for a particular use. Data is information for a particular work system if it can be used directly within the work system or is produced by the work system. Since the description or analysis of a work system focuses on whatever is truly relevant to that work system, the distinction between information and data is useful mainly as a reminder to identify the information that could be made more useful.

There are many methods for converting data into information:

- select the data pertinent to the situation and remove the irrelevant data
- combine the data to bring it to a useful level of summarization
- highlight exceptions that may bias the results
- explain more clearly what the data really say
- display the data in an understandable way
- develop models that convert data and assumptions into explanations of past results or projections of future results.

Except when work system participants are fluent with computer technology and analytical methods, the conversion from data into information must be pre-programmed in some way.

Figure 4.2 illustrated the relationship between data, information, and knowledge. Knowledge is a third essential element for understanding the role and use of information. Knowledge is a combination of instincts, ideas, rules, and procedures that guide actions and decisions. People use knowledge about how to format, filter, and summarize data as part of the process of converting data into information useful in a situation. They use knowledge to interpret that information, make decisions, and take actions. The results of these decisions and actions help in accumulating knowledge for use in later decisions. Thus, knowledge is necessary for using information effectively regardless of how brilliantly the information is gathered and combined. Unless a particular part of a work system is totally structured, system participants need knowledge to use the available information effectively.

In many work systems, lack of knowledge may be more of a problem than lack of information. Consider a drop-in clinic physician treating a patient whose medical records are unavailable. An experienced physician with broad knowledge might be able to treat most patients adequately even without the extensive information that the best care would call for. Similarly, an experienced manager or engineer with broad knowledge might be able to walk into a novel situation and make good decisions even though very little formal information is available.

Knowledge in work systems can be treated in a number of ways. At one extreme, it is possible to build knowledge into formal procedures as a way of structuring a business process. This approach forces the participants to conform to a particular way of doing things. A less coercive approach involves codifying expert knowledge in a specialized area as a set of computerized guidelines that help work system participants do their work. This is the way computerized expert systems were originally supposed to operate, although many systems currently called expert systems are more involved with automation of repetitive tasks and decisions. Unstructured situations call for a very different way to deal with knowledge. In these cases, the work system is designed under the assumption that participants must use their knowledge to figure out how the available information is related to the decisions at hand.

Types of Information and Knowledge

A hospital setting illustrates different types of information and knowledge that are often important:

- ***Computerized databases***. Computerized patient medical records and billing information are examples of databases in hospitals. The information in these databases is stored in pre-defined fields such as patient name, diagnosis code, lab test number, and care-provider ID. Information is collected when pre-defined events occur, such as examination of a particular patient by a particular physician, completion of a laboratory test on a specimen from a particular patient, or consumption of a particular pharmaceutical by a particular patient. The items collected when each event (or transaction) occurs are defined in advance as part of the design of a work system that collects the information in conjunction with performing other tasks. For example, examination of a patient by a physician typically generates billing information including the identification of the patient and physician, the date and time of the examination, and the type of examination. Other information related to the patient's condition and care might be recorded in a separate medical records database.

- ***Computerized documents***. These may include X-rays, medical notes, permission forms, insurance documents, email messages, and other information stored as complete documents, rather than as pre-defined fields in a database.
- ***Non-computerized documents and notes***. An example is hand-written notes that might be used as temporary instructions for the medical staff arriving on the next shift.
- ***Information in conversations and cooperative work***. Nurses, physicians, other health care providers, patients, and their families share information in face-to-face and phone conversations. At the end of nursing shifts, outgoing nurses inform incoming nurses about what happened on their shift. This type of information may never be recorded even though it may be quite important.
- ***Explicit knowledge***. This type of knowledge may be codified and recorded or may be conveyed verbally. Medical examples include diagnostic procedures and information in medical books and journals.
- ***Tacit knowledge***. This type of knowledge resides in people's heads but is not recorded or codified. For example, a quick glance at a patient might be sufficient for an experienced doctor to form a hypothesis about the diagnosis. The doctor might or might not be able to identify reasons for the hypothesis, especially if it comes from intuition based on years of experience.

These examples illustrate that important information in a work system may or may not be computerized. For example, X-rays and various documents may be computerized or may be stored as physical documents. Computerizing information is often helpful, but may not matter in some situations and may cause problems in others. For example, computerized medical records have many advantages, but not when the computers are down.[111] Similarly, attempts to computerize knowledge and knowledge distribution sometimes have little impact unless they are integrated into work systems.

Defining Information in a Database

We experience information in our everyday lives and usually don't have to be rigorous in the way we talk about it. Information must be treated differently when it is stored, retrieved, or manipulated through a computer system. To build, use, and maintain computerized databases it is necessary to know exactly what information the system contains, how the information is organized, and how users can obtain whatever information they need.

In a typical corporate database, the files and the items of information are all defined in advance, and the information is filled in with each transaction.[112] The definition of the information within a database determines what information it can store or retrieve for its users. For example, if all information about orders exists on a computer (or in a file cabinet, for that matter), it should be possible to retrieve order #752 if it exists or to verify that it does not exist. Similarly, if the marketing department needs to know what types of companies ordered a particular saleable item, a computer can find each order in a database, identify orders containing that particular saleable item, find the customers for those orders, and look up each customer's industry classification in the customer file. In contrast, a website provides whatever information its designers decide to provide, and there is no universally accessible definition of its content. The lack of universally accessible data definitions for web sites is a major reason why finding information by searching the Internet is a hit or miss proposition.

The following overview is designed to provide an appreciation of these issues for a manager or business professional who will probably rely on someone else to work out the details and perform the technical work of setting up a database. Anyone who will actually set up a database on a computer needs to learn the unique notation and details of the database software that will be used. This discussion uses the terminology *entity, entity type,* and *relationship.* These terms are a bit difficult to absorb, but possibly easier to absorb than the equivalent terminology of *object, class,* and *relationship*, which is used in UML (unified modeling language), a more current attempt to create a general modeling language based on an "object-oriented" paradigm. The meaning of object-orientation is discussed in many systems analysis textbooks for IT professionals, but is beyond the scope of this book.

To help in visualizing the concepts for talking about information in a computerized database, consider the example of a billing system in a consulting company. Every week the consultants submit timesheets containing the number of hours of their time that were devoted to different projects. Each project is related to a contract with a particular client. Several days after the end of the week, the consulting company submits an invoice to the client. To identify the information in the database, it is necessary to answer three questions shown below. A summary of the answers is sometimes produced in a graphical form such as the entity relationship diagram in Figure 12.1, which was generated using Microsoft Access.

1. What are the kinds of things this database contains information about? In database terminology, each client, contract, consultant, timesheet, and invoice is a separate *entity* (*object).* Each client, contract, project, consultant, timesheet, invoice, and invoice detail is a member of an *entity type* (*class* of objects). The definition of the database starts by identifying the entity types (or *classes*). In this particular case, the database contains information that is associated with the entity types (*classes*) clients, contracts, projects, consultants, timesheets, invoices, and invoice details.

Figure 12.1. Example of an entity-relationship diagram

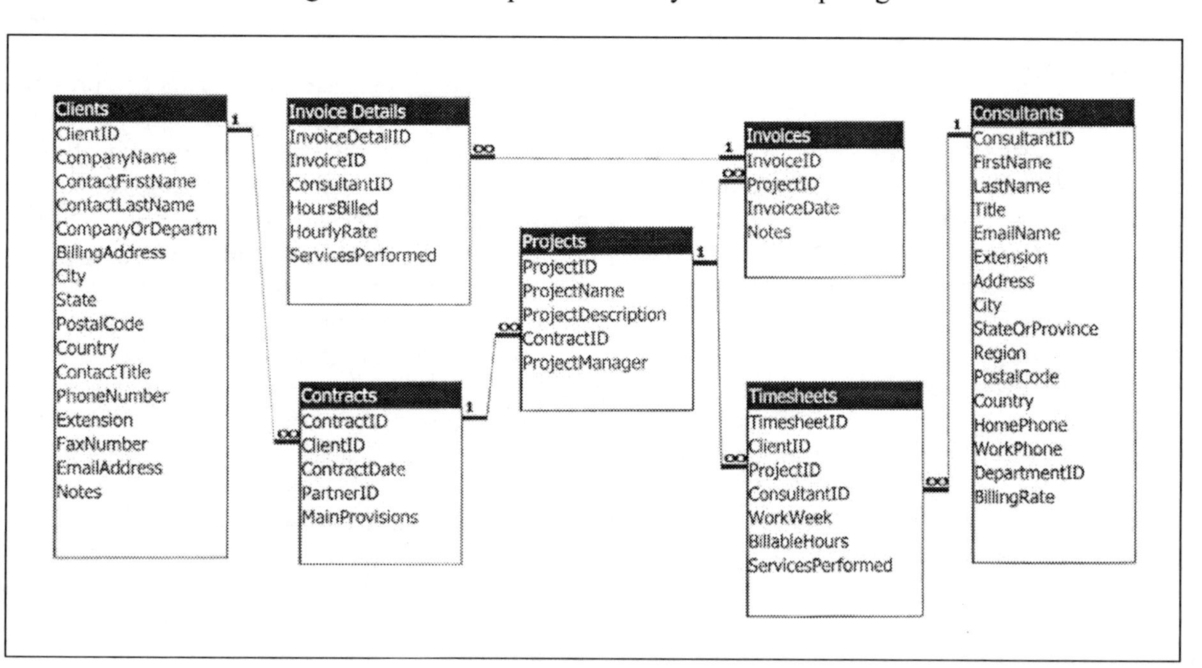

Each entity (*object*) within an entity type (*class*) is identified by an identifier, also called a *key*. For example, a unique identifier for people in the United States is social security number. Name is not a unique identifier because many individuals might have the name John Smith or Julio Sanchez. Each document in a document management system must have a unique identifier for the same reason. For example, a document management system storing letters from customers would need to identify each letter uniquely. Just using the customer number and the date would be inadequate because two different letters from the same customer could arrive on the same day.

2. What specific information does the database contain for each entity type? The database contains the same data items for each entity (*object*) within a particular entity type (*class*). For example, Figure 12.1 shows that the information for a particular timesheet includes timesheet ID, client ID, project ID, consultant ID, work week for which the timesheet is submitted, billable hours, and services performed. Each of these data items is called an attribute. The database treats each entity within an entity type consistently. Thus, the same set of attributes is stored for each timesheet, although the values of those attributes are different because different time sheets are submitted for different clients, different projects, and different consultants, and different workweeks.

3. What is the relationship between the entity types in the database? The relationship between two entity types is the way specific entities of one type are related to specific entities of the other type. Here are some of the relationships in this database:

- A client can have one or more contracts, but a particular contract is from a single client.
- A timesheet (for a particular week) is submitted by a particular consultant. A consultant may have many timesheets (for different weeks and different projects).
- A timesheet can report work on only one project. A specific project may have multiple timesheets (for different consultants in different weeks).

Identifying the entity types, the relationships between the different entity types, and the attributes for each entity type provides the basis for designing the relational databases that are used widely in today's transaction processing systems. A relational database consists of a set of tables of information, such as separate tables of information about individual consultants, clients, and invoices. The database software that operates a relational database makes it possible to use the structure of the database (the relationships between the tables, as illustrated in Figure 12.1) to perform computerized queries such as "List every consultant who has submitted more than three timesheets related to contracts from clients in France." Figure 12.1 identifies the specific data items that will be used in the query.

Unfortunately, the abstract terminology and inherent complexity of databases makes it inappropriate to go further with this discussion. To learn more about the structure of computerized databases, an interested reader will have to read the introductory chapters in a database manual or work with training demos provided with database software such as Microsoft Access.

Just technical details? Figure 12.1 summarizes consulting company's understanding of the source information that it intends to use in billing its clients. This is not just a bunch of technical details. Managers need to pay attention to this type of documentation because it may not represent the reality it is supposed to represent. For example, Figure 12.1 says that a given time sheet is for a given project. Consultants probably don't want to submit multiple timesheets if they work on multiple projects in a given week. It also says that an invoice is for a particular project. The client may want to monitor its use of consulting firms and may want to see billings for all projects on the same invoice. Thus, the relationships documented in Figure 12.1 should be discussed carefully before programming commences. Business professionals should participate fully in these discussions because they often have the best understanding of the reality that is being represented.

Small details can make a difference. Just identifying the information needed or used by a particular work system is often surprisingly difficult, starting with the definition of specific data items. For example, the development of an executive information system at Lockheed-Georgia discovered that the term "sign-up" had different meanings. To the marketing group, a "sign-up" was the receipt of a letter of intent to buy an aircraft. Legal services viewed "sign-up" as receipt of a contract. Finance interpreted it as receipt of a down payment. The standard definition eventually adopted was "a signed contract with a nonrefundable down payment."[113] In a similar example in a computer company, salespeople recorded a sale when the order was placed, manufacturing and logistics recorded it when the order was delivered, and finance recorded it when it was paid for. Likewise, a major railroad had disagreements about what a train is. It could be the locomotive, or all the cars pulled from one location to another, or an abstraction used for scheduling.[114]

Decisions about exactly what information to collect usually involve practical issues and may even involve political issues. Consider the way the United States Census treats information about race (ignoring, for our purposes, fundamental questions about the meaning of race). "In 1870, the U.S. Bureau of the Census divided up the American population into races: White, Colored (Blacks), Colored (Mulattoes), Chinese, and Indian. ... In 1950, the census categories reflected a different social understanding: White, Black, and Other."[115] The Census for 2000 used six categories: American Indian or Alaska Native; Asian; Black or African American; Native Hawaiian or Other Pacific Islander; White, and Other.[116] To recognize changes in the country's racial and ethnic make-up, for the first time, people could describe themselves as members of one or more categories, not just a single category. An earlier Office of Management and Budget proposal that people of mixed race declare themselves "multiracial" was opposed by civil rights organizations. They feared "that the grouping would dilute the numbers of blacks, Asian-Americans or American Indians [and] would make it harder to show discrimination patterns and to draw legislative districts where minorities constitute the majority of voters."[117] In other words, what might seem like a technicality about coding specific information can raise serious, controversial issues in some situations.

Data definitions and codes. In a computerized database, even the properties of each data item must be specified in a very detailed way. For example, properties that can be specified for a data item in Microsoft Access include data type (e.g., number, text, or currency), field size (number of characters), format, caption for reports, default value, and validation rule.

In some situations basic properties of a data item can have major impacts. For example, telephone numbers in the United States are 10 digit numbers, with three digits for the area code and seven additional digits. At one time, the area code identified the telephone's geographical region and the next three digits identified a more localized area. That scheme started to break down as the number of telephones multiplied. Soon it was necessary to split area codes, and some Los Angeles telephones with a 213 area code were switched to 310. Personal telephone lists had to be corrected and it was necessary to reprint telephone directories and business cards. The largest 10-digit number is 9,999,999,999. While it is unlikely that the United States will run out of phone numbers soon, the meaning of the three digit area code will become even hazier as cell phone owners bring their phone numbers with them to new locations.

A similar problem occurs when other coding schemes become obsolete. Consider a company with product lines 1, 2, and 3, whose products are numbered 1xxx, 2xxx, and 3xxx (in other words, product numbers such as 1654, 2377, and 3554). Adding two more product lines would result in products coded as 4xxx and 5xxx, but when the firm expands to its tenth product line, the first digit of a four-digit product number is no longer adequate for identifying the product line. If the product identification scheme is incorporated into the logic of computer programs, moving to a tenth product line might require programming changes related to the size of fields in the database, the operation of programs that access the database, and the format of reports that are generated. Technical tricks such as separating the data definitions from the programs make this kind of change easier. Even with technical tricks, the IT staff has to make sure that all of the details interact correctly. In other words, even small changes in the data definitions in a computer system may have many effects that someone has to check. The implications of changes become even more complicated in interorganizational systems such as supply chains that involve multiple companies, each of which may have its own conventions for numbering products and handling transactions.

Similar issues arise when the information being entered into a database is validated against predefined codes, such as error codes or service codes. The initial definition of the coding scheme is very important because changes in the codes could make past data impossible to reconcile with future data. For example, assume that a company had three service codes describing different types of service calls for the last five years. As the business develops, the service calls of types #4 and #5 become important. Later, managers find it necessary to subdivide each of those into several other codes. That type of change may be very useful for tracking current work, but now it will be impossible to use past data for trend analysis.

The discussion of data definitions and codes may seem like a lot of details, but details of this sort often determine whether information is useful. Information whose coding is designed poorly tends to become inconsistent or useless. In addition, industry initiatives for establishing standard formats for business-to-business transactions require a great deal of effort and negotiation about what might seem like small details. If the negotiated solution is similar to the way a particular company handles those details, it will enjoy the advantage of being able to support the new standard after fewer modifications of its internal information systems.

Databases and data warehouses. To make the use of computerized databases as efficient and convenient as possible, many companies separate their database capabilities into two parts. A transaction processing database is typically designed to make transaction processing as efficient and secure as possible. This type of processing focuses on controlling access and updating specific data items consistent with rules for specific types of transactions. In contrast, a data warehouse is designed to make the analysis of information, perhaps across an entire database, as convenient as possible without disrupting the processing of transactions. Information in the transaction database is copied and moved to the data warehouse daily or on some other consistent schedule.

The information in a data warehouse is usually a filtered and cleansed version of information in a database that feeds it. The difference results from different goals. For example, an order entry database typically includes the company name, address, names and phone numbers of customer contacts, and additional information for each customer. A data warehouse used for analyzing sales data typically does not need customer contact names and phone numbers, although it might contain customer ZIP code or other customer category codes that could be used in analyzing order information. Furthermore, the information in the order database is often at a finer level of **granularity** than is needed for analysis purposes. For example, the order database may contain each order that was received each day for several years, whereas the data warehouse might contain only weekly summaries of orders by product. If the corporation has thousands of products or product variations, the weekly totals in the data warehouse might be accumulated by product line, thereby providing more useable information for analysis. Further details of the "ETL" process of extracting, transforming, and loading data warehouse information are readily available on the Internet, but beyond the scope of this book.[118]

Linking Information to Action

Better information tends to have the greatest effect in work systems if it is used in a well-understood business process in which the information makes a difference, and in an environment in which system participants are encouraged to use that information to the utmost.

The ultimate question about creating value from information is to link that information to well-conceived decisions and actions. A useful initial premise is that better information may not help. There are many examples in which people with good information made bad decisions and people with little or no information made good decisions. The link between information and action is usually much clearer in highly structured situations. This link should also be examined in semi-structured and unstructured situations to avoid creating overload and to make sure that the information processing is worth the effort.

Value of Information

If better information would not change decisions or other outcomes, there is little reason to obtain the information or devote attention to it. Decision theorists use the idea of whether information makes a difference as the basis of a literal definition of the value of information - the monetary difference in the expected outcome when the information is available and when the information is not available. Assume, for example, that you want to buy a particular camera at the lowest possible price, and for some reason prices for this particular camera are not available on the Internet. An expert in camera prices will charge you $10 to identify the distributor offering the lowest price. By making assumptions about the probability of different purchase prices with or without this information, you could theoretically calculate whether on average you would receive enough of a price break to make the $10 purchase of the information worthwhile. If your calculations showed that knowing which distributor offered the lowest price would save you more than $24 on average, for example, it would be worthwhile to buy the information for any amount up to $24. This would be the value of the information to you regardless of whether the expert would charge $10 or $100.

Unfortunately, this theoretically elegant approach asks people to guess monetary values they have very little inclination or ability to guess, such as the average price across all distributors. Guessing values such as these is difficult enough when the work system is highly structured. The guesswork quickly becomes futile in semi-structured or unstructured work systems such as general management. Consider a weekly sales report that summarizes a company's sales by region. Managers use that report to monitor sales results and identify issues and opportunities. The monetary value of this information is difficult to assess because missing one week's report might cause a serious oversight or might have no consequences at all.

Information Needs in Semi-structured Work Systems

Highly structured work systems are designed to operate by using specific information in formulas or tightly specified procedures. In these situations, the information requirements are known and the methods for converting data into the required information are well articulated. Pre-defined information is an intrinsic part of the design of these systems. Human judgment plays a greater role in less structured work systems, where relationships between information, work practices, and outcomes are less clear.

Consider the authorization of credit card purchases. A computer can use a totally automated process to authorize a small purchase such as a $10 pizza. The decision factors for a small purchase include whether the credit card has been reported stolen and whether the recent account payment history is acceptable. The consequences of an incorrect authorization decision on a $10,000 purchase are much greater, and the relevant patterns and rules of thumb are much less clear. The authorization process for expensive items is semi-structured and the information requirements are not obvious. A great deal of information about the cardholder's recent credit activity, payment history, and financial resources may be available, but the decision in these cases requires human judgment.

In the late 1980s American Express developed an improved work system for authorizing purchases. Most purchases could be authorized automatically, but around 5% of authorization requests could not be granted automatically. These went to credit agents who had to decide what to do within 90 seconds, the guaranteed response time. Creating an effective work system required converting a mass of data from 12 different databases into coherent information that could be used for decision making. The new Authorizer's Assistant software used over 800 rules to simulate the decision process of experienced credit agents. Within 4 seconds, the credit agent making the decision about the $10,000 purchase received an automated recommendation plus an explanation of the reasoning behind it. The remaining 86 seconds was available to think about the situation and ask any necessary questions. Implementing the Authorizer's Assistant increased credit agents' productivity by 20%, reduced denials by 33%, and improved the accuracy of predictions of credit and fraud losses.[119] (Public discussion of this and similar systems is rare today because of competitive concerns and the possibility of giving hints that could abet fraud.)

In relation to information, the special characteristic of this system is that the requirements were not obvious and that a lot of experimentation was required to provide guidance while permitting authorizers to make decisions based on judgment. Many other semi-structured or unstructured work systems could be improved if someone could develop a way to convert data into actionable information. Typical decisions in such systems include which sales prospects to pursue, which product features to modify, or which investments to pursue.

Information for Management

General management is an important type of work that requires varied information from many sources and does not have reliable formulas or procedures for making decisions. Here are some of the general types of information used in the process of management:

- ***Comfort information***. Managers may not do a great deal each time they see the latest sales results or factory status information, but just having the information makes them feel more comfortable that they know what is going on and don't have major surprises on their hands.
- ***Exceptions and warnings***. Managers want to know what is out of the ordinary so they can take action before a problem becomes more serious or before an opportunity disappears.
- ***Metrics***. Managers track metrics related to key performance indicators to make sure the organization achieves its goals. Making it obvious to the organization that they care about measuring performance helps motivate people in the organization to attain those goals.
- ***Situational information***. Managers want current information about specific situations that currently require their attention. The challenge of including this type of information in formal information systems is that the situations change frequently.
- ***Gossip***. Managers need to know what people in an organization are thinking and saying, regardless of whether it is accurate or fair.
- ***External information***. Internal information may be easy to provide as a by-product of business operations, but managers should not allow themselves to fixate inward because many of their challenges and opportunities come from outside.

Given the semi-structured or unstructured character of management processes, the term *requirements* fails to capture the nature of management information needs. *Requirements* makes sense for structured processes because they can't operate without the information they require. For management activities, *desired information* might be a better term because managers can operate with much more latitude in terms of what information is used and how it is used.

Disconnects between Information and Decision Making

Better information helps only if it will be considered seriously. Psychologists and decision researchers have identified a series of common flaws in individual and group decision making. Each of these flaws can nullify the value of even the best information:

- ***Groupthink*** is a group's extreme aversion to disagreement among group members. The decision to go ahead with the 1962 Bay of Pigs invasion of Cuba is often cited as an example. Somehow, no one stood up and said that the attacking force would be outnumbered more than 10 to 1, making the plan exceedingly risky. After the invasion force was captured, President Kennedy is supposed to have said, "How could we have been so stupid?"[120]
- ***Poor framing***. Assume a business venture has an 80% chance of success, equivalent to a 20% chance of failure. One group was told the chance of success was 80%. The other group was told the chance of failure was 20%. The majority of the 80% group gave the go-ahead, but the majority of the 20% group did not. Logically the options are identical, but they were framed differently; the framing affected decisions.[121]
- ***Overconfidence*** is excessive belief in one's own opinion, often exaggerating the importance of favorable evidence and ignoring contrary evidence or opinions. Professional athletes who

guarantee victory sometimes learn they are overconfident.

- ***Poor probability estimation***. Overestimating the probability of familiar or dramatic events, while underestimating the probability of negative events. For example, one year before the space shuttle Challenger blew up, NASA estimated the probability of such an accident as 1 in 100,000 even though historically 1 in 57 booster rockets had blown up.[122]

- ***Association bias*** is an inappropriate belief that strategies used successfully in the past will succeed again in new situations that are different in important ways.

- ***Recency bias*** is a tendency to overemphasize the most recent information. If your boss seems susceptible to recency bias, try to be the last person to offer a suggestion before the decision is made.

- ***Primacy bias*** is a tendency to make a decision too soon by giving undue weight to the first relevant information. If your boss seems susceptible to primacy bias, try to be the first person to offer a suggestion.

Better information may not help in these situations because system participants are unprepared to receive and consider it. When there is a disconnect between information and decision making, even high quality information made readily accessible and presented effectively may not make much difference.

Evaluating Information in Work Systems

The quality and usefulness of information in a work system can be evaluated across many dimensions. **Information quality** can be defined as fitness for use by its users. Some aspects of fitness for use, such as accuracy and precision, are related to the information itself without regard to how it is used. Other aspects of fitness for use, such as timeliness and completeness, depend on how the information is used, and in some cases, who uses it. For example, some managers feel comfortable making decisions with much less information than other managers might say they need.

Dimensions for evaluating information. Table 12.1 identifies dimensions within five major categories for evaluating information.[123] Some dimensions within these categories can be assessed objectively by inspecting a sample of the information to identify inaccuracies, biases, and other shortcomings that can be measured. Other dimensions are subjective.

Although it is easy to say that all information should be accurate, precise, complete, and reliable, there are almost always tradeoffs involving the cost of providing the information and the benefits of using it. For example, the cost of each increment of accuracy and accessibility increases drastically as information quality approaches perfection. The additional cost in going from 98% accuracy to 99% accuracy might involve little more than additional follow-up and attention to detail, whereas going from 99% to 99.9% might require major changes in the entire work system's operation. Approaching 100% in timeliness and completeness has similar costs. For example, the ultimate of timeliness for a national retail chain would involve immediate updating of the central inventory database with each sale anywhere. This would be more expensive than using a nightly download of each store's total sales of each item for the day. Information in the nightly download would be several hours older on average, but the additional timeliness would probably have almost no effect on decisions. Similarly, the decision of whether to perform an expensive medical test to obtain more complete information about a patient's condition should depend on whether the possible beneficial effect on the treatment plan outweighs the additional cost and discomfort.

Table 12.1. Five dimensions of information quality

Dimension	*Typical metrics*
	Intrinsic quality of information
Accuracy	• percent correct • percent within a specified deviation (such as 1%) from a correct number • average bias
Precision	• number of significant digits in a number • pixels per square inch in an image • size of the units or components of information (also called granularity, e.g., whether sales totals are by product and day or by product line and week)
Age	• amount of time elapsed between when the source information was created and when the final information is used, possibly after further summarization
Believability	• subjective evaluation based on how information was produced or based on inaccuracy or bias in previous information from the same source
Traceability	• extent to which information can be traced back to its source, especially when many different groups contribute to different parts of the information
Reputation of the source	• percentage of bias (can sometimes be assessed by comparing information with similar information from other sources) • subjective evaluation based on inaccuracy or bias in previous information from the same source
Objectivity	• percentage of bias (can sometimes be assessed by comparing information with similar information from other sources) • subjective evaluation based on identifying points that are incorrect or exaggerated, or relevant information that is ignored or downplayed
	Accessibility of information
Access control	• extent to which only intended users have formal permission to access the information • amount of extra work and authorizations needed in order to access the information
Access time	• average time to retrieve information • average time to find an expert with the desired knowledge
Controllability by user	• controllability of data selection • controllability of data manipulation and summarization • controllability of formatting
	Contextual quality of information
Relevance	• subjective evaluation of the extent to which the information helps in understanding a situation, performing a task, or making a decision
Value-added	• subjective evaluation of the difference in the outcome resulting from the use of specific information
Timeliness	• extent to which information arrived on or before a deadline • subjective evaluation of the extent to which the availability and age of the information are appropriate for the task
Completeness	• extent to which stated requirements for specific information are fulfilled • subjective evaluation of the extent to which the information is complete for a particular task
Appropriateness	• subjective evaluation of the extent to which the information is appropriate for use in the context • legality (extent to which use of the information is legal in the situation)

Representational quality of information	
Conciseness	• amount of information presented versus amount of information available
Consistency of representation	• number or percentage of inconsistent codes or meanings within a body of information
Interpretability	• subjective evaluation of the extent to which information is in appropriate language and units, and data definitions are clear
Ease of understanding	• objective measure of the effort required to understand specific information • subjective evaluation of the extent to which the information is clear, unambiguous, and easily understood
Security and control of information	
Vulnerability to inappropriate access	• number of incidents of inappropriate access • subjective evaluation of the adequacy of information security
Vulnerability to inappropriate use	• number of incidents of inappropriate use • subjective evaluation of the adequacy of information security

Intrinsic Quality of Information

The intrinsic quality of information involves characteristics of the information that are independent of its use, such as its accuracy, precision, and age.

Accuracy and precision. Accuracy is the extent to which information is correct and free of error. Precision is the fineness of detail in expressing the information. For example, assume that my bank account contains $5430.94. Telling my wife that it contains $5430.94 would be 100% accurate, but saying that it contains around $5400 would be accurate enough for planning family expenditures for upcoming months. On the other hand, telling her it contains around $1000 would be suspiciously inaccurate. Telling her it contains $1153.71 would be more precise than telling her it contains around $1000, but would be about equally inaccurate (off by around 80%). In other words, it is possible for information to be extremely precise and extremely inaccurate, as happened in accounting statements by Enron, MCI, and other companies involved in criminal misstatements of financial information.

Computerized databases contain a surprising amount of incorrect or obsolete information. For example, the information in customer databases tends to become "stale" over time as customers change jobs, phone extensions, and addresses. Although it is difficult to find systematic research about the accuracy of information, several articles in the 1990s said, "between 1% and 10% of data items in critical organizational databases are estimated to be inaccurate."[124] In the 1980s, a study of computerized criminal records in several states and in the FBI found that records in some systems were 50% to 80% inaccurate, incomplete, or ambiguous.[125] More recently,

> "Gartner Inc. released a startling statistic: More than 25 percent of critical data used in large corporations is flawed, due to human data-entry error, customer profile changes (such as a change of address) and a lack of proper corporate data standards. The result: soiled statistics, fallacious forecasting and sagging sales. What's more, the research firm says that through 2007 'more than 50 percent of data-warehouse projects will experience limited acceptance, if not outright failure, because they will not proactively address data-quality issues.'" [126]

The accuracy of personal information in publicly available databases has significant impacts on many individuals, yet is often questionable. "A 2003 Government Accountability Office report, citing statistics from the Consumer Federation of America, found that 78% of credit-agency files omitted account information, 82% had inaccuracies regarding revolving accounts or collections, and 96% had bad credit-limit information."[127] One of the reasons for inaccuracies is that much of the information in publicly available databases is not verified. For example, the data broker ChoicePoint "does not check nor feel it is responsible for the accuracy of the estimated 17 billion files it has collected and stored. It assumes only that the facts it acquires are accurate when they arrive." According to its chief marketing officer, "We do not verify the factual basis of a record, but instead rely on the assertions of our data sources that created the record." [128] An example of the impact:

> When Brian Graham applied for a mortgage, he was surprised that "he was unable to qualify for a low adjustable rate. The culprit was a $72 bill for a cellphone service cancellation fee - one he had disputed with AT&T Wireless. Mr. Graham said he thought the matter had been resolved, but it had instead been placed on his credit report. … After months of letters and phone calls, he has hired a lawyer and filed a lawsuit to try to clear his name. … He learned how hard it can be to clean up one's credit history, even when it is soiled in error."[129]

"A study by the U.S. Public Interest Research Group determined that consumers spent an average of 32 weeks after initial contact negotiating with credit bureaus [trying] to get incorrect information removed, before they contacted federal officials."[130]

Finally, inaccuracy of verbal information is also important. For example, consider the accuracy that you expect when you call a help line to ask a question. A survey by the GAO (Government Accountability Office) found that the accuracy of tax-law information provided by the Internal Revenue Service's taxpayer assistance line increased from 79.5% in 2004 to 89.5% in 2005. That was a very good improvement, but it still means that over 10% of the callers received erroneous information.[131]

Granularity. This is the size of the units or components in which information exists. For example, sales totals stored by product and day have finer granularity than sales totals stored by product line and week.

Age of information. This is the amount of time between when the source information was created and when the final information is used, possibly after further summarization. As will be mentioned under timeliness, for many purposes it is quite acceptable to use information that is weeks, months, or even years old if nothing relevant to the use of the information has changed in the interim. For example, anyone who uses the latest census information should recognize how old it (especially later in each 10-year census cycle) and should think about whether anything relevant to its use has changed in the interim.

Believability. Important information sometimes has questionable believability because it may be biased or open to challenge for other reasons. For example, people providing information to formal information systems and planning systems may introduce bias based on their personal objectives and incentives (as was mentioned in the previous chapter). Project managers often exaggerate a project's likely duration if missing the scheduled completion date will have negative consequences for them. Sales people often try to negotiate low sales targets so that they will be more likely to receive bonuses for exceeding their targets. Even customers answering customer satisfaction surveys may provide biased information. For example, auto dealers for one of the Big Three auto manufacturers asked customers to fill out a customer satisfaction survey because marketing executives wanted to understand why satisfaction ratings for individual dealers were not closely related to dealer profits or growth. According to an article in *Harvard Business Review*, auto dealers saw the satisfaction survey as "a charade they play along with to remain in the good graces of the manufacturer and to ensure generous allocations of the hottest-selling models. The pressure they put on salespeople to boost scores often results in postsale pleading with customers to provide top ratings, even if they must offer something like free floor mats or oil changes in return."[132]

Traceability. In many circumstances it is important to be able to trace information to its original source and verify that its original source was valid. An example of the importance of traceability was cited by the Presidential Commission that investigated the reports of weapons of mass destruction in Iraq. Its report said,

> "One of the most painful errors concerned Iraq's biological weapons programs. Virtually all of the Intelligence Community's information on Iraq's alleged mobile biological weapons facilities was supplied by a source, code-named 'Curveball,' who was a fabricator … It is, at

> bottom, a story of Defense Department collectors who abdicated their responsibility to vet a critical source; of Central Intelligence Agency (CIA) analysts who placed undue emphasis on the source's reporting because the tales he told were consistent with what they already believed; and, ultimately, of Intelligence Community leaders who failed to tell policymakers about Curveball's flaws in the weeks before war."[133]

Traceability is also important in large projects in which requirements come from different sources. After extensive negotiations and compromises to whittle the project down to a more manageable size, the negotiated requirements may no longer satisfy anyone's initial intentions. Traceability back to the original requirements may help in deciding that the negotiated scope of a project is not worth pursuing.[134]

Reputation of the source. The believability of information is often related to the reputation of the source. For example, claims by politicians in political campaigns often have a believability gap, especially if their previous statements and actions reveal strong biases. Unfortunately, information from governments often has the same problem, such as when estimates of future budget deficits are frequently based on excessively optimistic assumptions.

Even auditors, whose job is to verify the validity of financial statements, have come under fire. The string of accounting scandals uncovered in 2001 and 2002 (e.g., Enron, WorldCom, Tyco) implicated major public accounting firms, whose audits were shown to be inadequate. As a result, the U.S. Congress passed the Sarbanes Oxley Act of 2002 to protect investors by improving the accuracy and reliability of corporate disclosures. Among other provisions it requires certification of financial reports by CEOs and CFOs and requires publicly traded companies to have an internal audit function that is certified by external auditors.

Objectivity. Although conscious misrepresentations by politicians may seem outrageous, expecting people to be totally objective and unbiased is often unrealistic. Incentives to hide and distort information exist in families, in businesses, and in government. Bias pervades informal systems in many organizations, especially in the way repeated filtering and sanitizing changes the meaning of verbal information and recommendations. Furthermore, even when people try to be as accurate and unbiased as possible, they see reality through the filters of their own personal experience. Since those experiences are different, the filters are different and people observing the same phenomena often see them differently. Even in companies that pride themselves on fact-based management, a total emphasis on objectivity and truth may miss something important. Businesses need honesty, but they also need passion and commitment. A manager proposing a new marketing program probably won't spend half of her presentation time on all the reasons it may not work. If she did so, one might wonder whether she really cared about leading that program.

Accessibility of Information

The accessibility of information is the ease with which authorized users can obtain the information. A file cabinet full of customer contracts in the form of paper documents illustrates the issue of accessibility. If the work system requires access to individual contracts by people who are in the same room as the file cabinets, information accessibility may be very good. But if the work system requires statistical summaries crossing many contracts, the manual task of compiling the information becomes daunting or impossible. Computerization may or may not solve this problem. Computerized information from different sources may contain incompatible data definitions or codes; technical interfaces between incompatible computers or networks may be difficult or costly to traverse.

The Internet provides information accessibility that would have seemed amazing to people in 1990. Although this level of accessibility is wonderful, it is also limited because accessibility involves a number of separate hurdles:

- ***Identifying what you want***. If you are not following a hypertext link from another site, and if you do not know the address (URL) of the web site you want, then you need to use a search engine such as those provided by Google, Yahoo, and MSN.

- ***Finding the relevant sources within the search engine's response.*** You may have a simple query in mind, but the search engine may identify thousands of possibly relevant sites, and the first few that are listed may not be the best match for what you need.
- ***Permission to access the information***. Some of the sites identified by a search engine may be available only to a restricted list of users.

Even information on your own personal computer may be difficult to access, such as when you want to find a document you produced last year but don't remember exactly where it was filed. Downloadable desktop search programs such as those from Google and Yahoo often make it easier to find documents stored on your computer, but those programs also have shortcomings.

Accessibility within a corporate database system can be simpler because the information in the database is defined in advance and because the query is specified using that data definition. On the other hand, the available information may not be at the level of granularity that is needed. For example, someone comparing health claims might need to know total sick days by division and month, but the available information may be for each sales office and week. Consolidating by division should not be difficult if there is a list that associates offices with divisions. The consolidation from weeks to months would be trickier because the number of days in a month is divisible by 7 only in February of non-leap years. In a more complex example, a hospital needed to estimate the cost of treating heart attacks when it switched from reimbursement by procedure performed to reimbursement by disease group. Information about different procedures and tests were scattered across several different databases, and there was no automatic way to perform the consolidation for heart attacks.

The above examples illustrate something about information accessibility that may be surprising. In practice, the accessibility of information depends partly on the way the information system is set up, and partly on the knowledge of the person who wants to access the information. More experienced Internet users can use search engines more effectively than novice users. Programmers and advanced users of databases are more able to perform database queries that are not defined in advance. Hence, even the ownership of information does not guarantee it will be accessible in practice.

Contextual Quality of Information

The contextual quality of information is the extent to which information fits the needs of the work practices within a work system. Contextual quality includes relevance, value-added, timeliness, completeness, and appropriateness of information.

Relevance. The information included within a work system snapshot is relevant by definition. Otherwise it wouldn't be included in the work system snapshot because it wouldn't be used or generated by the system's work practices. Information requirements of highly structured work systems tend to be defined clearly and intertwined with the way the work is done. Information requirements in semi-structured or unstructured work systems are less clear, and relevance is assessed more in terms of whether the information might make a difference. For example, when you are interviewing for a job, information such as the salary, benefits, job description, and location are obviously relevant, but other information about the company, industry, co-workers, and working conditions may be relevant to one person and irrelevant to another.

Value-added. The earlier discussion of the value of information explained that the value added by specific information is often difficult to assess. At minimum, it is easier to justify system changes that provide new information if that information can be linked directly to measurable improvements in results.

Timeliness. This is the extent to which the information's age and availability are appropriate for the task and user. Real time information is preferred for many tasks, but is not necessary for many other tasks. For example, an airline reservation system needs up to the minute information to avoid unplanned overselling of seats. On the other hand, a planning system that revises the airline's schedule needs to combine seasonality information with recent operational information that may be weeks or months old. Excessive attention to yesterday or today's information would be counterproductive because planning systems need to consider cycles and trends rather than daily blips that might misrepresent trends.

Completeness. This includes the breadth, depth, and scope of the information that is needed. Assume that your company needs to hire a new salesperson and all the information you have is a resume. That information might be enough for deciding whether to invite the candidate for an interview, but it is incomplete as a basis for hiring. A face-to-face interview would be required to observe the applicant's appearance, ability to interact, fit with the company, and other important characteristics.

A different problem related to completeness is the possibility of being swamped with too much information in relation to a task. Assume that a manufacturer of customized industrial equipment needs to produce a price quotation for a customer. Information might be available about past sales, about the customer, and about the status of competition. That information might be overwhelming and difficult to use unless the company has a model that uses specific information to generate at least a first cut at the quotation.

Appropriateness. There are many cases in which it is inappropriate or even illegal to use or discuss specific information. For example, large companies that compete in the same industry may be charged with price fixing if their employees discuss pricing strategies with competitors' employees. Also, as any U.S. job applicant should know, U.S. equal opportunity laws say that it is illegal to discriminate based on age, disability, gender, national origin, pregnancy, religion, or race.[135] Small talk and questions about these topics in job interviews are inappropriate because they may imply or represent illegal discrimination. In another example, advances in genetic engineering have raised issues about where genetic information can be used legitimately. A 2005 statement from IBM claimed that it is the first major corporation to adopt an explicit policy of not using the results of genetic tests to make employment decisions. Employment discrimination on the basis of genetics is barred in various forms in 33 states.[136]

In other situations, it is inappropriate to use certain information that might seem relevant but was developed for different purposes. For example, costs collected in an operational control system provide feedback about process efficiency and performance. Those costs are inappropriate for use in an activity-based-costing analysis, which uses standard costs (not actual costs) to describe the underlying economics of performing particular groups of activities.[137]

Representational Quality of Information

Information should be represented in an appropriate format that accentuates its meaning. Representational quality of information is described in terms of conciseness, consistency, interpretability, and ease of understanding.

Conciseness. Ideally, information for a particular task should be well organized and compactly represented. A major failing of many business documents is the lack of conciseness and clear direction. To help the reader, lengthy business documents usually start with an executive summary and outline. When the information is numerical, conciseness is accomplished by summarizing the information to a more aggregated level, by creating a model that computes results implied by the information, or by creating a graphical representation that converts information into a picture. Conciseness has its own problems, however, because excess summarization may hide information. For example, a manager with 500 products might want a one-page summary by product line, but that summary might hide the fact that the five new products introduced last month did badly.

Conciseness of information is also important on a personal level, where the flow of information through email, the Internet, and computerized databases has overloaded some individuals so much that they need information filtering and organization much more than they need access to additional information. On a work system level, providing participants ready access to massive amounts of tangential or ill-focused information may be counterproductive because it diffuses their attention.

Consistency of representation. This is the extent to which information is always presented in the same format and is compatible with previous information. The importance of consistent representations became apparent when NASA lost a $125 million Mars orbiter as it was about to go into Mars orbit. The engine fired after a 286-day journey, but the spacecraft came too close to the surface and failed to go into orbit. Engineers from Lockheed Martin had performed calculations using English units of measurement (feet and pounds), while engineers at the Jet Propulsion Lab had used metric units (meters and kilograms).[138] Inconsistent representation of information doomed the mission.

Lack of consistent representation in formats for numbers and dates in different countries around the world is sometimes a problem. For example, assume today is the 8th day of September 2006. Here are different formats that might be used around the world for displaying exactly the same information.

United States	*mm/dd/yyyy*	09/08/2006
Europe	*dd/mm/yyyy*	08/09/2006
China	*yyyy.mm.dd*	2006.09.08

It is easy enough to set a switch in a computer program to print these dates using whatever is the local notation. The more difficult issue involves making sure that someone in a United States office will not misinterpret a document produced in a foreign office, especially if the *yyyy* is abbreviated as *yy*, and the dates become 09/08/06 or 08/09/06 or 06.09.08.

Interpretability. This is the extent to which information is expressed in appropriate language and units, and data definitions are clear. For example, a speed limit of 60 means something very different to a person thinking in terms of kilometers per hour rather than miles per hour.

Interpretability is an important problem when non-experts receive or provide information that has a specific meaning to experts such as doctors, lawyers, or accountants. For example, when accountants present cost, profit, and depreciation on financial statements, the meaning of those terms is quite different from the meaning of the same terms when used in general conversation by non-accountants. On financial statements, depreciation is a calculated allowance based on formula (such as 20% of the original value per year). Depreciation is treated as an expense that reduces profit. Non-specialist conversation treats depreciation as a physical reality, such as "the value of my car depreciated due to the accident." The danger with terms such as depreciation is that non-specialists won't understand what they are supposed to mean in a specialist context, such as the financial statements that are used to manage companies and inform investors.

Interpretability is extremely important for medical information. For example, medical tests for detecting cancer and other diseases can have two types of error:

- ***False positives***. The result says the patient has cancer, but the result is wrong.
- ***False negatives***. The test misses a cancer that should have been detected.

Correct interpretation of medical test results requires an understanding of the rate of false positives and false negatives for any particular test.

Ease of understanding. This is the extent to which the information is clear, unambiguous, and easily understood. As an example, consider the description of Line 31 of the U.S. Government's Form 6251 for calculating alternative minimum tax as part of a 2004 tax return:

> "If you reported capital gain distributions directly on Form 1040, line 13; you reported qualified dividends on Form 1040, line 9b; or you had a gain on both lines 15 and 16 of Schedule D (Form 1040) (as refigured for the AMT, if necessary), complete Part III on page 2 and enter the amount from line 55 here.
>
> All others: If line 30 is $175,000 or less ($87,500 or less if married filing separately), multiply line 30 by 26% (.26). Otherwise, multiply line 30 by 28% (.28) and subtract $3,500 ($1,750 if married filing separately) from the result."

If one uses a tax preparation program, the software fills in the answer automatically based on previous choices and calculations. Looking at the excerpt from the form, it is obvious that ease of understanding involves more than just precise language. It seems unlikely that most people who are not accountants or tax experts would be able to read these instructions without moving their lips.

A very different example comes from a September 2003 press release by the Federal Open Market Committee (FOMC) of the U.S. Federal Reserve Board:

> "The Committee perceives that the upside and downside risks to the attainment of sustainable growth for the next few quarters are roughly equal. In contrast, the probability, though minor, of an unwelcome fall in inflation exceeds that of a rise in inflation from its already low level. The Committee judges that, on balance, the risk of inflation becoming undesirably low remains the predominant concern for the foreseeable future. In these circumstances, the Committee believes that policy accommodation can be maintained for a considerable period."

One commentator noted "all the murky phrases" such as unwelcome fall in inflation, the foreseeable future, and a considerable period, and said that the press release is "a mélange of stock phrases; it is imprecise, pregnant with haze."[139] Ease of understanding apparently was not one of the FOMC's goals in producing the press release.

Quantitative information in the form of tables and graphs is usually easier to understand than ambiguous prose, but even tables and graphs can be difficult to understand. For example, assume that a stock dropped from $75 per share to $72 per share over the course of a week. A graph showing the daily values will give a very different impression if the Y-axis goes from $0 to $75 versus $70 to $75. The y-axis from $0 to $75 will make the $3 drop appear negligible. The y-axis from $70 to $75 will make it look as though a serious drop has occurred.

Security and Control of Information

The first four categories of information quality dimensions (intrinsic quality, accessibility, contextual quality, and representational quality) are all about making information as useful as possible. The fifth category is about preventing misuse.

Information control starts with making sure that only authorized individuals have access. For example, it is important to control access to a customer list, to an employee's employment and health history, and to a salesperson's latest notes about progress in on-going negotiations with a customer. Employees of firms with highly computerized systems usually have explicit access privileges for specific information. This approach requires that a person or group in the organization must be in charge of setting up tables that specify which particular roles can have access to which information, and under which circumstances. For example, roles with the highest level of access can see information and can update it, roles with lower levels of access can see it or print it, but cannot update it.

Access restriction is all the more important when the information might be misused or stolen. Widespread use of computer networks and portable computing devices such as laptops has made it more difficult to know who is accessing sensitive information at any time.

The widespread use of the Internet and of commercial databases has greatly increased threats related to inappropriate access. For example, in October 2004 the data broker ChoicePoint discovered an identity thief's inappropriate access of its database. ChoicePoint has files "full of personal information about nearly every American adult. In minutes, it can produce a report listing someone's former addresses, old roommates, family members and neighbors. The company's computers can tell its clients if an insurance applicant has ever filed a claim and whether a job candidate has ever been sued or faced a tax lien." Months after the breach was discovered, ChoicePoint estimated that personal information about 145,000 people had been compromised.[140]

Causes of Information Risks and Stumbling Blocks

It is possible to discuss risks and stumbling blocks related to each dimension in Table 12.1. To avoid repetition, this section mentions some of the common causes of those risks and stumbling blocks.

Poor design. A widely cited study of accidents in complex systems such as nuclear plants, dams, tankers, and airplanes found that 60% to 80% of major accidents were attributed to operator error, but that many factors other than operator carelessness contributed to the problems. These factors included flawed system design, poor training, and poor quality control.[141]

A disastrous example occurred in 1995 when an American Airlines jet crashed into a mountain while approaching Cali, Columbia. The plane was taking an alternative route to make up for lost time. It was supposed to be guided by a radar beacon at an intermediate point, Rozo, but the crew entered only the letter "R," which represented a different beacon, Romeo, which was far from Cali. The plane turned to a different direction. The crew recognized that something was wrong, but continued the descent and the plane crashed."[142] The system had been designed to allow a person to designate a radar beacon with a single letter. The crew was supposed to verify the validity of the change, but failed to do so. A better-designed system would have reduced the likelihood of such as disastrous mistake. At minimum, it would have made sure that the change would be verified.

Sloppiness and lack of discipline. In a subsequent trial, the airliner crash was attributed partly to poor design, and partly to human error, which obviously occurred. In addition to entering incorrect information, the crew failed to follow procedures by verifying the new course. Sloppiness and lack of discipline is common in our everyday lives, but fortunately is rarely fatal.

Sloppiness and lack of discipline by work system participants affect the quality and usefulness of information quality. Accuracy of information may decline due to inattention to recognizable errors in information. Work system participants may not question the believability, source, or objectivity of information they receive. They may proceed with incomplete information even though more complete information might be available. They may make inappropriate assumptions about contextual or representational quality, thereby misinterpreting information or using information that does not fit the situation. They may do too little to maintain the security and control of information.

Efficiency pressure and overload. One of the reasons for sloppiness and lack of discipline is that care and discipline absorb time and effort. Unless people have clear incentives to identify and fix information problems, they tend to focus on task completion rather than information quality. Focus on producing output rather than identifying and fixing information quality problems is especially likely when people already feel overloaded with work and unable to take on new projects.

The conflict between efficiency needs and information quality is especially worrisome in relation to security and control of information. Information security requires systems for defining and enforcing access privileges, preventing unauthorized access to computers and networks, and monitoring data flows across networks. Resources devoted to those systems are not available for producing products and services for customers.

Changing contexts. Regardless of how well an information system was originally designed, changes in the context may make aspects of that system obsolete. For example, corporate mergers often lead to difficult information system transitions if the merged company is to operate as a single entity rather than two separate entities under one umbrella.

Changes in the external environment can also require that existing information systems handle new information. An example in the shipping industry is proposed anti-terrorism requirements that call for tracking both shipping containers and high security seals that assure nothing was added or removed while a container was in transit. Existing information systems track the containers, but do not separately track seals or their status over time. Every major shipping company bears the risk that new regulations will require expensive changes in information systems that support their basic value chain operations.

Lack of knowledge. Even correct information that was generated for one purpose may be highly inappropriate for other purposes. People sometimes perform their work mechanically and do not fully understand why they are doing work one way and not another. That lack of knowledge increases the chances of mistakes when novel situations call for new uses of existing information or uses of information that has not been used previously. For example, a sales forecast used for short term production planning may be totally inappropriate in a long term capacity planning analysis.

Crime. Information risks related to crime are large and growing, as is clear from the epidemic of network break-ins, identity theft, sabotage, and other criminal misuse of information by company insiders and by hackers and thieves on the outside. Here are several recent examples among many dozens that might have been cited:

> "A former Teledata Communications Inc. employee pleaded guilty in federal court to fraud and conspiracy charges in connection with what prosecutors have called the largest identity-theft case in U.S. history. ... Prosecutors said more than 30,000 people were ripped off as part of the scheme, in which credit reports were downloaded from three major credit-reporting agencies with stolen passwords and sold on the street for $60 each."[143]

> "Wells Fargo Home Mortgage, the nation's largest residential mortgage lender, on three occasions in the past year had to notify customers that confidential information was either stolen from its computers or viewed by unauthorized parties."
> ... A computer stolen from a consultant's office in California contained data about 200,000 customers.
> ... A laptop stolen from the trunk of car in St. Louis contained data on thousands of loan customers.
> ... Computers with student loan records were stolen in Georgia.[144]

The first example involved an employee who stole and sold personal information about customers. The second involved the theft of computers. The ChoicePoint example mentioned earlier involved the online sale of personal data to a person masquerading as a legitimate businessman.

Crime risks increase the challenges related to every other type of risk mentioned earlier. Poor design of systems often includes poor security. Many systems should be redesigned to improve security. Sloppiness and lack of discipline increase the likelihood that crimes will be perpetrated. Pressures related to efficiency and overload make it even more difficult to devote necessary resources to security and control. The increasing sophistication of criminals requires more knowledgeable employees. Taken together, the various risks related to information quality and usefulness present a substantial challenge for many important work systems.

Strategy Issues for Information

Many of the ideas about the identification and evaluation of information are also relevant to information strategies within work systems. These strategies can be grouped under information quality awareness, information quality assurance, ease of use for information, and information security. Information strategies will not be covered at length because a complete discussion of the topic would repeat parts of the previous sections. Instead a brief discussion will mention several points.

Quality awareness should be mentioned first because most strategies for information quality and security depend on widespread awareness of the importance of those goals. Following the passage of the Sarbanes Oxley Act mentioned earlier, CEOs and CFOs are surely more aware of information quality in financial statements because they are personally responsible for the validity of these documents. The requirement that companies have an internal audit function surely spreads those heightened concerns to other managers. Paying more attention is a starting point, but quality assurance needs to be systematized.

Quality assurance for information is basically a set of work practices and therefore needs to be built into work systems. Issues addressed in a quality assurance strategy for information include:

- checking and fixing the quality of incoming information instead of taking it for granted
- checking and fixing the quality of information produced by the work system, preferably performing the checks as soon as the information is produced
- tracking the quality of information as an integral part of the work practices for monitoring the work system.

The relevance of these aspects of quality assurance varies between work systems. Checking for detectable errors or omissions is easiest in transaction processing systems such as order entry and bill payment because these systems use pre-defined databases. For example, when someone enters an order using an ecommerce website, it is possible to check whether the address matches the zip code and whether the credit card number is valid and is not on a list of stolen or misused credit cards. Whether done by computer or manually, this type of checking is rule-driven and mechanical. In contrast, it is much more difficult to check information in the form of documents, conversations, commitments, or beliefs. The evaluation is largely subjective and rarely achievable through automation using pre-defined rules or criteria.

Ease of use. This is a key consideration in the design of databases and interfaces. The interplay between the various dimensions of information quality (see Table 12.1) illustrates why ease of use should not be taken for granted.

Ease of use for information starts with accessibility and with the amount of effort required for identifying and retrieving information. Standard daily and weekly reports for managers are designed to minimize that effort. When periodic reports and computerized queries are insufficient, ease of use calls for minimizing the amount of effort and delay in extracting, filtering, and formatting relevant information. In some cases, it is effective to provide retrieval and analysis tools for work system participants. In other cases, it is more effective to provide help from IT professionals or "super users" who are well versed in how to use a firm's data retrieval capabilities and who are willing to provide help needed by less knowledgeable users.

Security and control. Risks of misuse and theft of information generate a variety of tradeoffs between ease of use and control. For example, greater accessibility makes information easier to use, but often increases the risk of misuse. Conversely, security and control measures absorb time and effort that might otherwise be devoted to producing whatever the work system is designed to produce.

The need for security and control calls for procedures designed specifically to make sure that information is available only to authorized users who use the information appropriately and do not pass it on to unauthorized users. Unfortunately, convenience of using information often decreases as control increases. For example, it is possible to create lists of people who may access a particular file of information. It is also possible to have multiple passwords or other forms of control for logging on. Each security capability comes with a price, however, because time and effort related to security might be devoted to other things.

Too Little Information or Too Much?

This chapter's discussion of information began by noting that better information may not help even though information is essential for all work systems. As a final point, it is worthwhile to think about whether less information, rather than more, would be better in some situations. Even without the Internet, many of today's business people are swamped with so much information that they feel overwhelmed. Nobel Prize winner Herbert Simon described the phenomenon this way,

> "What information consumes is rather obvious: it consumes the attention of its recipients. Hence, a wealth of information creates a poverty of attention and a need to allocate that attention efficiently among the overabundance of information sources that might consume it."[145]

Simon's ironic point about information consuming attention applies to information in general, and it also applies to analyzing systems. It is always possible to extend the analysis of a system by looking for more information. Regardless of whether the task is selling a product, managing an organization, or analyzing a work system, the message throughout this chapter is that business people should pay attention to whether they have enough information and whether it is of adequate quality. They should be aware that information is absolutely essential but that more information may or may not help. Each section of the chapter discussed issues that help in making those assessments.

Chapter 13: Technology and Infrastructure

- Technology Basics
- Evaluating Technology
- Importance of Infrastructure
- Technical Infrastructure
- Information Infrastructure
- Human Infrastructure
- An Enterprise View of IT

This chapter expands on brief comments about technology and infrastructure in previous chapters. Unlike a chapter that might have been written 10 or 15 years ago, it assumes that most readers are familiar with common information technologies such as laptop computers, cell phones, and digital cameras. That familiarity makes it unnecessary to explain details of specific technologies, especially since those details are readily available on the Internet.[146]

This chapter focuses on IT and IT-related infrastructure even though the relevant technologies when analyzing a work system also include non-IT technologies. This chapter starts by summarizing IT basics, such as what IT does, what IT hides, why IT is important, and why better technology may not help. It summarizes key issues for evaluating technologies within work systems. It looks at technical, information, and human infrastructure. It summarizes management concerns within an enterprise view of IT.

Technology Basics

Previous points related to the basics of technology and infrastructure include:

- Technologies of interest when analyzing a work system include both information technologies (IT) and non-IT technologies. Technologies combine tools and techniques. Even when substantially computerized, technologies (such as cars) and techniques (such as checklists or sorting) may or may not be associated with IT.
- General-purpose technologies, such as search engines, cell phones, and spreadsheet software can be applied in a wide range of business situations. Situation-specific technologies are tailored to particular application areas, e.g., a spreadsheet model for calculating mortgage interest or a software program for designing kitchens.

- Technologies tailored to specific situations usually combine general-purpose tools, such as laptop computers, and specialized techniques, such as mortgage calculation formulas. The separation between tools and techniques is worth considering because it is often possible to use a different general-purpose tool (e.g., a better laptop) without changing the technique. Similarly, it is possible to change the technique (e.g., moving to a better mortgage calculation method) while using the same laptop. In other situations, different techniques require more powerful general-purpose tools. For example, early PCs were not fast enough to support voice recognition software and had no mechanical capabilities to record CD-ROMs or DVDs.
- Infrastructure includes human, informational, and technical resources that the work system relies on even though these resources are managed outside of it and are shared with other work systems. **Human infrastructure** consists of people and organizations that supply services shared by different work systems. **Information infrastructure** is information shared across various work systems. **Technical infrastructure** includes computer networks, programming languages, and other technologies shared by multiple work systems.
- Within the work system method, technologies whose surface details, interfaces, and affordances are visible to work system participants are treated as integral parts of a work system. In contrast, shared technical infrastructure is used by particular work systems, but often is largely invisible to work system participants.
- The choice between treating something as technology inside the work system or as technical infrastructure that the work system uses should depend on what is most helpful for performing a specific analysis. The Work System Framework distinguishes between technology and infrastructure because taking infrastructure for granted is often a mistake, as is apparent when a work system cannot operate due to problems in the infrastructure it relies on.

To extend those points further, it is useful to look at what IT does, what it hides, why it is important, and why better technology may not help.

What Does IT Do?

"Any sufficiently advanced technology is indistinguishable from magic."[147] This often repeated quote mirrors the way some business professionals treat IT as magic.[148] The many books that try to explain the intricacies of IT succeed to varying extents in peeling away one or two layers of magic only to reveal other layers of complexity and mystery. The difficulty is that technical aspects of technical innovations apply engineering knowledge that most business professionals don't understand. Just as they can drive a car without understanding the thermodynamics of combustion, most business professionals neither want nor need to understand much about how their computing engines operate internally. But they still want to make sure IT is applied to the right problems in the right way and genuinely supports work system performance.

Chapter 10 (Work Practices) noted that processing of information using IT can be boiled down to six basic functions. To see those functions in an everyday example, think about how a grocery store's information system supports customer checkout:

- It *captures* information about an item's identity using a bar code reader.
- It *transmits* that information to a computer that looks up the item's price and description.
- It *stores* information about the sale of the item.
- It *retrieves* price and description information in order to print each item on the bill.
- It *manipulates* information when it adds up the bill.
- It *displays* information when an LCD screen shows prices and a printer prints the receipt.

Looking at the six functions provides a direction for exploring possible work system improvements. Instead of starting with customer, product, or process characteristics, it is certainly possible to look at each of the six functions in turn and ask whether improvement in any area would improve work system results. When seen this way, information processing is not the headline, but rather, just another work system subcomponent.

What Does IT Hide?

All advanced technologies hide things from users. For example, most drivers don't know how an automobile engine operates, and don't care. Similarly, IT hides many aspects of its own operation that typical users don't care about. For example, the precise location of the computer that sends a web page to a user's personal computer is invisible to the user, as is the precise physical location of a database within a corporate computing system.

Hiding details from users is a two-edged sword. Hiding details helps users focus on the things they care about. However, users may not always benefit from what is hidden, such as:

- ***Surveillance***. Many IT applications are designed to perform monitoring of users. For example, every keystroke and utterance of a call center worker can be recorded. Likewise, many firms monitor usage of the Internet to make sure that employees are spending their work time productively.

- ***Unexpected storage of information***. There have been many cases in which people were surprised by the archiving of embarrassing emails on corporate or governmental computer systems. Deleting those emails from their PC hard drives had no effect on archived backups, and the emails were found years later during investigations. Similarly, authors of word processing documents are sometimes surprised that identification information is embedded in document metadata and corrections.[149]

- ***Unexpected use of information***. For example, the Drivers Privacy Protection Act was passed after criminals obtained victims' addresses from state departments of motor vehicles. Most car owners probably would never dream that states would sell the address of a person with a particular license plate number.[150] Similarly, links between disparate computer applications and databases allow many forms of information usage that were not anticipated when the applications or databases were originally designed. For example, new ways to match information across medical, insurance, and financial databases provide many possibilities for using personal information in unexpected ways. Some people worry that a diagnosis or even a symptom recorded in the privacy of a doctor's office will affect their future insurance rates and possibly their employment prospects.

- ***Lock-in***. Use of technology often creates a practical barrier that prevents switching to a different technology. For example, someone with 1000 documents written in Microsoft Word would need a significant impetus to move to a different word processing program if any of those documents could not be transferred automatically and invisibly.

- ***Hidden assumptions and logic***. All computer programs are based on assumptions and express a particular logic. Those assumptions and that logic are often inconsistent with assumptions and logic of users. For example, an elementary school's information system may be based on the assumption that a student has one mother and/or one father who should receive notifications about events and student progress. Some students whose parents remarried may have more than two parents who should be on the list. An information system that provides only two parent slots would not meet the school and students' needs regardless of whether the assumption of two parents seemed correct to the original designer.

IT hides essential technical details that users don't care about. Whether intentionally or not, it also may hide, obscure, or misrepresent threats to personal and organizational interests.

Why IT Is Important

A 2003 *Harvard Business Review* article with the inflammatory title "IT Doesn't Matter"[151] caused great consternation among IT professionals because it challenged the competitive value of IT investments. Anyone who has ever shopped at a large grocery store knows that the title is misleading. IT obviously matters because today's grocery stores cannot operate competitively without IT. If competitors in the retail food industry are using IT to record sales, track inventory, and make sure the shelves are stocked, a grocery chain that does not do so is at a great disadvantage. A more forthright title for the *HBR* article would have expressed something like: "IT brings competitive advantage only if a firm applies it to enable organizational capabilities that are unique and difficult to copy."

IT is important for work systems because it allows people to do work in ways that otherwise would have been difficult or impossible. Many relatively recent advances led to important capabilities that were previously uneconomical or totally unavailable. Consider three examples among dozens that might have been mentioned:

- ***Electronic invoices and data mining***. Until recently, the vast majority of invoices sent to businesses were generated using a computer, but mailed to the supplier in the form of paper, forcing the supplier to re-enter the same information into its computerized payables system. Liberty Mutual, a large insurance company, illustrated the potential benefits of electronic invoices when it began requiring electronic invoices from law firms that handle litigation. Aside from streamlining its clerical processes, Liberty Mutual created a database containing the details of the invoices it received. Analysis of the database revealed instances of duplicate invoices, charges for unauthorized services, and, in several instances, attorneys billing for more than 24 hours in a day.[152]
- ***Ecommerce***. Many aspects of the Internet, such as the use of packet switching[153] for transmitting data, have existed for over 30 years. Ecommerce became commercially viable during the 1990s due to a combination of additional advances including the invention of easy-to-use browsers, development of easily used conventions for web addresses, improvements in programming methods, and vast increases in data transmission capabilities.
- ***Digital photography***. The invention of photo-sensitive chips made digital photography possible. The first generations of those chips were a limited threat to traditional, film-based photography because the resolution was too low to produce high quality images. Subsequent improvements in chips made it possible to capture images whose quality equals or exceeds that of images captured on film. In turn, high-resolution images created a challenge for flash memory cards, because low capacity memory cards (e.g., 16 megabytes) could store only a few high-resolution images. Fortunately, the capacity of memory cards increased quickly to a gigabyte or more. Advances in flash memory technology also made it possible to store a large amount of information on iPods and other small devices.

Many other examples can be cited involving technologies such as inexpensive desktop and laptop computers, CD-ROMs, DVDs, PDAs, spreadsheets, and tools for building commercial application software. The main point is that advances in IT unquestionably matter a great deal because they enable activities and capabilities that were previously impossible.

A striking example of the impact of IT for organizational efficiency comes from the aftermath of the 1991 Persian Gulf War, when the U.S. Army sent 40,000 tractor-trailer sized containers to Saudi Arabia with no identifying information. It was necessary to open each container just to find out whether it contained tires, generators, or something else. In addition, there were many over-shipments because, according to *Fortune*, supply sergeants traditionally ordered everything three times in the expectation that two requisitions will go astray in unmarked containers. A review of logistics operations following the war led to a vastly improved approach in which bar codes on the containers are used to list their contents and global positioning satellites are used to signal their location.[154] In effect, newer technology made it feasible to change the information requirements, and the new accessibility of information made it possible to change the work system.

The Gulf War example also showed that simply having information is not enough because its quality, accessibility, and presentation all affect usefulness. The old system, over-shipments and all, was based on the now obsolete assumption that a globally accessible database identifying the destination and content of any container was impossible. The old system contained the right information at the point where the containers were packed, but could not make that information readily accessible at the receiving points in the War zone. The new technology of bar codes, databases, and global positioning satellites permitted new assumptions and made the existing information available where it was needed.

Why Better Technology May Not Help

Wonderful as current technology is, the assumption that better hardware or software technology will generate better results often leads to disappointment. A faster computer or more powerful software will not help an author who doesn't know what to say; it will not help an organization process transactions quickly if the staff is in disarray or if staff members don't know how to use the technology effectively. In other words, better hardware or software technology may not affect performance at all because it may not address the obstacle or bottleneck that limits performance. Changing only a specific technology has a positive impact mainly when the existing technology is too expensive, too slow, broken, or otherwise inappropriate.

Anyone faced with technology upgrade decisions should consider what a highway engineer said about the rebuilding of certain highway overpasses after the 1994 San Fernando Valley earthquake: "When we strengthen some of the older structures using the newest highway technology, all we are doing in many cases is moving the likely point of failure from one place to another."

In most cases, improving a specific technology in a work system makes sense only in conjunction with corresponding changes elsewhere in the work system. For example, years ago many observers believed that computer aided software engineering (CASE) software would solve productivity problems in producing application software. Firms that bought it discovered that it had only marginal impact unless work practices for producing software were changed to exploit the new CASE capabilities. The same issue occurred with many other hardware and software "solutions" that solved very little without corresponding changes elsewhere.

The benefits of better technology are also reduced when the potential users don't really know how to translate better technology into better work system results. This happens frequently, such as when people adopt new technology because they simply want to stay up to date or because a central IT organization told them they had to switch. Both reasons are common motives for continual upgrades from one version of software to the next. For many users, the additional training and learning time will not improve the product or the efficiency of the work system because they will continue to do the same work the same way.

Better technology may not help if existing technology is not a limiting factor in work system performance.

The productivity paradox. The fact that better technology may not help is part of the reason for the productivity paradox, the mystery of why it is so difficult to demonstrate the positive net benefit of computer expenditures on corporate success. Research on this topic usually analyzes the relationship between computer expenditures and economic success (profitability or market share) in a group of comparable firms. Although some correlations between IT spending and corporate results are beginning to emerge,[155] a technology upgrade is unlikely to yield results by itself unless the problem is a pure technology problem. Adding more or better computers or networks may improve a work system's performance but may also increase costs or expose weaknesses or points of failure elsewhere.

A final issue is that technology may absorb much more time than one would expect. A 1993 survey by the accounting software firm SBT Corp. asked customers to estimate how much time they spent "futzing with your PC." The average was 5.1 hours per week doing things such as waiting for computer runs, printouts, or help; checking and formatting documents; loading and learning new programs; helping co-workers; organizing and erasing old files; and other activities such as playing games.[156] Citing this estimate, a consultant with the consulting firm McKinsey quipped that if the estimate is correct, personal computers may have become the biggest destroyer of white-collar productivity since the management meeting was invented.[157] Many technological improvements have occurred since those quotations from the early 1990s. Better PC software and improved methods for installing and controlling networked PCs may have eliminated some of the "futzing around" that occurred then, but Internet use, email, instant messaging, virus protection, and many new capabilities provide new possibilities for absorbing time unproductively.

How might better technology help? Although better technology might not help, there are many situations in which an investment in new technology is justified. Typical situations in which better technology might help include, among many others:

- automating formerly manual activities to increase efficiency or reduce errors
- providing information that was not formerly available or accessible
- performing activities more quickly or efficiently
- structuring work to increase consistency or conformance with rules
- providing tools that help people use their intellects more effectively
- offloading unpleasant or dangerous work from people
- eliminating work system limitations caused by the previous technology
- helping people communicate while doing work.

Common indicators that better technology will help. Better technology is more likely to help under conditions such as:

- Work system participants can explain how the new technology will help them do their work better or will improve working conditions.
- Current technology is a recognized obstacle to doing work efficiently.
- Current technology is increasingly difficult to maintain.
- Current technology is incompatible with related or complementary technologies used in other relevant work systems.

Evaluating Technology

Strengths and weaknesses of the work system's technology can be assessed in terms of characteristics and performance indicators such as functional capabilities, price/performance, ease of use, uptime, reliability, compatibility with complementary technologies, maintainability, and technological obsolescence. Any technology user can appreciate issues in each of these areas even though each area includes many issues that require deep technical knowledge to understand fully.

Functional Capabilities

The first issue when evaluating technologies is whether they have the functional capabilities that are required. The history of PCs illustrates the importance of functional capabilities related to every aspect of information processing. Information about the historical development of IT components and capabilities is readily available through Wikipedia and other Internet sources,[158] but here are several examples: When PCs based on Intel's 286 chip were first used for word processing in 1983, typists often had to stop their fingers to let the chip catch up.[159] The lack of a mouse as an integral part of the first PCs made it clumsier to identify exactly which document was to be opened and to identify which location within the document was to be changed. Similarly, the first PCs used floppy disks (diskettes) for backup and for transferring information between computers. Even after an improvement from 5.25-inch diskettes to 3.5-inch diskettes, a diskette's storage capacity was only 1.44 megabytes. Today, a single high quality photo requires more storage than that. Fortunately, CD burners can store 700 megabytes on a CD-ROM and DVD burners can store 4.7 gigabytes.

Functional capabilities of computing devices were once mainly of interest to technical professionals, but the widespread use of personal computers, cellphones, PDAs, and iPods have made them evident to consumers. For example, the October 2005 announcement of the first video capabilities on Apple iPods prompted news articles that discussed:

- physical dimensions (4.2 by 2.4 by about 0.5 inches)
- screen size (2.5-inches)
- screen resolution (320 by 240 pixels)
- storage capacity (75 hours of video or 7,500 songs for a 30 gigabyte model)
- battery life (two hours for video and 14 hours for music)
- download time (12 minutes to download to a computer and 2 minutes to transfer to the iPod for a one hour TV program minus commercials)[160]

Functional capabilities in each area matter to users because they determine how the technology can be used effectively.

Price/performance

The history of IT is an amazing story of increases in performance relative to price. A desktop computer that might cost $500 today is more powerful than a desktop computer that would have cost $3000 ten years ago. And many capabilities such as recordable CD-ROMs and DVDs were not available then for any price.

In computer chips, the underlying phenomenon is often called Moore's Law, named after a 1965 prediction by Gordon Moore, a co-founder of Intel, that the number transistors per unit of area on semiconductor chips would increase dramatically every year for the foreseeable future due to improvements in manufacturing technology. A typical statement of Moore's Law is that the number of transistors on a chip will double every 18 months.[161] As illustrated in Table 13.1, which shows progress in the number of transistors in Intel microprocessors, this trend has been a reasonable approximation to reality for over 30 years. The Table does not mention other dimensions of progress involving processing speed, instruction set, and address space. The inventor and futurist Ray Kurzweil argues that similar rates of progress have occurred not just in integrated circuits, but in five generations of technology starting with mechanical computing devices around 1900.[162]

Table 13.1. Progress in Intel microprocessors[163]

Microprocessor	*Year of introduction*	*Number of transistors*
4004	1971	2,300
8080	1974	4,500
8086	1978	29,000
286	1982	134,000
386	1985	275,000
486	1989	1,200,000
Pentium	1993	3,100,000
Pentium II	1997	7,500,000
Pentium III	1999	9,500,000
Pentium 4	2000	42,000,000
Itanium	2001	25,000,000
Itanium 2	2003	220,000,000
Itanium 2 (9 MB cache)	2004	592,000,000

Doubling every 18 months is difficult to grasp because it defies our normal experience, in which back-to-back improvements of 10% or 15% starting from a respectable base are viewed as significant achievements. Even at a rate of 20%, the equivalent of hardware capabilities costing $100 in 1960 would have cost $10.74 in 1970, $1.15 in 1980, $0.12 in 1990, and little more than a penny in 2000. If automotive technology had improved at that rate since 1960, a new car would cost less than a meal at Burger King, and we could drive across the United States on less than a gallon of gasoline (if gasoline were still used). The exponentially increasing power of chips is just one of many aspects of the technical progress that is continuing. Rates of progress resembling or exceeding Moore's Law is continuing in other areas of IT including hard disk storage and data transmission.

Squeezing more memory bits into the same amount of space on an integrated circuit may not seem important until one thinks about what it means for software capabilities. VisiCalc, the first spreadsheet program, was released in 1980 and occupied 29,000 bytes of memory on an Apple II computer. The earliest word processing programs were similarly small because the PCs of the time could not run larger programs. The version of Microsoft Word that was available in 1995 was vastly more powerful, but it ran in a minimum of one megabyte of random access memory (DRAM) and needed twice that amount to run well. It contained hundreds of features that would have been infeasible with previous hardware capabilities. Ten years later, efficient operation of the Windows XP operating system and typical applications required 512 megabytes of DRAM. The successor Vista operating system requires one to two gigabytes.[164]

Ease of Use

Because every self-respecting vendor claims that its technology is user-friendly, it is worthwhile to look at how to validate this claim. Ideally, anything a person uses, ranging from utensils and vacuum cleaners to technically advanced products such as computers and copiers, should be user friendly. Cosmetic issues such as the aesthetics of computer screens are important, but genuine user friendliness involves more than just appearances. Technology is user friendly if most users can use it easily with minimal startup time and training, and if it contains features most users find useful. **User-friendly technology** is more productive because users waste less time and effort struggling with features and details that get in the way of doing work. **User-hostile technology** forces users to focus on the details and tricks of using the technology rather than on the work to be accomplished.

Business applications of IT are more user-friendly when their features help the user focus on the business problem rather than the computerized tool being used to help solve the problem. User-friendliness issues for IT cluster in three related areas: shielding the user from unnecessary details, providing starting points, and matching the user interface to the task at hand.

Shielding the user from unnecessary details. Assume that any use of computers divides the user's attention between two topics: the content of the work and the plumbing that is being used. The content is the work practices, know-how and information required to do the work, and the concerns of customers and management. The plumbing is the aspects of the hardware and software that might change without affecting the content of the work. Plumbing includes the internal operation of the computer and network hardware, internal operation of the application software, and location and internal format of the database.

People whose work includes processing information would like to maximize the percent of effort and attention they devote to the content of the work; they would like to minimize the effort and attention they devote to the plumbing. Accordingly, the user interfaces for PCs, cell phones or other types of equipment should help them focus on what they are trying to accomplish and not on the technology that is being used. An ideal interface should be simple and intuitive. Its use should require little or no thinking or memorization except for thinking and memorization related to the content of the work.

Ideally, plumbing should be invisible to its users because whatever attention goes to the plumbing does not go to the content of the work. In addition, attention focused on the plumbing is fraught with frustration and confusion because so many aspects of hardware, software, and databases are undecipherable to non-specialists. Even someone eager to master the intricacies of complex work may not be inclined to master plumbing details that seem arbitrary and ultimately impenetrable.

User-friendly technologies interact with the user in readily understood terms, never forcing the user to learn or pay attention to seemingly arbitrary or irrelevant details. The user must understand basic principles but does not have to remember meaningless details such as strange acronyms or the grammar for commands. Multiple applications have similar organization and appearance, and therefore are easier to learn. If software operates this way, the user manual is basically a reference. The users can figure out how the software works mostly by playing with and modifying pre-existing examples. These criteria for user-friendliness explain why typical business professionals often have difficulty using programming and query languages that require mastery of seemingly arbitrary terms, grammar, and other details.

Providing starting points. User-friendly software provides convenient ways to access and reuse work done earlier, thereby recognizing that users often prefer not to do everything from scratch. For example, instead of presenting a purchasing agent a blank data entry form, an information system for purchasing hospital supplies can display a list of the items that were purchased in the past or a list of items whose inventory level indicates reordering is necessary. A more general approach is to provide templates as potential starting points. For example, model building software can include a library of components or partially constructed models that can be combined or tailored to a particular situation instead of forcing model builders to start from scratch. Similarly, desktop applications provide sample word processing documents, presentations, and spreadsheets as usable starting points for many situations.

Matching the user interface to the task. The user interface in user-friendly technology is matched to the nature of the task. Different input methods are combined to make the work efficient. The menu system is well structured, easy to understand, and consistent with menu systems in other applications the user sees frequently. The flexibility of the user interface reflects the flexibility of the task, permitting the user to do the task in whatever way the user finds easiest. Regardless of the order of the inputs, the software is designed to identify likely errors and make it easy to fix any user errors that occur. Ideally, the software adjusts to what the user knows, with novices seeing and using only basic features while experts can use the software in a more advanced way.

Importance of simple physical features. Although aspects of user friendliness can be a bit ethereal, the actual circumstances of using any particular technology often reveals that simple physical features determine the usefulness of the technology just as much as sophisticated technical features. For example, anyone who has carried a laptop knows that weight makes a difference. People who want a lighter laptop often need to spend more and sacrifice features. Similarly, one of the limits of the video iPod is that "it can get annoying to hold the new iPod in a good viewing position for long enough to watch a TV episode, because it doesn't come with a stand."[165] Both examples serve as a reminder that technology is always part of a larger work system. Genuine user-friendliness is determined partly by "technical" features such as the nature of the user interface, and partly by the way physical characteristics such as size, weight, shape, physical configuration, and operating conditions fit into the larger work system.

Uptime and Reliability

Imagine going to a doctor's office for an appointment and learning that your medical records are not available because a computer went down. Or imagine working in a highly automated factory when the computers that control manufacturing go down. Unless a backup system springs to action immediately, the consequences range from inconvenience, on the one hand, to medical errors or lost production on the other. The implication for analyzing systems is that the analysis is incomplete unless issues related to uptime and reliability are considered. Even if the technology is operational 99% of the time, the analysis should consider what will happen during the 1% of the time when the technology is unavailable.

Compatibility

We encounter compatibility issues in our everyday lives, and often manage to find reasonable solutions or workarounds. For example, when people who speak different languages need to communicate they may call upon workarounds such as using sign language or travelers' dictionaries, drawing pictures, or employing a translator. In relation to technology, compatibility is the extent to which the characteristics and features of a particular technology fit with those of other technologies relevant to the situation. IT incompatibilities often sound mysterious and impenetrable because IT brings unforgiving requirements for completeness involving technical details. The practical issues involve standards, backward compatibility, and the ability to combine technologies that provide complementary functions.

Standards as the basis of compatibility. Standards are the basis of technology compatibility. The significance of standards is clear if you have ever tried to use metric wrenches on nuts and bolts measured in inches, or if you ever tried to plug an American hair dryer into a typical wall socket in Europe or Asia. The metric wrench slips. The plug's shape is not compatible with the outlet's shape, but special plug adapters provide a workaround. Compatibility issues related to IT standards range from "technical" details, such as internal machine languages and data coding methods, to mundane issues such as the size of paper in copy machines and the shape of plugs. The metric versus English tools and the different shapes of wall sockets in different countries demonstrate that inconsistent standards often coexist. In the world of IT, these are often called **competing standards**.

Although the value of having standards is apparent, tradeoffs related to standards should not be overlooked. Standards may disallow certain features or capabilities that are valuable in a particular situation but are inconsistent with the standard. Standards may also contradict the features vendors have built into their own proprietary products. Standards are an important competitive issue because vendors often rely on proprietary data and product architectures as strategies to lock out competitors. With the proliferation of different types of computers and other devices, purchasers of large computer systems often look for conformance to agreed-upon standards.

Effectiveness of information technologies usually requires compatibility with other technologies in the setting.

The bottom line issue related to standards is the ability of hardware and/or software products to operate together conveniently and inexpensively whether or not they are designed and produced by the same manufacturer for the same situations. For example, the interoperability of two different enterprise software packages is determined by the extent to which their files contain the same information and handle it the same way. In this case, interoperability requires compatible internal coding of data, compatible program logic, compatible user interfaces, and compatible communication with storage and output devices. It is no surprise that cobbling together information systems based on software from different vendors can be a nightmare.

Backward compatibility. The need to operate in conjunction with previous versions of related technology is a persistent bugaboo for technology vendors. For example, this is the reason why Windows 95 was designed to be able to run programs written for the DOS operating system from the early 1980s. Likewise, Pentium chips from Intel had to handle application code written for earlier Intel microprocessors. Backward compatibility of this type makes each successive step forward more complicated. As backward compatibility requirements expand, each new version of a technology becomes more like a dinosaur that has to drag around a heavier and heavier tail as it grows. Backward compatibility is also part of the price of progress, however, since few IT users are willing to abandon their earlier work just because a newer release of hardware or software provides greater potential for the future.

Components that snap together. Compatibility is an important concern when thinking about the technology in a work system because most technology is genuinely useful only in combination with other technologies. Incompatibilities can make it exceedingly difficult to build, operate, and maintain technical systems. Complementary components should "snap together." The less compatible they are, the more effort must go into workarounds. Sometimes these are simple, somewhat like the adapter for a hair dryer's electric plug, but in other instances the workaround itself absorbs a lot of resources.

As a loose analogy, assume you need a research report translated from Japanese to English, but do not have access to a translator who can do the job directly. Instead you hire three translators, one who knows Japanese and French, one who knows French and German, and one who knows German and English. You can get the translation done, but the cost will be much higher and the quality probably lower than if you could hire just one translator. Business professionals may not be able to assess most technical compatibility issues, but they need to appreciate these issues because plunging ahead without considering them may lead to unanticipated development or maintenance problems.

Maintainability

Anyone who has tried to add a bathroom to an existing house understands that maintaining and upgrading a system can be complicated and frustrating. The problem is the need to work around everything that has already been done, which is often more difficult than working from scratch. For example if you don't have architectural plans you may not know what is inside a wall until you cut into it. The analogy between remodeling a house and maintaining an information system illustrates some of the issues but doesn't communicate the complexities of changing hardware and/or software used in organizations.

Building information systems for maintainability. Business professionals whose main exposure to computer use involves spreadsheets and word processing often wonder why it takes IT professionals so long to produce what seem like minor changes in an information system. They sometimes fail to realize that personal work creating calculations on a spreadsheet is quite different from development or maintenance of software meant to be used over extended periods by an organization. Their personal work on the spreadsheet is about calculations they control completely and may not need to explain to anyone else. A week or month later it may not matter if they cannot remember exactly how or why they did those calculations.

In contrast, IT professionals building or maintaining a system need to work in a way that makes their work products understandable to users and other IT staff members months or years in the future. An information system designed to be maintainable is built with a modular structure that makes it comparatively easy to visualize the impact of changes in any component. The entire information system is documented from both user and technical perspectives. The programs and documentation go through careful inspection for completeness, accuracy, and comprehensibility. Development shortcuts that try to bypass this lengthy process often generate problems that will haunt the people who maintain the information system in the future. A typical shortcut of this type is placing a computerized system into operation as soon as the software is debugged, even if documentation and training materials are incomplete. The likely result is that changes during the shakedown phase will never be reflected properly in the documentation, which will gradually become less and less accurate with additional changes to the hardware and software.

Keeping technology in operation. Computer hardware and software are far from infallible. That reality is often ignored when people tout the advantages of computers as stoical and totally reliable workers that never become tired or go on a coffee break. There are many situations in which people cannot make reservations or obtain information "because the computer is down" or because a database backup is in progress.

While it is obvious that non-operating computers and networks do nothing to make information processing effortless, it is important to remember that someone needs to keep the plumbing running. If possible, that someone should be dedicated to maintaining technology and infrastructure. That someone should not be "Joe down the hall" who is a finance whiz but also happens to understand how computers work. Joe should be doing financial work, not unplanned and unbudgeted computer maintenance.

Standards and maintainability. The impact of standards on maintainability is apparent. Why should a firm support ten different brands of personal computers or five email programs when one or two cover most of the real requirements? The advantages of standardization are substantial, but the disadvantages should not be ignored. Adherence to standards may disallow certain features or capabilities that are valuable in a particular situation but are inconsistent with the standard.

Technological Obsolescence

Rapid technological progress provides better technology for system builders but also complicates the challenge of maintenance. In addition to creating new tools and methods that must be mastered, nonstop technological change implies that today's leading edge will soon be bypassed by something else. Given the current pace of change, the hardware and software used comfortably to write a book today may be obsolete within five years because they may be incompatible with new requirements in the future. Unfortunately, those requirements are unknown.

Every organization faces inevitable obsolescence of whatever technology it currently uses. Even if hardware technology itself continues to meet operational requirements, at some point it will be old enough that repairs will be impossible for lack of replacement parts. The U. S. Government faces this problem with some of the voluminous tape libraries of atmospheric and other research data it collected in the 1960s. The tapes are still there, but the hardware and software for reading those tapes is no longer available.

Conformance with Basic Principles of Information Processing Using IT

Chapter 5 identified 24 principles that can be applied to any work system. Table 13.2 presents an additional set of principles that apply to processing of information using IT. Although guidelines such as "capture a given item of data only once" are repeated frequently and may seem obvious, many current information systems do not conform to these guidelines. In some instances, technical or economic considerations dictated that system design would not conform to the guidelines even though design flaws were recognized. In other cases, software was designed without considering the guidelines but the effort to change it is not cost-justified. In yet other cases, the guidelines provide ideas that can help in identifying potential improvements. Like any other investment, those potential improvements should be analyzed for both costs and benefits.

Importance of Infrastructure

It is tempting to take infrastructure for granted because someone else will manage it. Infrastructure is relevant to almost any analysis of a work system because it is essential for work system operation. Attention to infrastructure may identify risks and dependencies. It may also identify ways to use infrastructure more effectively or achieve a better balance between what is treated as infrastructure and what is treated as part of the work system.

Infrastructure consists of essential resources shared and relied upon by multiple work systems. A region's physical infrastructure includes its roads, public buses, power lines, sewers, and snow removal equipment. Its human and service infrastructure includes police, fire, hospital, and school personnel. Telephone or electricity outages and work stoppages by police or transportation workers demonstrate that infrastructure is at least partially beyond the control of people and work systems that rely on it. Consider what happened as a result of a simple power outage at an AT&T switching station that provided the telephone infrastructure for three airports.

> A simple power outage occurred at a New York City AT&T switching station at 10 a.m. on Sept. 19, 1991. Workers activated backup power, but a power surge and an overly sensitive safety device prevented backup generators from providing power to the telephone equipment, which started drawing power from emergency batteries. Workers disobeyed standard procedures by not checking that the generators were working. Operating on battery power was an emergency situation, but over 100 people in the building that day did not notice the emergency alarms. Some alarm lights did not work; others were placed where they could not be seen; alarm bells had been unplugged due to previous false alarms; technicians were off-site at a training course. At 4:50 p.m. the batteries gave out, shutting down the hub's 2.1 million call per hour capacity. Because communication between the region's airports went through this hub, regional airport operations came to a standstill, grounding 85,000 air passengers.[166]

Table 13.2. Principles for Processing Information Using IT

Capturing information

- Capture information once. (Assume that a customer's address and contact information is captured in several different information systems. It is unlikely that the information will be consistent.)
- Where possible, capture information automatically to minimize errors introduced by human intermediaries. (This is a goal of bar code and RFID technologies.)
- Where possible, apply error detection techniques at the point of capture. (Finding errors immediately makes it more likely that erroneous information will be corrected before it is used.)

Transmitting information

- Verify that the information arrived at the destination. (You assume the addressee received your email unless it bounced back to you. Ideally, verification that it was opened should eliminate doubts about whether it was actually seen.)
- Identify and fix any information degradation that occurs in transit. (When information is digitized for transmission, it is possible to use coding schemes that enable automatic identification and correction of errors due to lost bits.)

Storing information

- Store only one "official" version ("version of record") for each item of information. That official version might be replicated in several places to maximize the ease of recovery if a computer goes down.
- Store information using data structures that expedite retrieval of the information. (These data structures may be different from data structures that expedite data capture and fast processing of transactions.)
- Treat the authorization to change information as separate from the authorization to access information. (Authorization to change information should be more stringent than authorization to see information.)

Retrieving information

- Enforce access controls to minimize unauthorized access.
- Facilitate access to information by providing pre-programmed queries. If appropriate, also provide readily usable methods for creating new queries.
- Hide the technical details of information storage and retrieval from users.

Manipulate information

- Provide pre-programmed calculations, reports, and models that facilitate effective use of information.
- If feasible, enable users to transform information into other forms that they need.
- Hide the technical details of information storage and retrieval from users.
- If possible, warn information users that certain calculations are likely to cause errors or misinterpretations. (For example, adding quantities makes sense, but adding percentages usually does not make sense. 45 apples plus 64 apples equals 109 apples, but 45% of one group plus 64% of another group does not produce 109% of the combined group.)

Display information

- Display information in a manner that is meaningful for the user and the situation. (Regardless of how information is stored, display only the information that is needed for the user's purpose, and display the information in a meaningful format.)
- If possible, provide pre-programmed queries and reports.
- If feasible and appropriate, enable users to generate novel queries and reports.

Virtually every important work system in industrial society relies on a human and technical infrastructure. Consequences of a simple power outage in the AT&T example inconvenienced or endangered thousands of people due to a combination of power equipment failure, alarm system failure, and management failure. Infrastructure failures like this are fortunately rare in advanced industrial countries. Chronic inadequacies that limit the average performance of work systems are much more common.

The AT&T example also illustrates that the infrastructure for one work system is the work system for someone else. The air traffic system used the telephone system as infrastructure. Maintaining phone service was the primary work system at the New York City switching station. Like any other work system, a work system that produces or maintains shared infrastructure creates value by producing products and services that its customers use.

Aside from demonstrating the havoc that can result from infrastructure problems, this example illustrates the need to look at various aspects of infrastructure. Technical infrastructure is often the most visible because failures in this area often affect work systems in immediate and dramatic ways. Also important in many situations is the way the organization's information infrastructure provides information generated elsewhere in the organization. Human infrastructure includes the human resources staff, trainers, and others who work in the background to keep the human side of the enterprise operating effectively.

Focusing on IT infrastructure. Since we are primarily concerned with IT-reliant work systems, the discussion of infrastructure will focus on IT resources that are shared among many different work systems. An organization's technical IT infrastructure consists of computer and telecommunications hardware and software shared across multiple business units or functions. The software includes operating systems, database management systems, programming languages, and system development tools used for building information systems. The human side of IT infrastructure includes technical expertise related to specific technical components and managerial expertise in areas such as planning, staffing, and negotiating. IT infrastructure is typically budgeted and provided by an organization's IT function. IT infrastructure investments are usually long term in nature, take advantage of economies of scale, and are difficult to change once they are made.[167]

Technical Infrastructure

Overlap between technology and technical infrastructure. It is worthwhile to distinguish between technology within a work system and the technical infrastructure that the work system uses. For example, consider the way many insurance salespeople use laptops during sales presentations. These computers can be viewed as the technology used within a sales work system or, alternatively, as a part of the firm's technical infrastructure for receiving e-mail, linking to the Internet, and performing other general communication and data processing activities.

Whether the laptop computers should be viewed as shared infrastructure or as technology within a work system depends on the purpose of the analysis. The relevant factors include:

- ***Dedicated use or shared use.*** Technology within a work system is dedicated to that particular work system and is viewed as part of it, such as hand held terminals used by UPS drivers. Shared infrastructure is viewed as something outside a work system that the work system relies on to operate, such as a telephone network.
- ***Ownership and control.*** Technology within a work system tends to be controlled by the work system's managers. Infrastructure is shared with other work systems. Infrastructure tends to be owned and managed by a centralized group in charge of infrastructure.
- ***Users' depth of understanding.*** Technology users have more need to understand details of how the technology works for them. Details of infrastructure are more hidden, and differences between brands or sources are less consequential to users.

Platforms, tools, and methods. Decisions about technical infrastructure start with choices among alternative **platforms**, specific families of computers, networks, and software that are the technical basis of information systems. Since different applications fit best with different platforms, a key challenge in this area is to select one or more platforms that support today's needs, are efficient to maintain, and provide a migration path that is likely to satisfy future needs.

Many firms have found themselves weighed down by the need to support multiple platforms for similar purposes. For example, many universities support both IBM-style PCs and Apple Macs. Each platform has certain advantages, but sharing information and applications while using different platforms often wastes resources. Uncoordinated hardware and software purchases and upgrades have wasted huge amounts of time and effort in most large companies and many small ones as well. For example Nynex, which later became part of Verizon, discovered it could save $25 million per year simply by standardizing to two brands of personal computers.[168] To avoid situations of this type, even companies with highly decentralized information system activities often maintain central authority over brands and versions of hardware and software that can be purchased and used.

Another part of IT infrastructure is the tools and methods used for developing, operating, and maintaining information systems. The basis of most business data processing systems built before the 1980s was the programming language COBOL. Database management systems brought another layer of capabilities. Developments related to the Internet and private networks added programming tools related to building web sites, transferring data, and producing new technical interfaces to existing software modules.

Building information systems using new tools and approaches is often challenging due to a combination of newness, lack of experienced personnel, and lack of mature tools that support the programming and debugging effort. Business professionals tend to view these issues as someone else's problem, but they are important to the business because the adequacy of technical IT tools and methods affects an IT group's ability to create and maintain information systems. Although better technology may make little difference in the current operation of a particular work system, the argument for better technical infrastructure may be powerful if it supports many work systems more successfully, or if it provides the basis for future changes in those work systems.

Migrating functionality into the infrastructure. The continuing migration of data processing capabilities out of specific applications and into technical infrastructure makes the development and maintenance of a robust IT infrastructure even more important. Thirty years ago a computer program inside of a financial accounting information system might have controlled far-flung details such as how the database is defined and how the printer is controlled. Today, these capabilities are increasingly controlled by IT infrastructure components performing functions such as:

- providing corporate access to the Internet
- providing a corporate communication network
- storing information on centrally controlled computers
- providing centrally controlled backups and standard recovery procedures
- using shared resources such as printers and communications networks
- performing software updates using a network instead of individual installations on PCs.

Investing in IT infrastructure for performing these functions differs from improving a specific work system along a firm's value chain. Direct investment in a work system tends to have more immediate impact on that work system's performance. In contrast, an infrastructure investment is usually a long-term investment whose value is felt across many work systems, often after long delays while attaining critical mass. The nature of infrastructure as a shared resource makes it more difficult to justify infrastructure investments. Managers in charge of individual work systems often wonder whether they receive enough direct benefit to justify their department's contribution to the cost.

Security and control standards. A crucial part of IT infrastructure involves the creation and maintenance of corporate standards for IT security and control. This part of IT infrastructure tries to minimize risks related to threats such as software bugs, unauthorized changes to software, attack by malicious software, unauthorized access to computers and information, theft of information, and damage to physical facilities.

Obstacle or opportunity? Aside from supporting current work systems, technical infrastructure provides the basis on which new computer applications can be built. It creates opportunities by making it easier to build or improve software that reduces internal costs or helps provide better products and services for customers.

Eliminating infrastructure obstacles is usually a big, difficult project. The Bank of America required over 5 million lines of computer code to consolidate a set of separate databases into an integrated database of customer information for its 20 million accounts. Establishing the unified customer database made it possible for branches to sell and service bank products and to do cross selling among different products.[169] This new marketing approach was part of the bank's strategy.

Many of the obstacles and opportunities related to infrastructure occur at a less visible level, however. For both business and technical professionals, ineffective infrastructure causes duplication of effort, unnecessary interfacing of incompatible databases or technologies, and unnecessary training. Clumsy infrastructure often forces creation of awkward workarounds that make it possible to do work despite the infrastructure that is supposed to be helpful. A well-designed, well-operated technical infrastructure makes it easier for people in an organization to achieve their goals.

Information Infrastructure

Examples of shared information infrastructure include:

- an intranet that makes phone and email directories and other corporate data available
- a shared corporate database for convenient access to production schedules, orders, inventories, employee skills, and other information that is needed across functions
- external market research data that is purchased and made available across business functions.

Cross-functional ERP initiatives exemplify the creation of information infrastructure. ERP is a complex form of software that supports transactions within and across business functions such as sales and finance. It maintains an integrated database that can be used for different functions across the organization. Although the acronym ERP stands for enterprise resource planning, ERP software is much more about performing transactions in a consistent and controlled manner. The acronym ERP is actually an expanded reincarnation of MRP II, manufacturing resource planning; MRP II was an expanded reincarnation of MRP, material requirements planning.

An explanation of enterprise systems[170] in *Harvard Business Review* illustrates the goal and potential value of ERP using the following example:

- A French sales representative of a U.S. manufacturer uses a laptop computer to prepare a quotation for a customer order.
- A formal contract is produced automatically in French.
- If the customer accepts the contract, and if the customer passes a computerized credit check, the order is entered.
- The shipment is scheduled.
- Required materials are reserved from inventory or ordered from suppliers.
- Factory capacity is allocated across several different factories in Taiwan that will manufacture parts of the order.
- The timing of each stage is verified for feasibility.
- All commissions and costs are calculated and allocated appropriately.

That degree of automatic integration is possible only if the company has an integrated database that can be used across the various transactions. Twenty years ago that degree of integration across a large company was infeasible due to the enormous amount of work required to define all of the data and program all of the modules. With the advent of information infrastructure supported by commercial ERP software, more companies have created a corporate information infrastructure that allows them to move closer to the degree of integration imagined in this example.

Human Infrastructure

A city's human infrastructure includes its municipal government, its courts, its police and fire departments, its schools, its garbage collection, its street maintenance, and a myriad of other functions. These infrastructure functions support the city's primary reasons for existing, such as providing an environment for residential and commercial activities. In businesses, most work systems depend on the surrounding organization to provide similar forms of infrastructure and support services, thereby making it far easier for work systems to produce what their customers want.

Technical IT infrastructure cannot be effective without support from human infrastructure. The AT&T story earlier in this chapter showed how a minor failure in part of an organization's technical infrastructure was allowed to become a major problem because the human part of the infrastructure was lax. Going the other direction, the best-intentioned human infrastructure will be ineffective if it lacks the appropriate technical tools for doing its job. For example, inadequate programming tools may prevent programmers from working efficiently. Similarly, inadequate computer system monitoring tools may prevent computer system managers from making well-informed decisions when they try to keep computer system components in balance.

The human infrastructure for information systems may be distributed across a number of locations, including line divisions, a centralized information system department, and outsourcing vendors hired to run computer centers and perform other functions. This human infrastructure often includes:

- shared training staff
- shared consulting staff for application configuration and other help in making technology as useful as possible
- shared help desk for support related to details of using technology
- shared operational IT staff that controls passwords, security, standards, and networks.

Consider the human infrastructure aspects of creating and maintaining a data warehouse that stores information used by work systems across many departments. Someone must make sure that the data files are defined properly, that the network performs efficiently, and that the data warehouse and network are secure from misuse. Someone must make sure the software is updated appropriately and that users know when it is updated. As users gain experience using the data warehouse, someone should make sure they are using it effectively and should obtain and implement their suggestions for improvements.

Under-appreciated and under-funded. The human and service side of infrastructure is often under-appreciated and under-funded. For example, a study of computing and organizational practices in a large number of leading companies concluded, "Despite widespread recognition among IT planners that most systems are underutilized because of inadequate training, the deficiency is repeated in systems implementation after systems implementation."[171] Even when discussions of new systems or system enhancements touch on infrastructure requirements, they tend to over-emphasize the technical and capital-intensive components of infrastructure. Conversely they tend to downplay the human and service infrastructure required to achieve the desired benefits from the proposed investments.

IT staff members charged with keeping technical infrastructure operational often believe they have a thankless job because of the expectation that everything will always work. No one says anything until a computer or network goes down, and then the people who are inconvenienced complain vociferously. Since infrastructure is shared among applications, managers in the application areas often try to reduce the infrastructure expenditures they must absorb. They seem to want perfection, but at a lower cost.

The tendency to underestimate the importance of human and service infrastructure related to IT may also result partly from a general tendency to spend on tangible things such as buildings and computers, and then to skimp on intangibles, such as training and support. Over-optimism about the users' ability to figure out how to use infrastructure effectively may also be a factor. Another possible cause is surprise at the amount of effort and expense needed to train and counsel users and to keep technical infrastructure operating. Regardless of the cause, the result is often ineffective use of infrastructure. The trend toward organizational decentralization and outsourcing of many IT-related functions make it even more important to include human infrastructure in the analysis of new systems.

An Enterprise View of IT

Almost all important work systems in today's organizations rely on IT in order to operate. Many of their capabilities depend on corporate IT capabilities. For completeness, it is worthwhile to provide a brief overview of key IT-related decisions at a corporate or enterprise level. According to a 2004 book[172] by researchers at MIT's Center for Information Systems Research, every enterprise needs to make five types of decisions about how IT will be governed. These decisions concern:

- IT principles
- IT architecture
- IT infrastructure
- business application needs
- IT investment and prioritization.

IT Principles

A firm should establish a set of strategy-driven IT principles that provide a rationale for high-level decisions about IT. These IT principles should clarify enterprise expectations in three areas:[173]

- ***What is the enterprise's desired operating model?*** Issues here concern topics such as the extent to which the firm is centralized or decentralized, the extent to which different units share products and customers, and the extent to which it is beneficial to establish and enforce centralized services. For example, a centralized purchasing function should be part of an enterprise's operating model if centralized purchasing could negotiate better prices and service for the enterprise without causing significant delays or other major problems.
- ***How will the IT function support the desired operating model?*** The IT function should be organized to mirror the firm's operating model. If the firm has important capabilities that must be shared across units, IT should be deployed to contribute to that effort. For example, if stores should be able to swap inventories to meet customer demand, information systems need to support that capability. Similar opportunities to benefit from standardized data and highly coordinated processes are an important reason to use ERP software.
- ***How will IT be funded?*** Shared IT infrastructure can be funded totally by a central IT organization or its costs can be allocated to business units. Cost allocation is often a highly political issue because some units will benefit more than others and because the allocation appears as an expense on divisional income statements. In other words, these allocated costs are not controllable by the divisions but might swing a division from profit to a loss.

An article[174] in *Sloan Management Review* in 1997 presented a business perspective on IT principles by contrasting what it called business maxims and IT maxims. **Business maxims** summarize an enterprise's desired operating model, including its competitive stance, its approach for coordinating business units, and its approach for managing information and information technology. **IT maxims** summarize how IT should be deployed to support the desired operating model and how IT decisions should be made. The desired deployment of IT is expressed in relation to the firm's approach to transaction processing (e.g., degree of standardization, common interfaces, and local tailoring) and access and use of different types of information (e.g., financial, product, or customer.)

Table 13.3 identifies typical categories[175] of business and IT maxims and identifies one business maxim and one IT maxim in each category for TRBG, the bank in the example in Chapter 8. To emphasize that business and IT maxims are choices, possible maxims that were not chosen are also shown.

IT Architecture Decisions

"At the enterprise level, an IT architecture is the organizing logic for applications, data, and infrastructure technologies, captured in a set of policies, relationships, and technical choices to achieve desired business and technical standardization and integration." ... "The key to process integration is data standardization – providing a single definition and a single set of characteristics to be captured with a data element."[176] A company's desired approach to data and process standardization depends on the diversification in its various units. Even highly diversified companies may benefit from technical standardization for cost-effectiveness, negotiated vendor agreements, and enterprisewide security.

IT architecture can be an important source of competitive advantage, or can prevent a company from pursuing possible strategies. For example, Johnson & Johnson had traditionally operated through over 100 largely independent operating units that produced products ranging from bandages to pharmaceuticals, and more recently, heart stents. The decentralized approach was successful for decades, but by 1995 customers complained that they "had to deal with multiple sales calls, multiple invoices, and multiple contracts with J&J operating companies. They wanted a single point of contact." The first step toward a more centralized architecture involved installing a single global network and desktop configuration. Next was a shared services organization to attain efficiencies of technical standardization. In sectors in which different units shared customers and markets, J&J standardized critical data across operating companies.[177] The goal was to attain competitive benefits of standardization without losing the competitive benefits of operating as autonomous units.

Firms seem to move through four stages of "architecture competency."[178] Firms can generate significant benefits at each stage. Today's leading firms tend to be at least at the second of the following stages:

- ***Application silo stage***. The firm's IT architecture is no more than the architectures of the various applications.
- ***Standardized technology stage***. The firm attains efficiencies through technology standardization. During this stage firms shift resources away from developing silo applications and toward building shared infrastructure. The primary benefit is IT efficiency. Important risks during this stage include resistance to standards and resistance to whatever approach is used for enforcing standards.
- ***Rationalized data stage***. The firm standardizes important data and processes rather than just technology. During this stage, resources are shifted away from application development and toward data management and infrastructure development with special emphasis on data that supports core activities across the organization. This shift requires centralized coordination of major processes that produce data.
- ***Modular architecture stage***. The firm preserves global standards while enabling local differences. Currently this stage is more a goal than a reality.

Table 13.3. Business and IT Maxims for TRBG (a bank described in Chapter 8)

Note: Business and IT maxims are big-picture intentions about how a business will operate and how IT will contribute. To illustrate the basic idea, one maxim under each category of business or IT maxim is presented to summarize TRBG's intentions. Also shown are other possible maxims that were not chosen.

Business maxims	***IT maxims***
Cost focus Accept relatively high personnel and marketing costs in order to expand profitability and market share. • Not chosen: Become a low cost provider. • Not chosen: Maintain industry-average cost structure but charge costs to customers.	***Expectations for IT investments*** Evaluate IT investments based on whether they reduce costs and/or improve decision making. • Not chosen: Use IT to provide better service to customers. • Not chosen: Use IT to create new cross-selling opportunities and new financial products and services.
Value differentiation perceived by customers Provide relatively aggressive terms for loans that some other banks might find risky. • Not chosen: Become a conservative lender that provides top-notch service in other areas. • Not chosen: Focus on established clients with long histories of stability and success.	***Data access and use*** Provide easy access to whatever information is needed to create and evaluate loan applications. • Not chosen: Maintain tight control over data access to minimize risk of data theft or sabotage. • Not chosen: Establish consistent customer information available to all functional areas of the bank. • Not chosen: Provide identical information access to mobile workers and office workers.
Flexibility and agility Gradually adopt new business methods within existing markets and new markets. • Not chosen: Maintain current business methods and stay within existing markets. • Not chosen: Provide loan officers more flexibility in obtaining approvals different ways.	***Hardware and software resources*** Perform all customer and accounting transactions through an integrated information system. • Not chosen: Maintain existing legacy systems that are not integrated. • Not chosen: Acquire sophisticated analysis programs from vendors and integrate those programs with existing information systems.
Growth Pursue aggressive growth of the size and profitability of the loan portfolio by expanding penetration of current and new geographical regions, client groups, and loan types. • Not chosen: Stay within current geographical regions, industry groups, and types of loans. • Not chosen: Expand by merging with other banks.	***Communication capabilities and services*** Maintain simplest possible communication capabilities consistent with the bank's size. • Not chosen: Use the Internet extensively as a medium for internal and external communication. • Not chosen: Exploit high bandwidth communication capabilities such as videoconferencing.
Human resources Hire ambitious sales people who are willing to work hard. • Not chosen: Hire more experienced loan officers who will demand higher compensation but may be more productive. • Not chosen: Reduce the number of employees through greater automation.	***Architecture and standards approach*** Central IT group controls all computing and IT-related services and purchasing for the entire bank. • Not chosen: Establish a single, integrated IT architecture for all branches and offices. • Not chosen: Integrate the most cost-effective application software from different vendors.
Management orientation Retain centralized decision making for loan approvals and other important decisions. • Not chosen: Decentralize decision making on loan approvals by supporting greater discretion for loan officers. • Not chosen: Promote fact-oriented management.	

IT Infrastructure Decisions

IT infrastructure decisions have an important impact on a firm's strategic agility, especially its ability to pursue business initiatives that cross the entire firm or that link with customers or suppliers. Building IT infrastructure takes a long time. Therefore any distinctive competitive advantage that results from IT infrastructure improvements is difficult to match in the short term.

Many IT infrastructure decisions fall within 10 clusters of infrastructure services.[179] The first six clusters are the enterprise's technical infrastructure.

- ***Channel management services*** provide electronic links to customers and business partners.
- ***Security and risk management services*** include firewalls, policies for remote access, encryption, use of passwords, and disaster planning and recovery.
- ***Communication services*** provide electronic communication, both internally and externally.
- ***Data management services*** involve storing and distributing corporate information using data centers, data warehouses and Internet-based services.
- ***Application infrastructure services*** attempt to reduce costs, increase reliability, and increase standardization of IT applications using tools such as ERP.
- ***IT facilities management services*** coordinate and manage physical facilities that perform large-scale data processing.

In contrast with the enterprise's technical infrastructure, four clusters represent management-oriented IT capabilities:

- ***IT management services*** include information system planning, project management, service-level agreements, and negotiations with vendors.
- ***IT architecture-and-standards services*** govern the way IT will be deployed across the enterprise.
- ***IT education services*** provide IT- and system-related education and training across the enterprise.
- ***IT R&D services*** perform research about new ways to use IT to add value.

Decisions about IT infrastructure are often difficult because it affects so many functions but operates in the background. Some of the key issues concern deploying IT infrastructure to achieve business objectives, locating infrastructure services so that they can be effective, pricing services fairly, deciding when to update services, and deciding what to outsource.

Business Application Decisions

The direct benefit of IT infrastructure decisions occurs when business application needs are met. Business application decisions are most strongly related to the operation of specific work systems, and often involve significant trade-offs and changes in the way work is done. This bulk of this book focuses on the work system analysis that is required for good business application decisions.

IT Investment and Prioritization

IT investment makes up over 50% of the capital budget of most large companies. Research at MIT's Center for Information Systems Research identified four broad objectives for IT investments. In their sample of 147 firms, 54% of IT investment was for infrastructure objectives, 13% for transactional objectives, 20% for informational objectives, and 13% for strategic objectives.[180]

- ***Infrastructure objectives.*** Reduce costs or create a flexible base for future initiatives through a combination of business integration, business flexibility, reduced cost of IT within separate business units, and standardization.

- ***Transactional objectives***. Reduce costs and/or increase speed in performing transactions.

- ***Informational objectives***. Provide information for accounting, management, control, analysis, communication, collaboration, or compliance with government reporting requirements. Ideal results include increased control, better information, better integration, improved quality, and faster cycle time.

- ***Strategic objectives***. Try to achieve increased sales, competitive advantage, response to competitive necessities, and/or better market positioning.

Many firms look at IT investments in portfolio terms, with a number of investments spread across different types of assets with different risk-return profiles. It is important to take a portfolio approach because success is uncertain and even successful investments do not pay off immediately. The MIT research estimated that around 50% of strategic IT investments fail. Other types of IT investments are less likely to fail. Also, significant IT infrastructure investments may have a several year lag between the time of the investment and effective use. These investments often have a negative impact on firm profits in the year after the investment due to the cash outlay and the significant disruption that they cause.

Part of the payoff from many significant IT investments comes after subsequent investments that build on the first investment. This implies that the investment decision should not be made based on the immediate outcome of the investment, but rather should consider a decision tree of options that might occur after the initially uncertain results of the investment. A technique called **real options** allows the analysis of an investment decision to work through a decision tree of possibilities, and to recognize that subsequent projects can be deferred, modified, expanded, or abandoned based on initial accomplishments and problems.

The MIT research concluded that firms with greater IT savvy actually attain better investment results across all four types of objectives. Those firms tended to be the ones that used IT heavily for communication, that had a high level of digital transactions, that used the Internet extensively, that had firm-wide IT skills, and that had strong top management involvement in IT decisions.

Chapter 14: Environment and Strategy

- Related Work Systems
- Organizational History and Distribution of Power
- Organizational Culture
- Organizational Policies and Procedures
- Legal and Regulatory Requirements
- Competitive Environment
- Alignment of IT with Corporate Strategy

A work system's environment includes the organizational, cultural, competitive, technical, and regulatory environment within which the work system operates. It also includes other work systems that it affects or that affect it directly or indirectly. The organization's general norms of behavior are part of the culture in the environment that surrounds the work system, whereas behavioral norms and expectations about activities within the work system are considered part of its work practices.

Factors in a work system's environment affect its performance even though the work system does not rely on its environment in the way it relies on the infrastructure it uses. Accordingly, it may be a mistake to assume that practices and people outside of a particular work system are irrelevant factors, externalities, or simply someone else's problem. In particular, work practices that are seriously at odds with work practices in nearby work systems may cause problems in the larger organization. For example, a sales department's innovation of awarding vacations in Hawaii to top performers may generate resentment in a customer service department that uses t-shirts and coffee mugs to recognize excellence. In general, a new or modified work system that seems to satisfy needs for performing a particular function may be rejected like a foreign body if does not fit with the organization's history and culture, and with the needs of other systems and non-participant stakeholders.

Issues related to a work system's environment sometimes invalidate the assumption that a work system will be successful if it is defined clearly and rationally enough. A work system's environment may include organizational and cultural forces that create obstacles and drain resources and attention away from the work system's operation. In other cases, forces in a work system's environment may lead to win-win alliances that provide additional resources and remove obstacles.

This chapter uses brief examples to illustrate issues related to aspects of a work system's environment. As listed above, its sections cover related work systems, organizational history and distribution of power, organizational culture, policies and procedures, legal and regulatory requirements, and competitive environment. It closes with issues related to the alignment between a firm's strategy and its IT function.

Related Work Systems

Work systems should be aligned with other work systems that receive or use what they produce, or that are affected in other ways. Common types of misalignment include:

Mismatches related to inputs and outputs. The customers of a given work system usually include other work systems that receive and use the products and services it produces. In addition to thinking about customers as they were discussed in Chapter 9 (Customers and Products & Services), it is worthwhile to consider customer work systems as part of the environment surrounding a work system. For example, assume that the production department's output is the input to the packaging department. Near month-end, the production department is often substantially behind its monthly schedule and uses overtime during the last week to catch up. Whether or not the production department meets its schedule, the packaging department may find itself swamped with an unusually large amount of finished goods. It may need to use overtime in order to meet its goal of a low average cycle time for shipping finished goods. Thus, the production department's attempts to meet its schedule may have ripple affects that make the packaging department's performance look bad. The production department should try to meet its own goals, but in terms of the larger organization it should not merely shift its problems to another department.

Work systems never exist in a vacuum. They always affect and are affected by other work systems.

Policy inconsistencies between work systems. In some cases the policies and procedures in one work system operate at cross-purposes with policies and procedures in other work systems. In a supply chain example cited in *Harvard Business Review*, "the required merchandise was often already in factory stockyards, packed and ready to ship, but it couldn't be moved until each container was full." The policy of shipping only full containers "delayed shipments by a week or more, forcing stocked-out stores to turn away customers. .... When manufacturers eventually deliver additional merchandise, it results in excess inventory because most distributors don't need a container load to satisfy the increased demand."[181]

Poorly aligned incentives. In some situations the incentives of different work systems operate at cross-purposes. During the 1980s, Campbell Soup's work system for sales promotion offered distributors discounts several times a year. Distributors responded by buying more soup than they sold during those periods. For example, 40% of the company's chicken soup for an entire year was sold during the six-week promotional periods. The result was costly fluctuations in the company's supply chain activities. When Campbell recognized the mismatch between its sales promotion system and its operational systems, it started to collect information about distributors' sales rather than just their purchases, and started discounting based on distributors' sales, thereby eliminating the large forward-buy quantities.[182]

Organizational History and Distribution of Power

An organization's history and distribution of power is part of the background that should be considered before trying to change work systems in significant ways. A typical example involves a change in technology within one department that affected the power of another department.[183] A machine shop used computer numerical control (CNC) machine tools to cut, grind, and shape metal parts. Its many types of orders included frequent emergency replacements and small test lots. Its response to customers was often delayed by external departments that scheduled its resources and performed numerical control programming. New CNC machines allowed the machine shop to schedule its own work and to program the machines. Even inexperienced machinists could program simple cuts. The machinists found the new equipment a welcome challenge, but external departments including resource planning, NC programming, and manufacturing R&D were far from enthusiastic. Representatives of these groups protested to divisional management because they had been eased out of the loop.

This case illustrates the possible importance of non-participant stakeholders even in deciding how a work system will operate internally. Machining could be done using a variety of technical options. The selection of the new technology was based partly on the way the machines could help in doing the work and partly on the way these machines could promote the machining group's goals for autonomy and noninterference from outside. The external departments responded by complaining about the new technology not because the machining was being done inadequately, but because the new technology reduced their power and influence. In this case the external concerns had little impact, but with different divisional managers the outcome might have been different.

Organizational Culture

Discussions of systems often focus on formal procedures for performing tasks even though the organization's culture can be as important as the system's official rationale in determining system success. Corporate culture can be defined as "a pattern of shared basic beliefs that the group learned as it solves its problems of external adaptation and internal integration, that has worked well enough to be considered valid, and therefore to be taught to new members as the correct way to perceive, think, and feel in relation to those problems."[184] This definition expands into three mutually reinforcing layers.

- Most readily evident to outsiders is the top level, the visible features of the work environment.
- The second level is the espoused values and explicitly articulated beliefs often found in the organization's mission statement, strategic plan, or corporate credo.
- At the heart of an organization's culture are the underlying beliefs -- the largely unexpressed and unconscious assumptions shared by its members. These beliefs guide thinking, suggest the ways in which problems are addressed, and define notions of equity and justice.[185] They influence daily routines related to organizational hierarchies, communication, expression of dissent, treatment of employees, and many other aspects of work life.

The slow implementation and acceptance of many CASE (computer aided software engineering) tools illustrates the importance of culture in adopting new technology within work systems. (CASE was mentioned earlier to illustrate the ineffectiveness of changing technology without changing other parts of a work system.) CASE products are basically programming toolkits used to systematize the process of defining software requirements and converting those requirements into computer programs. These toolkits are purchased based on the assumption that performing programming tasks in a consistent manner will improve coordination, increase productivity, and create uniformly organized programs that can be modified easily over time as business requirements change.

Despite this plausible rationale, the implementation of CASE systems has often been disappointing. Although CASE products certainly have limitations, their technical shortcomings explain only part of the problem. The culture in many programming groups explains much of the rest, especially if programmers have been treated as prima donnas who are proud of their own hard-earned skills and have been allowed to do their work however they want to do it. Whether or not a high degree of autonomy for programmers is a good idea in general, programmer autonomy doesn't fit with tools whose benefits come from standardization around pre-determined formats and procedures. Half-hearted implementations resulted from contradictions between the assumptions of the culture and the assumptions of CASE. The technical staff used parts of the toolkit, but did not achieve the integration into work practices needed for major improvements in productivity and coordination.[186]

Rules and Taboos

Culture is subtle because it combines explicit and implicit rules and practices that govern how people behave. For example, contrary to the way other law firms operate, the culture in a West Coast law firm includes a policy of not recording and measuring the level of business origination by individual attorneys. The firm's internal systems operate consistent with this explicit policy. For instance, its marketing support systems do not focus as much as those of other firms on questions related to who brought in each client.

In contrast to this explicit policy, the CASE example (mentioned above) showed how important parts of organizational culture reside in unwritten rules. A programming group might have formal, well-documented processes for doing many things, but it wouldn't have an official statement that "we treat programmers like prima donnas." Similarly, written rules might say that service technicians are supposed to fill out a form describing each service call, but the unwritten rules about the spirit of using the information would determine how thoroughly the written rules are followed.

An organization's culture may determine whether even a technically brilliant work system will succeed.

Is management walking its own talk? A relevant, but often uncomfortable part of analyzing a system is asking whether management is walking its own talk. Yes, the organization espouses its desire to balance home and work life, but it requires that everyone arrive at work just before schools open. Yes, the organization espouses the value of everyone's contribution, but the only way to get an idea accepted is to convince the boss that she invented it. Yes, the organization espouses customer focus, but its engineers are more motivated by sophisticated engineering than by solving simple customer problems. Contradictions such as these may be readily apparent to work system participants, but being the messenger who documents management inconsistency is often uncomfortable and sometimes risky.

A key difficulty in dealing with organizational culture is that discussing certain cultural issues is virtually taboo in many organizations. Even departments that hold wide-ranging discussions about strategies and tactics may still avoid public conversations about cultural assumptions. For example, you probably wouldn't hear many public conversations about one department's widely held belief that managers have the right to berate poor performers in public or a different department's belief that managers should never say anything confrontational because the CEO hates conflict.

Euphemisms, spin control, and political correctness sometimes make it more difficult to address culture issues cleanly. Even everyday language contains strange distortions, such as saying "I won't take any more of your time" to express the thought "I don't want you to take any more of my time," and "thanks for the criticism," to express the thought "I found that criticism ridiculous and it offended me deeply." Having a genuine discussion about work system participants can be especially difficult when the issues involve topics such as the skills, seriousness, and potential contributions of specific individuals.

Organizational culture encompasses many topics, but we will say a bit more about three that affect work systems: trust, privacy, and information sharing.

Trust

Who hasn't heard executives claim publicly that they respect and trust their employees, that people come first, and that they have the best, most loyal employees? Hearing executives say these things leads one to wonder whether the organization's work systems actually reflect the same respect, loyalty, and trust. Consider what happened when a consultant gave a group of managers the exercise of designing an organization that would produce the lowest possible level of trust. People in this management audience fidgeted and acted embarrassed as they saw their own organizational practices in the extensive rules and control systems that they proposed as a way to produce low trust.[187] It is easier to talk about trust than to build it into the way an organization works.

Instead of using extensive rules and procedures, some leading companies such as the department store chain Nordstrom try to maintain both control and trust through a well-articulated culture and careful attention to whether employees fit the culture. Nordstrom's employee handbook consists of a single five-by-eight card containing seven short sentences. After saying that outstanding customer service is the company's one goal and that employees should set their personal and professional goals high, the card says that there is only one rule at Nordstrom: "Use your judgment in all situations. There will be no additional rules." Nordstrom demands conformity to a gung-ho, highly customer-oriented culture. Its employee training about the Nordstrom Way includes selected stories about how employees extend themselves and use judgment to provide outstanding customer service.[188] An attempt to install systems with extensive rules governing customer-related decisions about adjustments or returns would probably be rejected in this culture.

Organizations with less committed cultures often attempt to develop systems that compensate for low levels of trust. For example, a study of software design found an information system that limited the information available to bank tellers and tried to limit their discretion in making decisions. The software was designed to freeze the teller's computer screen and require a supervisor's approval when unusual account activity or other unusual conditions occurred. According to the lead developer of software used by this bank and other mid-sized banks, "The first guideline is that banks purposely and deliberately keep tellers ignorant of what the bank is try to do [because they] believe that their greatest protection is in keeping the teller and everyone else, including the branch managers, purposely ignorant."[189] Ironically, some of this system's restrictions were too awkward to enforce in practice. For example, the information system permitted only a limited number of separate transactions by one customer during a day, but a long-time customer came to the branch quite frequently. To make it unnecessary to call a supervisor for every unusual but well understood occurrence, branch management adopted a workaround that bypassed the security provisions in the information system. The supervisor's codes were made available to the tellers so that they could unfreeze the computer screens themselves in order get their work done despite the security provisions that had been programmed.[190]

Privacy

I came to appreciate the significance of privacy as an aspect of organizational culture when I visited a Japanese office in Tokyo 20 years ago. The office was a rectangular room, whose main feature was two rows of 10 desks, each extending from near the entry door toward the back of the room. The manager's desk was furthest from the door. The two rows of desks faced each other, and each desk touched both the desks next to it and the desk in the facing row. The workers appeared to have almost no personal privacy in this office, but keeping these 20 people coordinated was certainly much easier than it might have been if they had individual offices or even cubicles in an open floor plan. A formal coordination system that might have been helpful in an American setting would have been meaningless in this office because the entire department sat together every day.

The traditional Japanese language has no word for privacy[191], but privacy raises many issues for work systems in Western organizational cultures. Consider the privacy of personal communications within a business setting. While training 700 employees to create and send e-mail messages to fellow workers, the e-mail administrator at Epsom Computer assured them that e-mail communications would be totally private. Later she discovered that her boss was copying and reading employees' e-mail. When she was fired after complaining, she sued her former employer on behalf of the employees whose e-mail had been opened. Epsom argued that state privacy statutes make no mention of e-mail. The judge agreed with Epsom and dismissed the case.[192] To clarify expectations and avoid lawsuits, most American firms now inform their employees that e-mail is company property and is subject to inspection.

Information Sharing

Computer network proponents and vendors often proclaim that information is an organizational resource to be shared. They usually don't say that information sharing may be inconsistent with personal, organizational, and regulatory realities. As is clear from everyday life, reasons for reluctance to share information include desire for privacy, need to maintain control, and desire to avoid embarrassment or even prosecution. The related pitfall in designing information systems is the assumption that people within an organization should want to share information with others.

Successful use of technology for information sharing depends on technology and on having a culture of sharing information.

The issue of local control delayed the international consulting firm McKinsey & Co. when it set out to develop PeopleNet, a company-wide human resources information system for keeping track of employees. Since consultants are a consulting firm's most important resource, it was important to make personnel information more of a corporate resource. The project was delayed by two factors: glitches related to inadequate technical infrastructure and worries at McKinsey's highly independent branch offices that PeopleNet would reduce their traditional local control over staffing. After many meetings, control of the part of PeopleNet related to tracking staff assignments was given to local offices, which agreed to share with headquarters only a limited amount of personnel data, such as languages spoken, skills, and experience.[193]

A similar result occurred when Andersen Worldwide (previous parent of Accenture) implemented an electronic network designed to promote information sharing and collaboration within its highly skilled but widely dispersed staff. The network reached 82,000 people in 360 offices in 76 countries through data, voice, and video links. Its electronic bulletin boards permitted posting of requests for help on problems and provided methods for accessing company information related to subjects, customers, and resources. Investments in the network and employee training yielded disappointing results until Andersen management started considering evidence of information sharing in promotion and compensation decisions.[194]

These brief examples related to trust, privacy, and information sharing illustrate the importance of alignment between work system operation and organizational culture. Over-emphasis on how the computer operates or on details of work procedures may obscure cultural misalignments that can be showstoppers.

Organizational Policies and Procedures

Every mature organization has written or unwritten policies and practices. Employees are expected to follow these policies and procedures to satisfy the organization's need to be coherent and follow its own strategy. For example, a national consulting firm's expectations of all of professional employees include:

- They will submit a weekly progress report.
- They will use the company's templates for all presentations.
- They will use the company's templates for developing project plans and reporting progress against goals.
- They will complete certain levels of prescribed training.
- They will follow company policies related to employee complaints.
- They will follow company policies regarding use of email, computers, and other resources.

Policies and procedures such as these seem totally mundane, but they create constraints on how work systems can operate and how employees spend their time. In some cases it is possible to work around or ignore organizational policies and procedures, but it is usually easier to abide by corporate policies and procedures unless they create significant obstacles.

The examples above concern very general policies. Other corporate policies are more focused and attempt to solve specific problems. For example, the pharmaceutical company Syntex (later acquired by Hoffman La Roche) found that essential information systems for monitoring laboratory procedures or communicating critical data to the Food and Drug Administration had been developed by end users and contained no security, backup, or recovery capabilities. A task force including business managers and IT staff members created guidelines for mission-critical information systems. One of the policies said that business managers would be responsible for identifying mission-critical information systems and making sure they complied with corporate standards.[195]

Organizational policies and procedures often affect information systems through technical standards and policies. Uncoordinated hardware and software purchases and upgrades have wasted huge amounts of time and effort in most large companies and many small ones as well. To avoid this type of situation, even companies with highly decentralized information system activities often maintain central authority over which brands and versions of hardware and software that can be purchased and used. Topics such as these fall under an enterprise view of IT, which was discussed at the end of Chapter 13.

Legal and Regulatory Requirements

"You can't ask me that." This was a new employee's response when a human resources officer asked her whether she was married. Her response would have been appropriate at her job interview, since marital status had nothing to do with the job she had applied for, but she had already been hired and the human resources officer was trying to find out who should be covered by her new health insurance policy.

Whether or not particular information can be used for a particular purpose is one of many areas in which the legal and regulatory part of the external environment affects work systems. Inquiries about marital status are illegal in the work system of hiring new employees for many jobs because the information might be used to discriminate against married or unmarried individuals. The same information is not only acceptable, but expected when an employee enrolls for company-sponsored insurance. Similar issues arise concerning race, age, medical conditions, and other types of information that might confer an advantage or disadvantage to particular individuals in particular circumstances.

HIPAA, the Health Insurance Portability and Accountability Act, is an example of a law directed at personal privacy. Among many provisions, it dictates that patients can have access to their medical records and must be informed of how their personal information will be used. It also requires that health care providers establish procedures for identifying which employees have access to protected health information and for permitting access only by those employees who need the information.

The legal and regulatory environment often forces expansion of work systems to collect information that might not otherwise seem important or productive. For example, banks are required to collect information that demonstrates they do not engage in redlining, the practice of discriminating again people in certain areas when offering loans or insurance. Similarly, federal regulations about the work hours of pilots and flight attendants force airlines to collect and report specific types of information about their actual work hours regardless of whether that information is needed for internal staffing analysis.

The Sarbanes Oxley Act of 2002 had much wider impact. A response to management fraud at major companies such as Enron and WorldCom, it imposed extensive requirements related to corporate governance, internal controls, and record keeping by publicly traded corporations. The rules include:

- CEOs and CFOs must certify corporate financial reports, and face significantly longer jail terms if financial statements are knowingly misstated.
- Corporations must provide independent annual audit reports on the existence of internal controls related to financial reporting.
- A corporation's audit firms cannot provide consulting, actuarial, and legal services for the corporation.
- Corporate loans to executive officers and directors are banned.

Sarbanes Oxley required changes in work systems related to financial reporting systems and internal auditing. Many corporations had difficulty bringing their work systems in these areas into compliance. For example, according to a 2005 survey, nearly half of all companies affected by the second round of Sarbanes Oxley deadlines did not expect to meet a July 2006 target for retaining and archiving any messages and records, including e-mail and instant messaging, that may affect financial decisions or public disclosures."[196]

Competitive Environment

The external competitive environment creates requirements of a different sort. Competitive pressures call for faster response time, higher quality, and greater productivity with fewer people doing more work after corporate downsizings. These circumstances force changes in work systems because fewer people often can't do the work as it was once done.

In some cases competitive pressures lead to complete automation of activities that were previously performed by people, especially in information-related tasks. Thirty years ago, receptionists or telephone operators answered telephone calls to major corporations. Today, computers answer those calls and use artificial speech synthesis to ask the caller to specify an information request or connect to a phone extension. Corporate web sites serve a similar function by allowing people to obtain information without talking to a person. In both cases, the cost drops from dollars to pennies per call or inquiry.

Competition in the brokerage industry exemplifies the way external competitive pressures led to new, IT-enabled business models. Traditionally, stock sale and purchase transactions for individual investors were handled through stockbrokers, often with high commissions. The ability for individuals to enter buy and sell orders using the Internet bypassed brokers and reduced brokerage commissions to $15 or less for small trades. The possibility of using the Internet for entering trades changed the competitive environment, and each brokerage had to decide on its approach for competing based on its unique combination of services offered and fee structure.

Competition led to IT-related changes in work systems even before the Internet. Automobile manufacturers and large retailers discovered that handling inconsistent paper invoices from suppliers could cost $50 per invoice. They responded by requiring that their suppliers send invoices in specific electronic formats using a technique called electronic data interchange (EDI). For many smaller suppliers, the conversion from their previous order entry and invoicing systems to EDI brought substantial costs and comparatively few internal benefits. They were forced to convert to EDI, however, because their large customers would not do business with them if they did not convert.

Alignment of IT with Corporate Strategy

A final aspect of environment and strategy is the extent to which a firm's IT function is aligned with corporate strategies. The need for this alignment seems obvious. Today's businesses simply couldn't operate without current IT capabilities. Typical business transactions could not occur efficiently, recordkeeping using real time databases would be impossible, and managers would not receive the information they need to monitor and guide business activities. From that starting point, it might seem surprising that there should be any issue at all about aligning a firm's IT efforts with its business needs and strategies.

Despite the need for alignment, the lack of alignment is often viewed as a serious issue by both business and IT executives. Table 14.1 summarizes key results of a 1992-1997 survey of over 1000 executives from over 500 firms in 15 industries at IBM's Advanced Business Institute. Listed in rank order are the primary enablers and inhibitors of alignment between business and IT. The results were nearly identical for business and IT executives, who were equally represented in the sample.[197] In the survey only about half of the respondents believed that their firm's business and IT strategies were aligned.

Table 14.1. Enablers and inhibitors of alignment between business and IT

Enablers	***Inhibitors***
Senior executive support for IT	IT/business lack close relationships
IT involved in strategy development	IT does not prioritize well
IT understands the business	IT fails to meet its commitments
Business - IT partnership	IT does not understand business
Well-prioritized IT projects	Senior executives do not support IT
IT demonstrates leadership	IT management lacks leadership

Around a decade later, 2003 and 2004 surveys[198] of members of the Society for Information Management, an organization of IT executives, still identified alignment between business and IT as the top concern among 22 managerial issues. The 2003 SIM survey identified inhibitors of alignment similar to those in Table 14.1, including lack of senior management support, lack of influence at headquarters, lack of communication, and lack of business commitment to budgets for IT investments. In 2004 the rest of the top ten included:

- Attracting, developing, and retaining IT professionals
- Security and privacy
- IT strategic planning
- Speed and agility
- Government regulations
- Complexity reduction
- Measuring the performance of the IT organization
- Creating an information architecture
- IT governance

It might seem surprising that alignment would rank above the other nine as a key issue. However, using a 5-point scale (where 5 indicates the highest level of alignment), around 70% of the respondents assessed their organization at level 2 or 3. The enablers and inhibitors in Table 14.1 imply that statements such as the following applied to many of the firms:

- Executive support for the IT function is sometimes tepid or worse.
- The partnership between business and IT, and the involvement of IT leaders in strategy development is often lacking.
- The prioritization of projects is sometimes inadequate.
- The IT group's understanding of the business is sometimes inadequate.
- The leadership of the IT group is sometimes inadequate.

Although the surveys paint a disappointing picture, it is obvious that today's businesses could not operate as they do without extensive contributions from IT groups. The real message from the surveys is that much more can be done to improve the alignment between IT and the business.

This page is blank

Chapter 15: Work System Ideas in a Broader Context

- Goals and Limitations
- Antidote to the Seven Temptations
- Recognizing Missing Links in IT Success Stories
- Interpreting Jargon
- Interpreting Information System Categories
- Conclusion: Focusing on Fundamentals

This final chapter discusses work system ideas in a broader context. After reiterating goals and limitations of the work system approach, it summarizes how the work system method (WSM) addresses the seven temptations mentioned in Chapter 1. It illustrates how work system ideas can be used to identify missing links in success stories about IT and systems, and to interpret IT- and system-related jargon. Next it shows how a work system approach helps in interpreting common information system categories whose meaning changes continually as new capabilities are incorporated into real world systems. It closes with a reminder that this book identifies fundamental ideas that are useful now and should be useful in the future, even as technology continues to progress in scope and power.

Goals and Limitations

This book's primary goal is to explain practical ideas that business and IT professionals can use for understanding and analyzing systems in organizations. I wrote the book because I believe that most business professionals need an organized but flexible method for thinking about systems in organizations with or without the help of IT professionals or consultants. I hope WSM users will be more able to evaluate systems from a business perspective and more able to participate effectively in system-related projects. I hope WSM will help business and IT professionals communicate more effectively about the realities addressed by new or improved information systems.

"If you have a hammer, everything is a nail." This book's hammer is ideas about work systems. I hope you are convinced that these ideas are valuable, but I also hope you have no illusion that they are supposed to solve all important problems related to systems and IT. The work system method was designed for thinking about situations that can be conceptualized as work systems. A work system approach can be used, but far less effectively, for discussing families, small groups, organizations, firms, or even industries. Before using a work system approach to discuss situations or phenomena at those levels, you should consider using other approaches that are designed specifically for those situations.

It is especially important to recognize that focusing on work systems does not fully address a number of important topics that are associated with systems and IT, especially the following four:

Infrastructure decisions. Decisions about IT infrastructure may have positive or negative effects on specific work systems that rely on shared infrastructure capabilities. Some of the ideas in WSM may be useful in making these enterprise-level decisions, but often it is impractical to analyze every affected work system before making infrastructure decisions. Those decisions should be analyzed from an enterprise perspective, not just a work system perspective.

Programming concepts and methods. IT professionals use analysis and programming methods that are not covered here. The brief comments about techniques for specifying business processes and databases (e.g., Figures 10.2 and 12.1) only scratch the surface for topics that are covered in entire books on systems analysis and design methods for IT professionals. Nonetheless, IT professionals may find WSM useful for clarifying system-related issues and goals before they embark on their technical work. They may also find WSM useful for analyzing their own work systems. However, their technical work should use concepts and methods that are tailored to the specific type of work they are doing.

Six Sigma and statistical process control. Six Sigma began as a statistical approach for analyzing repetitive processes and finding and fixing the causes of variation and defects.[199] It has expanded into numerous versions of a general approach for attaining and maintaining high levels of quality across an organization. The core techniques of Six Sigma focus on collection and analysis of process measurements over weeks or months. Consulting and training companies have set up educational programs that certify process analysts as Six Sigma greenbelts or blackbelts in applying statistical analysis and improvement techniques based on documentation, analysis, and experimentation. In contrast, WSM is designed to attain a balanced understanding of a situation that encompasses the various elements of the Work System Framework. In many cases, the goal of the exercise does not require the amount of time or analysis involved with Six Sigma, especially if no statistical information is available and if statistical variability is a relatively small factor in the situation. In a more extensive analysis where statistical patterns exist or statistical data could be collected, it would be extremely useful to enlist Six Sigma expertise to measure and analyze the business process.

Organizational politics. WSM mentions personal and political issues, but does not say how to analyze these issues in depth. These issues can be just as important as seemingly more rational issues related to efficiency and output. Systems approaches are less useful when a situation seems to be driven primarily by personal interests and political rivalries.

Antidote to the Seven Temptations

This book's goal is to encourage organized, balanced thinking about IT-reliant systems in organizations. Using WSM at almost any level of detail should reduce the effects of the seven temptations mentioned in Chapter 1.

Temptation #1: Assuming technology is a magic bullet. WSM's balanced emphasis on all nine work system elements makes it less likely that anyone will see technology as a magic bullet. A work system's success depends on how the parts of a work system operate together.

Temptation #2: Viewing technology as the system. Once again, the Work System Framework makes it clear that technology is just one part of a work system. Technology is just one of nine elements for understanding a work system.

Temptation #3: Abdicating responsibility for systems. Work systems are systems of doing work in organizations. The inclusion of work practices, participants, products and services, and customers should make it apparent that business managers and executives are responsible for work systems whether or not IT is used extensively.

Temptation #4: Avoiding performance measurement. The use of metrics is central to WSM. Work system recommendations are much more convincing if they are supported by estimates of how much performance will improve from its current level.

Temptation #5: Accepting superficial analysis. WSM provides guidelines about the main questions that should be pursued and about many common topics under each question. It permits flexibility in terms of the depth of analysis. Its structure should raise awareness of issues that are not explored.

Temptation #6: Accepting one-dimensional thinking. WSM looks at more than the work practices, participants, information, and technologies inside the work system. It also looks at the products and services produced, customers, the surrounding environment, and relevant infrastructure and strategies. Including so many elements outside of the system makes it unlikely that a work system will be viewed as though a single dimension is adequate for the analysis.

Temptation #7: Assuming desired changes will implement themselves. The implementation phase of the work system life cycle model identifies major activities that are usually required for implementation success. The recommendation and justification phase of WSM asks for a rough project plan to verify that the recommendation is feasible.

Recognizing Missing Links in IT Success Stories

Assume you are reading an article that discusses a great IT success, a promising ecommerce system, or a proposed supply chain. Or assume that a vendor or consultant identifies a problem or proposes a change in a system in your organization. In each case, you need to interpret a story that includes some topics, skims over other topics, and omits others completely. Stories always stress some things and deemphasize others in order attain their own goals and maintain their own coherence, but that does not mean that you need to ignore what is left out.

The Work System Framework establishes a balanced view of a work system and recognizes that all nine elements are usually important. The typical issues for each element provide more depth. Together, the framework and the issues provide an instant checklist for trying to identify the important topics that were not mentioned in an IT- or system-related story. Table 15.1 identifies typical omissions from system- and IT-related success stories, cases, and newspaper articles. Omissions from stories usually are not meant as misrepresentations. People have a story to tell and they focus on what is important for telling it. Recognition of omissions is important, however, because it helps you understand how to interpret a story, especially in terms of which elements were emphasized, downplayed, or ignored. It also helps you decide what questions to ask. Instead of asking a question about a detail of the technology, for instance, it may be more useful to ask a question about the work system that uses the technology.

Interpreting Jargon

Every field of knowledge has its particular vocabulary of specialized terms that identify the important ideas within the field. Parts of that vocabulary are sometimes viewed as jargon, especially when they sound pretentious or are overused mindlessly, as in "At the end of the day our empowered associates leverage seamless IT solutions by operationalizing scalable learnings of best practices and incenting actionable mindshare." Joking aside, what may be one person's jargon may be another person's standard vocabulary of a professional field. (Admittedly, some of the ideas in this book can be considered jargon even though the main ideas are defined and used carefully.)

Jargon is especially problematic in relation to IT-reliant systems because many of the terms have taken on multiple meanings that are only somewhat related. For example, assume you heard that a firm's sales *system* is down, its manufacturing *system* is backlogged, and it bought five new Pentium *systems* from Dell. The sales system is probably software used in the sales department; the manufacturing system is the firm's system of manufacturing products for customers; the systems bought from Dell are personal computers. The context in each case would help in interpreting the term *system*, but there are many situations in which the meaning of jargon terms is unclear.

Work system ideas are often useful in interpreting jargon related to systems and IT. The key is to ask yourself a simple question:

> *Is a work system being discussed, and if so, what are its elements?*

Table 15.1. Recognizing common missing links in IT success stories

Work system element	***Typical omission in IT success stories***	***Significance of omission***
Customer	• The story doesn't mention the system's customers	• An evaluation of a work system is incomplete without some evidence of customer satisfaction.
Products and services	• The story mentions what the system produces but focuses on how the system operates internally.	• An evaluation of a work system is incomplete without some discussion of what it produces.
Work practices	• The story treats an idealized business process as though it is the system. • The story assumes participants will perform the work exactly as specified. • The story talks about the steps in a business process but ignores awkward issues related to decision making and communication. • The story talks about decision making or communication, but omits business process steps that make it possible for the decision making or communication to occur.	• Workarounds, errors, and sloppiness often violate idealized business process. • Participants may not perform the work as specified. • The heart of the matter may reside in decision making and/or communication.
Participants.	• The story ignores participant incentives and individual differences in skill, ambition, and attentiveness. • The story speaks about technology users as though using technology is the most important thing they do.	• Incentives and individual differences may determine the outcome. • Being a work system participant is more important to most technology users.
Information	• The story ignores the information in the system or it assumes that the information is close to flawless. • The story focuses on the information that is processed but says little or nothing about the knowledge needed to use that information correctly.	• The presence or absence of useful information is often a key determinant of system success. • The knowledge of work system participants is often a key determinant of work system success.
Technology	• The story focuses on technology but says little about how the technology or its features matter to the work system's success.	• The features and benefits may seem nice for technology users, but those advantages may not help users produce better results.
Environment	• The story focuses on the internal operation of a work system and ignores important aspects of the environment within which it exists, such as the organization's unique characteristics and culture.	• The unique environment, rather than the features of the technology, may be the main determinant of success.
Infrastructure	• The story focuses on how the work system operates and assumes that infrastructure is invisible and unproblematic.	• Infrastructure may prove problematic.
Strategies	• The story explains the brilliance of the IT strategy.	• Often the strategy is a plausible explanation of what happened in the past, rather than a pre-existing plan for what should happen.
Work system as a whole	• The story focuses on an information system and makes claims about its important impacts, but may say little about how that information system is an integral part of a larger work system.	• The business results come from the work system, not the information system that supports it.

Consider a few examples:

IT system. Systems that use IT are sometimes called IT systems. Chapter 1 mentioned the example of "The IT System that Couldn't Deliver," which turned out to be an improved sales system that had not yet been implemented in the sales force. In other words, something called an IT system might be software that runs on computers or might be any system that happens to use IT. When someone else uses the term IT system, you often don't know what it means until you ask for clarifications.

IT solution. This might mean whatever software is being proposed for use in a work system, or it might mean the application of IT to solve a problem that is basically about IT. The need for balance between the elements of a work system implies that changing only the technology in a particular work system will be effective mainly if the problem is a purely technical problem. Otherwise, IT solution is a bit of an oxymoron because changes related to IT are only part of the solution to the problem faced by the organization.

To interpret the meaning of terms such as IT system, IT solution, and ebusiness, ask yourself whether a work system is being discussed, and if so, what are its elements.

Business process. This is usually defined as a structured set of steps with a beginning and end, and with each step triggered by an event or condition such as the completion of a previous step or the detection of a new problem. In some situations business process is treated as an idealized set of steps that will be executed accurately by work system participants. In other situations business process may be treated as whatever activities are performed, even if those activities are unstructured. WSM uses the terms work practices and business process carefully. Work practices include all activities within the work system. Business process is one of many perspectives for thinking about work practices. Among other perspectives are decision making, communication, and coordination. The goal is to deal with reality instead of just theorizing about how things should be.

Best practices. This may mean a set of frequently effective practices that a software firm has encoded into its application software, or it may mean the best work practices for a particular situation in a particular firm. The Work System Framework shows why changing from current practices to best practices defined somewhere else is often infeasible. The problem is that the work practices within a work system need to be consistent with the other elements of the work system. For example, a hiring method that might be viewed as best practices for a manufacturing firm with 50,000 employees might be impossible for a public relations firm with 12 employees.

The paperless office. A book called *The Myth of the Paperless Office* asks, "Why, when we have all the latest technology to allow us to work in the digital world, do we depend on paper so heavily? Indeed, why are most workplaces so dependent on paper? It seems that the promised 'paperless office' is as much a mythical ideal today as it was thirty years ago."[200] When that book looks at paper as a technology, it finds that paper allows flexible navigation, facilitates cross-referencing, is easy to annotate, and allows interweaving of reading and writing. When it looks at work systems that use paper, it finds that paper remains very useful for tasks including writing documents, reviewing documents, planning, collaborating, and working effectively in organizations. In other words, there are good reasons to question the value of paperlessness for its own sake. There are many reasons why paperless office projects may not get rid of all the paper.

System development. This may mean the creation of software or it may mean the creation of an information system (a type of work system) that uses IT. In WSM, the work system life cycle model identifies development as one of four major phases. Development is the process of defining, creating, or obtaining the tools, documentation, procedures, facilities, and any other physical and informational resources needed before the change can be implemented successfully in the organization.

Implementation. Computer scientists think of implementation as the creation of software that performs specific data manipulations on a computer. Business professionals think of implementation as establishing a desired change in an organization. The work system life cycle model identifies implementation as the process of making a new or modified work system operational in the organization, including planning for the rollout, training work system participants, and converting from the old way of doing things to the new way.

Ebusiness. Although this term is a bit outdated, it is worthwhile to look at how it became visible. EDP (electronic data processing), email, and EDI (electronic data interchange) had all existed for many years when IBM started an ebusiness advertising campaign. The day before IBM's ebusiness ad campaign appeared on Oct. 7, 1997, the *Wall Street Journal* said Louis Gerstner, IBM's CEO, wanted "to position IBM as a cutting edge company and shake off for good its image as a stodgy, if reliable, supplier of computers to giant corporations."[201] Soon "e" was seen commonly as a prefix in ecommerce, e-retailing, e-advertising, and so on. Definitions of ebusiness in the next several years included:

- "By connecting your traditional IT systems to the Web you become an ebusiness."[202]
- "Ebusiness is the complex fusion of business processes, enterprise applications, and organizational structure necessary to create a high performance business model."[203]
- "Electronic business" … "includes everything having to do with the application of information and communication technologies (ICT) to the conduct of business between organizations or from company to consumer."[204]

Differences between these definitions illustrate why it is unclear how to decide whether a particular business is or is not an ebusiness. Much more important is whether the business uses IT effectively and whether its work systems accomplish its goals.

Digital enterprise, digital economy, digital society. The range of definitions for each of these terms would probably mirror the range of definitions of ebusiness. Although companies that deal primarily in information (such as eBay, Yahoo, or Google) are poster children for the digital enterprise, most companies that use computers and communication technology extensively still have to deal with the physical world. Consequently, the term digital enterprise could cover almost any enterprise that makes extensive use of information and communication technology, even if the enterprise itself does something quite physical, such as producing oil and refining it into gasoline. The broader terms digital economy and digital society seem to encompass even more pervasive uses of information and communication technologies.

Ebusiness, digital enterprise, digital economy, and digital society were included above as an illustration of terms whose connotations may deserve more attention than their various definitions. In contrast, terms for discussing specific systems should be clear enough that they can be used for understanding real situations.

Interpreting Information System Categories

Information systems are a type of work system that exists to process information. Some information systems, such as a dispatching system in a factory or transportation company, exist to support other work systems and are often seen as subsystems within those larger systems. In other situations, information systems are seen as independent work systems that produce valuable products and services directly.

Information system categories, such as management information system (MIS) or decision support system (DSS) are used frequently, but are often ambiguous because each category was coined at a particular time and subsequently took on additional meanings and connotations. Most real world systems combine capabilities and characteristics from the various categories because system designers care much more about producing valuable systems than about conforming to a ten or twenty year old definition of an information system category.

A work system approach for understanding information system categories such as MIS or DSS starts by assuming that you don't know what someone else means by one of these categories. When one of these terms is used, ask yourself whether a work system is being discussed and what it does. Sometimes it will turn out that a work system is being discussed, such as a particular way to provide management information or to make a decision in a particular situation. In other cases, the topic will be a type of software that is being touted, sold, or installed. For example, a software vendor selling what it calls a DSS actually is selling software that might be used in the future. In a literal sense nothing is wrong with this view of DSS software because software can always be viewed as a system that transforms inputs into outputs. However, viewing DSS software as the system is problematic if it leads one to forget that software is installed as part of a larger work system in which business professionals perform work and happen to use DSS software. The success of the software should be evaluated in terms of how well it helps people do their work, not in terms of its theoretical capabilities or how well it operates on a computer.

We will look at a number of common information system categories in order to introduce the essence of each category and also to explain how work system ideas can be used to visualize important issues within each category. Most of the categories mentioned here can be used in any functional area of business. A different set of categories applies to types of information systems that are associated with specific functional areas, such as computer aided design (CAD) systems, corporate planning systems, and financial accounting systems. The categories related to functional area information systems will not be discussed here. Suffice it to say that these have many of the same ambiguities as the more general categories we will mention.

Office automation system is a general category for tools (technologies) such as word processors, spreadsheets, and databases that can be used in typing and modifying documents, performing calculations, and storing and retrieving information. Contrary to their category name, these tools help in performing office work but do not automate offices. Today, office automation tools are so commonplace that they are taken for granted.

Some uses of word processors, spreadsheets, and small databases are too inconsequential and too isolated to be analyzed as a work system. For example, even though some work system ideas might help in thinking about a spreadsheet calculation that will never be re-used, a full work system analysis of a one-time situation would not be worth the effort.

In contrast, the typing of memos by lawyers and legal assistants in a law office might be analyzed as a work system because the creation of memos occurs frequently in law offices and absorbs a substantial amount of time. Although the work system analysis might pay attention to how the word processor is used, it would give greater emphasis to work practices related to taking notes, drafting memos, producing documents in a form for distribution, and storing and protecting the documents. The analysis would pay attention to the appropriate division of labor between lawyers and less expensive personnel. It would also pay attention to how the memos would be indexed and stored in a document database for retrieval.

The taken-for-granted nature of word processors and spreadsheets sometimes results in uses that are ineffective. An example is a large bank's budgeting system that relied heavily on spreadsheets produced by different departments and consolidated by a central budget group. The use of departmental spreadsheets reinforced a silo mentality in which each department did its budgeting separately, consolidations occurred at several different levels of aggregation, and the results were returned to the departments for correction and further iteration. Some of the budget analysts were frustrated about being in spreadsheet hell, focusing a great deal of attention on whether the spreadsheets were handled consistently, and less attention on whether the budget met corporate objectives. Use of specialized corporate budgeting software rather than spreadsheets might have afforded a more efficient budgeting process and better results.

The widespread availability of word processors and spreadsheets also results in their being treated as default tools when nothing else is readily available. Many businesses use spreadsheets for recording inventory levels, receipts, and other transactions that should be recorded in database applications designed for reliability. The use of spreadsheets as a pseudo-database may seem cheap and simple compared to buying and implementing specialized software, but it brings many risks. An example mentioned earlier involved a pharmaceutical company with high vulnerability to accidental loss of mission-critical information because their employees used spreadsheets to store information about clinical trials. Spreadsheets are designed as calculation tools, not data storage tools. Many people who set up spreadsheet databases have little knowledge of how to design databases for usefulness, reliability, and security. Their minimal knowledge of debugging techniques is one of the reasons for the high rate of error in spreadsheets.[205]

Communication system is another category that often refers to electronic tools, in this case communication tools such as traditional telephones, cell phones, pagers, e-mail, voice mail, instant messaging, videoconferencing, intranets, extranets, and web sites. Each tool potentially affords a different type of convenience. For example, instant messaging supports brief interactions via text, whereas email is rarely used for interactive communication but handles lengthier messages conveniently. Similarly, intranets and extranets provide Internet-like interfaces to information for company employees or for customers and suppliers.

From a work system viewpoint, a communication system involves much more than the capabilities of the tools, which can be considered part of the organization's shared communication infrastructure. For example, a communication system for police officers can include different technologies that are used differently depending on whether the officer is in a car, on foot patrol, at a police station, or at home. As with many communication systems, an important aspect of a police communication system is the guidelines and expectations for using particular technologies in particular situations. Without such guidelines, people don't know which capabilities to use when, and communication becomes chaotic.

Transaction processing systems provide the basic information for keeping track of what is going on in a business. The information is collected one event (transaction) at a time, stored in a database, and later used to control operations and to provide management information. A transaction processing system (TPS) collects information about specific events such as creation of an order, receipt of a shipment, or completion of a manufacturing step. Other examples of transactions include:

- Providing a password to an employee
- Withdrawing money from an ATM
- Correcting an error in a customer order
- Placing an order to buy stock
- Signing up for telephone service
- Purchasing an airline ticket from a web site
- Making a hotel reservation using a web site.

Notice how some of the transactions involve movement of money whereas others, such as providing a password to an employee or making a hotel reservation, are non-financial operational events that should be recorded.

Over the years, transaction processing has seen a number of significant innovations. The first TPSs collected information on paper forms. In a separate step, information from those forms was entered into a computer system, often hours later. Real time computing made it possible to enter transactions almost immediately as they occurred. Automatic data collection tools such as barcode readers minimized data entry errors and delays by automating the previously manual entry of information. A recent extension along this path is the use of RFID tags, which use radio waves to enable automatic collection of information about individual items that move past a reader, thereby making manual scanning of individual items unnecessary. For example, by using radio waves instead optics, RFID makes it technically possible to identify each of 20 separately tagged shirts inside a cardboard box. Ongoing projects such as one at Wal-Mart are testing the economic feasibility of using RFID for inventory management.[206]

TPSs are often viewed as highly mechanical, but looking at them from a work system viewpoint reveals that human participants often play important roles. For example, before entering an appointment into a doctor's schedule, the appointment scheduler makes judgments about how quickly the patient needs to see the doctor and about how to accommodate scheduling needs of both the patient and the doctor. Performing the data entry transaction is a relatively minor part of the job compared to making good judgments and communicating effectively with the patient.

Real time TPSs rely on pre-defined information stored in a database and processed in a predictable way. For example, a TPS for entering orders typically uses database information such as:

- for individual customers: customer ID, name, address, and possibly whether or not the customer's payments are up to date
- for individual orders: order ID, date, customer ID, name of customer contact for the order, shipping arrangement
- for individual line items on an order: item ID, unit price, quantity, and any applicable discount

Performing the transaction encompasses filling in a computerized form and automatically storing the data in the database. As the transaction is entered, the computer uses information in the database to help minimize mistakes and to fill in existing information such as customer address.

Management information systems. MIS has taken on several different meanings. It might be an information system that provides information for managers, or it might be a TPS that controls transactions and also provides information to management. Seeing MIS as a system of providing information for managers leads to questions about whether work practices for consolidating and displaying management information meet the needs of management. Seeing MIS as a TPS that has good reporting capabilities leads to the same questions, plus other questions about whether the transactions are controlled appropriately and performed accurately.

MIS is one of the areas where the distinction between data and information is important. An MIS that controls transactions effectively and collects a huge amount of transaction data may do so without providing useful management information. Again, this is a reason to use a work system view to see whether it produces the types of information that are directly helpful in management work practices. The analysis of management information is less direct than the analysis of transactions because management work is far less structured than transaction processing work. Although management work is not totally repetitive, it usually has a lot of regularity. For example, most managers like to have consistent, repeatable methods for monitoring the work done by their subordinates. This is why managers receive standard reports at the end of shifts, days, weeks, months or quarters.

Decision support systems and decision automation. The term decision support system (DSS) was coined as a reaction to optimization modeling done by operations researchers in the early 1970s. Optimization models perform complex calculations that determine the best possible decision under certain highly restrictive mathematical conditions.[207] For totally structured situations, such as allocating shipments to trucks in a way that minimizes cost, the model may determine the final decision. In many other situations, an optimization model may be used as the basis for a sensitivity analysis that compares the best decision under different assumptions.

The original idea of decision support was that interactive use of computerized data and models could help managers make semi-structured decisions that are not susceptible to optimization models. Managers could examine information in databases and could use simulation models to test alternatives. This approach seemed more practical than optimization because most management decisions are not structured enough to be automated. That was 30 years ago, before personal computers existed and while there was some doubt about whether computers were too difficult for most managers to use. Personal computers made interactive computing readily available, but the original idea of DSS frequently bumped into the reality that even clever tools have little impact unless they are used as part of a work system.[208]

Recently some vendors have started to replace the term DSS with the term business intelligence (BI). Formerly BI was associated with secretly obtaining and digesting information from the external environment, rather than analyzing internal operations. The capabilities of BI software are much more about analyzing operational data, as illustrated by the way Microstrategy, a provider of BI software, identifies five styles of business intelligence:[209]

- enterprise reporting: static MIS reports programmed and delivered using current technology that is much more convenient than the tools of 30 years ago
- cube analysis: "slicing-and-dicing" information to examine questions such as which products had the greatest sales improvements, which regions sold those products most successfully, and what types of promotions were used in those regions to attain those successes
- ad hoc query and analysis: use of convenient data analysis tools to generate reports that were not programmed in advance
- statistical analysis and data mining: professional data analysts using advanced statistical techniques for correlation analysis, trend analysis, financial analysis, and projections
- alerting and report delivery: proactive delivery of event-triggered "alerts" to large populations of users.

A trend toward increasing use of decision automation is another important departure from the original idea of DSS. Formula-based decisions have been automated ever since people started using computers, but increasing power and pervasiveness of computing technology has enabled many new applications of decision automation. Whereas the original idea of DSS focused on semi-structured decisions, "today's automated decision systems are best suited for decisions that must be made frequently and rapidly, using information that is available electronically. The knowledge and decision criteria used in these systems need to be highly structured, and the factors that must be taken into account must be well understood."[210] Decision automation has been applied to a number of decisions previously made by knowledgeable individuals. Examples include yield optimization in airlines (frequent revisions of prices of airline tickets based on seat availability), configuration of products to meet customer needs, allocation decisions among customers or suppliers, fraud detection for credit card companies and government agencies, and operational control decisions for physical environments such as farms or power grids.

Knowledge management systems. The last decade saw substantial interest in issues related to knowledge management systems (KMS). Different people define knowledge management differently, but the basic idea is to make sure that people within an organization have access to the knowledge that they need at the time when they need it. This is easier said than done. Even in organizations with vast collections of documents and databases of transaction information, much of the knowledge in organizations is never written down. Instead, it resides in the minds of the employees as tacit knowledge. This knowledge is of no use to employees who don't have it, and it disappears from the company when employees leave.

There are two basic approaches to knowledge management. The first approach is to make knowledge explicit by recording it, cataloging it, and making it accessible when needed. In effect, the knowledge is stored in a document database or other type of database and then retrieved through structured queries or keyword searches. The second approach is to provide access to people who have the appropriate knowledge. One way to find people with knowledge is to broadcast an email message requesting help. Another approach is to maintain a database identifying special knowledge that individuals have, such as experience in construction management or fluency in Italian.

As with the other types of information systems, the success of KMS involves much more than technology. From a work system viewpoint, a KMS is real only when particular people are doing a particular type of work that makes knowledge available to potential users of that knowledge. The success of many KMS applications depends on the ability and willingness to share knowledge. Personal incentives are an obstacle if people believe that sharing their knowledge will erode their unique personal advantages within the organization.

ERP. Chapter 12 explained that ERP (enterprise resource planning) is a complex form of software that establishes an integrated database used for transactions across functional areas of business, such as manufacturing, sales, customer service, and accounting. Although called ERP systems, ERP software is more accurately described as an integrated technical and information infrastructure that is configured to support transaction processing within local work systems.

ERP is actually a misnomer. The acronym ERP stands for enterprise resource planning, but it is more accurate to call these systems enterprise information systems or enterprise systems because they are actually much more concerned with recordkeeping and coordination than with planning. The history of the name goes back to MRP (Manufacturing Requirements Planning) software developed in the 1960s for ordering components required by manufacturers to meet factory output schedules and to replenish their inventories. MRP was succeeded by MRP II (manufacturing resource planning), which added capabilities related to plant capacity, sales, and other functions. ERP added accounting, invoicing, shipping, logistics, human resources, and other functions. The way ERP emerged from MRP II, which emerged from MRP is typical of the way various types of information systems evolved and added new capabilities with or without changing their names.

ERP systems are not work systems. Rather, ERP is infrastructure shared by many different work systems. Various parameters of the software are configured to make the software as useful as possible for specific work systems that use it. A major challenge in setting up ERP software is that the same configuration decisions affect multiple work systems. There are many situations in which a configuration decision that is good for one work system is quite awkward for another.

CRM, customer relationship management, is another example of a misleading name for a category of software. CRM emerged as a category of software related to customer-facing transactions. Various CRM products support sales cycles, order entry, customer service, and other customer-facing activities. With its transactional focus, CRM software is not really about relationships in any genuine sense of the term. CRM overlaps to some extent with ERP because ERP vendors provide software for some customer-facing functions such as order entry.

The unusually high failure rate of CRM has been attributed to many factors, including the immaturity of CRM software, difficulty in integrating CRM software with other software the organization, and confusion about what CRM is supposed to do. A work system perspective explains part of the problem. The real issue in using CRM is creating better work systems for selling to customers, entering orders, providing customer service, and performing other customer-facing work. Starting from the premise that those are the goals, rather than implementing CRM per se, might have avoided some of the confusion that occurred when companies launched CRM initiatives.

Conclusion: Focusing on Fundamentals

This final chapter completes an arc that started when Chapters 1 and 2 introduced basic ideas about work systems. Chapter 3 showed how those ideas could be woven into a flexible analysis method that can be used at various levels of detail. The top layers are useful for attaining a big picture understanding of a system in an organization. The more detailed layers help in drilling down to identify specific issues that may be important. Chapters 4 through 7 presented clarifications about the basic ideas, including a chapter about how work systems change over time. Chapters 8 through 14 identified a large number of topics and issues that arise frequently when designing, evaluating, and improving work systems. Certain issues and topics may or may not be important in specific systems, but the organized compilation of these issues and topics provides guidelines for increasing the likelihood that common issues are addressed in an analysis effort.

This chapter started by mentioning goals and limitations of the work system approach, and then illustrated several ways in which a work system perspective can help even when a complete analysis is not needed. Keeping the Work System Framework in mind helps in identifying omissions from success stories about IT and systems in organizations. For example, stories that focus on features and benefits of new technology often omit characteristics of participants and culture that may have been just as important as the technology in generating the results. Similarly, IT and management jargon terms, including categories of information systems, have taken on many meanings that have changed over time. Simply asking whether a work system is being discussed often helps in understanding what these terms mean within a current context.

In addition to completing the arc, this final chapter was a reminder of themes that pervade the book. Aside from the usefulness of the work system approach, one of the main themes is that business and IT professionals need organized and non-overwhelming ways to think about systems in a variety of situations. Sometimes they need a picture to see how things fit together and to see what is missing. Sometimes they need a common vocabulary to make it easier to work together. Sometimes they need an outline for digging in and increasing the likelihood that major issues will be uncovered. Sometimes they need a type of analysis that is rigorous enough to be clear, but nontechnical enough to encourage genuine agreement between business and IT professionals before technical programming work begins. The work system approach addresses all of these needs.

Another major theme is the limits of techno-centric thinking. Even when IT is absolutely essential for work system operation, a techno-centric focus can be misleading because the operation and success of systems in organizations relies on much more than having the right technology. New terms related to systems in organizations emerge continually and absorb new meanings as technology and work practices evolve. With so many aspects of business and technology changing rapidly, it is easy to be fascinated by the latest developments. The work system approach fully respects the power of technology, while also providing a lens for keeping a moving target in perspective.

To my knowledge this is the first book to explain a work system approach of the type presented here. It tried to present fundamental ideas that could have been applied 100 years ago, before digital computers existed, and might be applied 100 years from now, when technology will surely be beyond anything we can imagine today. As author, I would settle for the less grandiose goal of genuine usefulness to you, next week or next year. I hope it meets that goal.

Appendix: Example Illustrating the Work System Method

- Level One -- Summary
- Level Two -- Important Questions
- Level Three - Analysis Checklists
- Level Three - Templates

This Appendix is organized to illustrate different levels of detail for using the work system method (WSM) introduced in Chapter 3. WSM is designed to provide flexibility for users while maintaining clarity about what is being omitted. Users may want nothing more than a work system snapshot identifying the work system that someone else is analyzing. In some cases, they may want a summary related to a problem and recommendation. In other cases, they may want a deeper analysis and more detailed recommendation and justification. Chapter 3 explained how WSM's structure is designed to support a range of goals and to make it convenient for users to proceed at any level they choose.

To provide an example, this Appendix shows how different levels of WSM might be applied to the loan approval work system introduced in Chapter 8. The purpose is to show what some of WSM's outputs might look like. The specific checklists and templates that are shown are not cast in concrete. To the contrary, they are starting points that can be modified to suit the user's situation and needs.

Level One of WSM can be used as an executive summary of an analysis and recommendation. For example, in the loan approval case, a one-page version of Level One would briefly identify the system and problem, would say very little about the analysis, and would summarize the recommendation. It would go further than a work system snapshot that only identifies the work system, but would provide almost no justification for the recommendation. The lengthier version of Level One shown in this Appendix gives a more complete list of issues and identifies important points in the analysis, but still does not provide enough information about the analysis to evaluate the recommendation fully.

Level Two of WSM is organized as 25 questions, 5 about the system and problem, 10 about the analysis and possibilities, and 10 about the recommendation and justification. Chapter 8 introduced the example by summarizing the bank's situation and then providing hypothetical answers to the first 15 questions. Chapter 10 added three diagrams that summarized the business process. Those answers and diagrams will not be repeated here. Only answers to the 10 Level Two questions about the recommendation and justification will be shown.

Each of these answers might have been pursued in even more detail. In a real world situation that matters, business and IT professionals should discuss these questions and should identify issues that need further clarification or analysis. Level Two may seem long to someone not involved in the situation, but it represents only a fraction of what should be considered by a project team before recommending a significant investment.

Level Three of WSM consists of checklists and templates that might be used at various points in the Level Two analysis. The purpose of the checklists is to help in identifying topics, issues, and possibilities that might be important to the analysis, but might be overlooked. Each checklist identifies frequently relevant topics and provides room for comments.

- The principles checklist may trigger realizations that the original problem statement ignored important issues, or that certain possible changes would cause other problems.
- The work system scorecard reminds analysts of a range of performance indicators.
- The strategy decisions checklist encourages consideration of big picture strategies that go beyond fixing details and local symptoms.
- The possibilities checklist helps in remembering that changes in one part of a work system often must be accompanied by related changes in other parts of the system.
- The risk factors and stumbling blocks checklist provides reminders of things that often go wrong.

The user of a checklist decides whether each topic is important enough to merit further consideration for the work system that is being analyzed. If a topic is important, the user jots notes about why it is important. If not, the user indicates that it isn't important and goes on to the next topic. Use of the checklists helps in making sure that important issues are not being ignored, but certainly does not guarantee that all relevant issues have been considered.

The five checklists shown in this Appendix are based on Figures 5.1 through 5.5 in Chapter 5. Each of these checklists uses a specific theme (principles, performance indicators, and so on) to scan across all of the work system elements. A practical difficulty of using these checklists is that several different checklists may point to the same issue. For example, a stumbling block may be directly related to a possibility for change or an important performance indicator. Consequently, using all five of the checklists in the same analysis would probably seem redundant even though each checklist poses issues organized around a different theme.

The checklists shown at the end of Chapter 5 took a different form. Those checklists were related to specific work system elements and were designed to help in answering the analysis and possibilities (AP) questions related to specific elements. For example, someone answering AP3, the question related to work practices, would look at the associated checklist to identify potentially relevant work system principles, performance indicators, strategies, stumbling blocks, and possibilities for change. Ideas from the checklists at the end of Chapter 5 were incorporated into the answers to the Level Two questions that were used to organize Chapter 8.

The Level Three templates can be used for typical project activities such as developing a preliminary project plan, estimating likely costs and benefits, and identifying project risks.

IMPORTANT: This Appendix illustrates an analysis method; it would not be submitted as a final report to management. A report for management would probably start with an executive summary that might (or might not) look like Level One. The remainder of the report would be organized in whatever way is appropriate for the situation. The Level Two and Level Three details would be discussed as needed. Questions and topics that are listed in checklists and templates but are not important for that particular work system would not be mentioned. The filled out checklists and templates might be made available as supplementary material, but would not be included in the main report. Managers want to know how the project team made sense of the situation; typically they are much less interested in seeing all of the details that the project team considered.

IMPORTANT: As mentioned in Chapter 8, this example was constructed to show the potential significance of many topics covered in this book. For purposes of illustrating the work system method, a constructed example seemed more appropriate than a narrower real world example that would have touched fewer issues. You may question the rationale for some aspects of the situation in the example, and you may or may not agree with the recommendations. That type of questioning is exactly what business and IT professionals need to do when they collaborate to create or improve work systems. The purpose of the example is to illustrate the topics that might appear in the analysis and the level of detail that would just be a starting point for producing a carefully designed and well-justified replacement for the current work system.

Level One -- Summary

System and Problem

The system. The work system starts when a loan officer and a new client discuss the possibility of obtaining a loan from TRBG. The loan officer helps the client assemble a loan application including financial statements and tax returns. The loan officer submits the loan application to a credit analyst, who prepares a "loan write-up" summarizing the applicant's financial history, providing projections explaining sources of funds for loan payments, and discussing market conditions and the applicant's reputation. Each loan is ranked for riskiness based on history and projections. The loan officer presents the loan write-up to a senior credit officer or loan committee. Senior credit officers approve or deny loans of less than $400,000; two loan committees and an executive loan committee approve or deny larger loans. The loan officer may appeal their decisions. The loan officer notifies the client. Loan documents are produced if the client accepts the loan.

Problems. The current loan approval system is not generating the level of interest payments needed to meet the bank's profitability targets. Over the last five years the current work system has produced a substandard loan portfolio for the bank, i.e., too many loans were approved that should have been denied, and some denials probably should have been approved.

- The current loan evaluation model provides inadequate guidance for approval or denial decisions, for setting risk-adjusted interest rates, and for setting terms and conditions for the loan.
- The current loan approval system is too expensive because it absorbs too much time of too many employees.
- Senior credit officers are extremely overloaded, especially after two of them quit and moved to other banks last month after complaining about overload for a year.
- Credit analysts are frustrated by a combination of awkward technology, inexperience, frequent interruptions, and many starts and stops in their work as they gather information and clarifications from various sources.
- Some of the market and financial projections provided by clients are questionable.
- Some borrowers believe the approval process takes too long.
- Some loan officers believe that the approval process is partly political and shows favoritism to certain loan officers.
- Some loan committee members believe they have too little time to review loan applications before loan committee meetings.
- Some loan write-ups are incomplete, wasting time and causing delays.
- Some loan applicants complained that they were led to believe loans would be approved, only to find out that a senior credit officer or loan committee required unrealistic terms in loan covenants.

Analysis and Possibilities

- Decision-making in approving or denying loans is inadequate. Too many risky loans are approved, and some loans are denied that should have been approved. There is no formula for approving loan applications, but the current decision process is too subjective.
- The existing loan evaluation model does not provide adequate guidance. A validated loan evaluation model is needed that will provide better guidance, especially by comparing applicant's situation with past results from comparable applicants.

- When the salaries and benefits of employees are considered, the time devoted to processing and approving loan applications costs between $800 and $3000, even with the clients paying for real estate appraisals. Reducing the amount of time devoted to decisions would reduce costs.

- Credit analysts produce loan write-ups using an awkward and error-prone combination of three tools: a complex spreadsheet, a loan evaluation model, and a loan write-up template in Microsoft Word.

- Around 30% of applicants complained that the fees or interest rates were too high, or that excessively stringent loan covenants made the loan impractical to pursue.

- For applications that received denials or especially stringent loan covenants, around 25% of the applicants complained that the explanation for the denial or the stringent covenants was inadequate.

- The time from receiving a completed loan application to making the approval or denial decision averages around one month; two weeks is the absolute minimum. Although it might be possible to reduce turnaround time, most clients believe turnaround time is less important than terms and conditions of a loan.

- Currently there is no formal process for tracking the reasons for approvals or denials. It is possible to track parameters of each loan, the guidance provided by the models, and the decision that is taken. Periodic analysis of this information might clarify whether the decision process is biased in favor of or against particular loan officers.

- The loan portfolio is substandard, partly due to the quality of the current loan evaluation model's guidance, and partly due to poor decisions, especially where procedures were bypassed.

- The loan write-ups may be ready 3-4 business days before a weekly loan committee meeting. Some write-ups could be sent to loan committee members earlier than is currently done.

Recommendation and Justification

- Enforce creditworthiness standards.

- Provide a better loan evaluation model and better guidance for setting interest rates and fees. Three companies provide such models. The preferred model, subject to further analysis, is a commercial product sold by Spreadcomp. It has been used in many banks. The fee per analysis is $75 plus .005% of the amount by which the loan exceeds $1,000,000. Thus, the fee for a $2,400,000 loan would be $145. This expense is justified based on two factors. First, it seems likely that loan decisions will improve. Second, use of the model could provide guidance that will reduce the amount of high cost of employee time that is devoted to credit analysis and decision making.

- Integrate the improved loan evaluation model into a single tool for creating a loan write-up. Increase efficiency and eliminate errors by providing an integrated tool through which credit analysts enter loan information once, thereby eliminating manual copying between documents.

- Monitor whether unwarranted approvals or denials occur. Do this by archiving the model's probability estimates and using them to evaluate the actual decisions that were taken.

- Provide a loan evaluation model for use with clients. If the new loan evaluation model proves effective for decision making, negotiate with the model provider to establish a basic version of the model that loan officers can use with clients to expedite their work, make them more efficient, and allow them to deal with more clients and find more opportunities.

- Provide more timely information for loan committees. Increase their efficiency by making information available earlier before meetings and by providing better guidance from a loan evaluation model.

- Do not reduce the size of the loan committees.

- Investigate possible patterns of bias in loan approval decisions. Use of a new loan evaluation model should make it possible to monitor patterns of loan approvals and denials to see any systematic bias, as revealed in excess approvals or denials for certain loan officers.

- Produce and use better explanations of approvals and denials.
- Faster turnaround is low priority. Focus on consistency, rationale, and decision making rather than on the fastest possible turnaround for loan applications. Faster turnaround would be nice, but making appropriate decisions and treating customers professionally is more important.

A project to evaluate three alternative models, select one of them, and implement its use will cost $200,000 including training and consulting. This investment, plus the cost of running the model, should pay for itself within six months of implementation. These models have been used in 15 banks and are mature enough to operate reliably. There is some risk that bank personnel will ignore the models when making certain loan decisions. (Note: In a realistic situation, a cost/benefit study and risk analysis would be based on a tentative project plan and other analyses not included here.)

Level Two – Important Questions

System and Problem

NOTE: Answers to the five SP questions for Level Two were presented in Chapter 8 and will not be repeated here.

Analysis and Possibilities

NOTE: Answers to the five AP questions for Level Two were presented in Chapter 8 and will not be repeated here.

Recommendation and Justification

NOTE: The following response to RJ1 (including the work system snapshot) is a more detailed version of the Level One recommendation and justification (shown previously). The other RJ questions at Level Two look at the recommendation from other perspectives.

RJ1: What are the recommended changes to the work system?

The main recommendations include:

Enforce creditworthiness standards. To improve the quality of the loan portfolio, our creditworthiness standards must be enforced, not overridden at the whim of senior credit officers or loan committees.

Provide a better loan evaluation model and better guidance for setting interest rates and fees. The three commercially available loan evaluation models are from Spreadcomp, Winter Meadow Software, and Bayesian Parameters, Inc. The Spreadcomp model is the preferred option, subject to further analysis. It has been used in many banks. The fee per analysis is $75 plus .005% of the amount by which the loan exceeds $1,000,000. This model should be used to generate a recommended decision for each loan. The loan committee should use the recommended decision as guidance, but should be able to reach its own conclusions. The fee for running the model is justified based on two factors. First, loan decisions will probably improve even though the vendor's claims for the model's value are not fully substantiated. Second, use of the model could provide guidance that will reduce the amount of high cost of employee time that is devoted to credit analysis and decision making.

The italicized and bolded phrases within brackets highlight the main ways in which the work system snapshot in Figure 2.2 would change if the recommendations were adopted.

Customers	Products & Services
• Loan applicant • Loan officer • Bank's Risk Management Department and top management • Federal Deposit Insurance Corporation (FDIC) (a secondary customer)	• Loan application • Loan write-up • Approval or denial of the loan application • Explanation of the decision • Loan documents

Work Practices (Major Activities or Processes)
• Loan officer identifies businesses that might need a commercial loan. • Loan officer and client discuss the client's financing needs and discuss possible terms of the proposed loan. ***[If it is possible to provide a simplified loan evaluation model for use by the loan officer, it may be possible for the loan officer to establish more realistic expectations.]*** • Loan officer helps client compile a loan application including financial history and projections. • Loan officer and senior credit officer meet to verify that the loan application has no glaring flaws. • Credit analyst prepares a "loan write-up" summarizing the applicant's financial history, providing projections explaining sources of funds for loan payments, and discussing market conditions and applicant's reputation. Each loan is ranked for riskiness based on history and projections. Real estate loans all require an appraisal by a licensed appraiser. (This task is outsourced to an appraisal company.) ***[Using an integrated software tool instead of three separate tools for the write-up will improve credit analyst efficiency and eliminate errors.]*** • Loan officer presents the loan write-up to a senior credit officer or loan committee. • Senior credit officers approve or deny loans of less than $400,000; a loan committee or executive loan committee approves larger loans. ***[A new loan evaluation model would provide guidance that might help the senior credit officers and loan committees make better decisions. Having better guidance from a model might also help them do their work more efficiently.]*** • Loan officers may appeal a loan denial or an approval with extremely stringent loan covenants. Depending on the size of the loan, the appeal may go to a committee of senior credit officers, or to a loan committee other than the one that made the original decision. ***[A new loan evaluation model might make it more difficult to appeal loans that fall outside of the model's guidance.]*** • Loan officer informs loan applicant of the decision. • Loan administration clerk produces loan documents for an approved loan that the client accepts.

Participants	Information	Technologies
• Loan officer • Loan applicant • Credit analyst • Senior credit officer • Loan committee and executive loan committee • Loan administration clerk • Real estate appraiser	• Applicant's financial statements for last three years • Applicant's financial and market projections • Loan application • Loan write-up • Explanation of decision • Loan documents	• Spreadsheet for consolidating information • Loan evaluation model • MS Word template • ***[Integrated software application will combine a new spreadsheet, a loan evaluation model, and template.]*** • Internet and telephones

Integrate the improved loan evaluation model into a single tool for creating a loan write-up. Increase credit analyst efficiency and minimize errors by replacing the separate spreadsheet, loan evaluation model, and write-up template with what appears to credit analysts as a single software application. The analyst will enter loan information, the model will run automatically, and the credit analyst will fill in only the additional information that was not in the spreadsheet or model. Whenever any number changes, the software will automatically regenerate the write-up, including all ratio analysis and risk category analysis that is non-subjective.

Monitor whether unwarranted approvals or denials occur. Archive the new loan evaluation model's probability estimates in order to analyze whether unwarranted approvals or denials occur, and if so, under what circumstances. Use that analysis to develop rules about when it is appropriate to override the guidance from the model. Do similar analysis to answer questions about whether loan applications submitted by certain loan officers receive positively or negatively biased treatment.

Provide a loan evaluation model for use with clients. If the new loan evaluation model proves effective for decision making, it would be helpful to provide a basic version of the model that loan officers can use with clients to expedite their work, make them more efficient, and allow them to deal with more clients and find more opportunities. The simplified version would be used to provide quick feedback to clients about the likelihood of loan approval. A run of the model would not be a guarantee, but would help the loan officer establish realistic expectations. To avoid creating premature negotiations about fees and interest rates, the loan officer would say no more than describing general guidelines that will govern decisions to be made at headquarters.

Provide more timely information for loan committees. Increase their efficiency by making information available earlier before meetings and by providing better guidance from a loan evaluation model.

Do not reduce committee size. Given the weakness of our loan portfolio, we should not reduce the number of experienced viewpoints that will be applied to loan decisions. (On the other hand, after obtaining a better loan evaluation model we should take the guidance from that model more seriously.)

Investigate possible patterns of bias in loan approval decisions. Monitor patterns of loan approvals and denials to find any existing bias related to loan officers, clients, or information from the models.

Produce and use better explanations of approvals and denials. Applicants need to receive better explanations of denials and stringent loan covenants. For internal use, explanations of approvals and denials should help in tracking the quality of our decisions.

Faster turnaround is low priority. Focus on consistency, rationale, and decision making rather than on the fastest possible turnaround for loan applications. Customers would like faster turnaround, but making better decisions and treating customers professionally is more important.

RJ2: How does the preferred alternative compare to other alternatives?

Noteworthy alternatives were considered in three areas.

Loan analysis model. Alternatives include keeping the current loan evaluation model, developing a new homegrown model, or adopting a model provided by one of three software vendors. Keeping the current model was ruled out because almost no one is satisfied with the guidance the current model provides. Although one staff member was interested in developing a homegrown model, that alternative seemed risky because of the lack of prior experience with developing such a model and lack of other staff members with sufficient statistical knowledge to maintain that model if its developer moves to another position inside or outside of the bank. The preferred vendor is SpreadComp, Inc., a software company whose loan software has been on the market for over 10 years. Instead of purchasing the software, we should install the SpreadComp software on a server at headquarters and pay a fee for each analysis. The fee per analysis is $75 plus .005% of the amount by which the loan exceeds $1,000,000. Other software vendors such as Winter Meadow Software and Bayesian Parameters, Inc. were considered but those options cost 35% and 55% more, respectively. They each provide more extensive analysis capabilities, but the additional cost outweighs the additional value (pending further analysis).

Model for use with clients. A loan evaluation model for use in discussions with clients might help establish realistic expectations about the likelihood of loan approval. The alternatives are to provide no model for loan officers (as is currently done), to provide a simplified model, and to provide a complete model that would also be used by credit analysts. The preferred option is to provide a simplified model for loan officers, assuming that is feasible. (This approach has not been used anywhere else, but seems possible in theory.) The alternative of providing the complete model to loan officers might have two problems. First, access to the complete model might encourage loan officers to game the system by making a less qualified loan applicants seem more attractive based on the calculations in the software. Second, clients might interpret favorable outputs of the model as a commitment for loan approval. The appearance of a commitment would make it more difficult for the bank to deny applications that the software would approve because it does not take into account factors that human decision makers might recognize.

Configuration of the loan committees. A variety of options were considered, but the recommendation is to stay with the existing five-member loan committees. The idea of retaining loan committees of more than a few people seemed beneficial because most committee members believe that discussions within the committees sometimes identify issues that a smaller committee would not have found.

RJ3: How does the recommended system compare to an ideal system in this area?

An ideal system would make the decision automatically based on the best objective information possible. It is not clear that the subjective evaluations by the loan committees lead to substantially better decisions. Adoption of the new loan analysis model was chosen is a step toward the ideal, but maintains involvement by the loan committees because no one trusts existing software to be reliable enough to make the best decisions, especially since forecasts of future conditions are highly subjective.

An ideal system would also reduce the amount of time required for credit analysts to transfer information from financial statements, tax returns, and other sources into a loan evaluation model. Integrating the three tools (a new spreadsheet, new loan evaluation model, and new Word template) is a step in the right direction.

RJ4: How well do the recommended changes address the original problems and opportunities?

TRBG's loan approval process is not generating the level of completed loans that are required to meet the bank's profitability targets. There is substantial disagreement about the extent to which approval standards should be tightened or relaxed, but there is general agreement that TRBG is not generating enough current revenue from its loan portfolio and that its competitors are taking business that it might have taken itself. The following table summarizes the extent to which the recommendations address the original problems and opportunities.

Problem or opportunity	*Extent to which the recommendations address this problem or opportunity*
Failure to generate the level of completed loans required to meet the bank's profitability targets	The expenses related to implementing the new loan evaluation models will hurt profitability in the short term. The use of a better loan evaluation model and integrated interface for credit analysts may lead to fewer errors in loan approval decisions. The expense of running the models should be outweighed by increased revenues from additional correct approvals and decreased losses from incorrect approvals. However, there is no reliable way to estimate how many decisions will be affected, especially given that the models provide guidance for decisions that will still incorporate subjective criteria. Also, it is possible that attaining profitability targets will require a larger number of prospects and loan officers. That possibility was not part of this analysis.
The current work system has produced a substandard loan portfolio for the bank.	The use of a better loan evaluation model may lead to fewer errors in loan approval decisions. However, the error reduction will appear only in new loans. The existing substandard loans will remain in the portfolio until they are paid off or foreclosed. Also, there is no reliable way to estimate how many decisions will be affected (i.e., changed from what they would have been), especially given that the models provide guidance for decisions that will still incorporate subjective criteria.
The current loan evaluation model provides inadequate guidance approvals or denials, for setting risk-adjusted interest rates, and for setting terms and conditions for the loan.	Existing loan evaluation models from three vendors would provide better guidance than the existing loan evaluation model that is filled out by credit analysts.
The current loan approval system is too expensive because it absorbs too much time of too many employees.	Guidance from a better loan evaluation model should reduce the amount of time spent by senior credit officers and loan committees. The integrated interface for credit analysts should increase their efficiency and eliminate errors. A simplified version provided to loan officers would increase their efficiency by helping them set expectations with clients. There is no definitive estimate of how much time of which employees would be saved.
Senior credit officers are extremely overloaded, especially after two of them quit and moved to other banks last month after complaining about overload for a year.	This analysis did not discuss how to replace the two senior credit officers who quit. Adopting a new loan evaluation model should increase efficiency, but certainly not enough to make it unnecessary to find replacements for the two who left.
Credit analysts are frustrated by a combination of awkward technology, inexperience, and frequent interruptions.	The integrated interface for credit analysts should increase their efficiency and eliminate errors.
Some of the market and financial projections provided by clients are questionable.	The recommendation does not address this issue directly, although some of the comparisons by the loan evaluation model might reveal unrealistic projections.

Some borrowers believe the approval process takes too long.	The proposal may hasten the loan approval process, but the clients' highest priority is better interest rates and terms, rather than a faster approval process.
Some loan officers believe that the approval process is partly political and shows favoritism to certain loan officers.	Use of a new loan evaluation model should make it possible to monitor patterns of loan approvals and denials to see whether there is any systematic bias or favoritism, as revealed in excess approvals or denials for certain loan officers.
Some loan committee members believe they have too little time to review loan applications before loan committee meetings.	Loan committees meet once a week. Therefore loan write-ups may be available as much as six days in advance. Instead of providing all loan write-ups one day in advance, provide them as soon as they are available and let the loan officers schedule their own time for reading them.
Some loan write-ups are incomplete, wasting time and causing delays.	Consistent use of a new loan evaluation model that is integrated into a single interface for credit analysts should establish more consistent loan write-ups.
Some loan applicants complained that they were led to believe loans would be approved, only to find out that a senior credit officer or loan committee required unrealistic terms in loan covenants.	A simplified version of a new loan evaluation model should help loan officers establish more realistic expectations. If a simplified version cannot be provided, the information from the new loan evaluation model used by credit analysts could be made available to loan officers to help them decide how to handle clients with questionable loan applications.

RJ5: What new problems or costs might be caused by the recommended changes?

Enforcement of guidelines. TRBG has not clarified the appropriate tradeoff between increasing short-term profitability by granting marginal loans versus long-term quality of the loan portfolio. Creditworthiness standards were overridden occasionally in the past. Enforcement of guidelines may cause a short-term profitability crunch. Use of a new loan evaluation model may have many benefits, but it will not guarantee enforcement of guidelines.

New loan analysis model. Although there is some experience in the banking industry with loan evaluation models that are more advanced than the ones used currently, there is some possibility that those models will not fit well at TRBG. It is possible that our credit analysts are not experienced enough to use those models properly. We may not be able to obtain simplified versions of the models to help the loans officers. Also, the guidelines from a model-based loan analysis will be more powerful than the guidelines from the current model, but may still be overridden by senior credit officers or loan committees. The models are not reliable enough to make loan approval decisions automatically. Overall, the adoption of new loan evaluation models is likely to be a learning process, and we may encounter some unanticipated glitches.

Use of the model with clients. The idea of providing a simplified loan evaluation model for the loan officers seems attractive, but it could encounter a number of problems. First, it may not be feasible, given that no one has provided an appropriate model for use by a loan officer who is not a trained credit analyst. Assuming a simplified model could be made available, there is the risk that using this type of model with clients might short circuit the loan approval process. Clients would have even more grounds to complain about denials if they saw a loan model and believed they were be pre-qualified. Also, providing the model to the loan officers could increase the chances that they will present loan applications that are stretched and possibly misrepresented to meet the criteria in the model.

RJ6: How well does the proposed work system conform to work system principles?

Ideally the proposed work system should be as good or better than the current system in terms of all 24 principles. At minimum, areas where it is better should be more important than areas where it is worse.

The five Level Three checklists in the next section include a checklist for conformance to the 24 work system principles. That checklist is filled out for the existing work system. The same checklist could be filled out for the proposed work system. (The second version is not included here because it the example in the next section illustrates enough about the use of the checklist.)

RJ7: How can the recommendations be implemented?

Templates RJ7-1, RJ7-2, and RJ7-3 provide a starting point for thinking about whether the recommendation is practical to implement.

- RJ7- 1: Ownership and management of the project
- RJ7-2: Tentative project plan
- RJ7-3: Resource requirements

These are shown with the Level Three templates at the end of this Appendix. They are not filled out here because of the constructed nature of the example. A complete project plan would involve extensive information about TRBG's organization and its IT group's capabilities.

RJ8: How might perspectives or interests of different stakeholders influence the project's success?

The main stakeholders are clients, loan officers, credit analysts, senior credit officers, members of loan committees, and the bank's Risk Management Department. Clients should be pleased or neutral about the improved work system. In some cases, they will receive quicker service and clearer explanations. In other cases, their interactions with loan officers and TRBG will be similar to those with the current work system.

Work system participants within TRBG should be pleased because the improvements will provide better guidance for decisions, increase efficiency, and reduce the rate of errors and rework. Anyone who actually benefits from favoritism may not like the improvements because it will be easier to monitor for favoritism.

RJ9: Are the recommended changes justified in terms of costs, benefits, and risks?

Templates RJ9-1 through RJ9-5 illustrate the type of cost/benefit/risk analysis that could be done in a real situation.

- RJ9- 1: Summary of project justification
- RJ9-2: Direct costs of the project
- RJ9-3: Indirect costs of the project
- RJ9-4 General risks applicable to this project
- RJ9-5: Project risks applicable to specific phases of this project

These are shown with the Level Three templates at the end of this Appendix. They are not filled out here because of the constructed nature of the example. A complete financial plan and project plan would involve extensive information about TRBG's organization and its IT group's capabilities.

RJ10: Which important assumptions within the analysis and justification are most questionable?

The most important assumption is that use of a better loan evaluation model will lead to better decisions. It is unclear whether that will happen, both because the software might not give better recommendations and because senior credit officers and loan committees could still override those recommendations due to competitive or other considerations (as sometimes happened in the past).

Level Three – Analysis Checklists

Work System Principles Checklist

This checklist is a reminder to consider each work system principle and to ask whether it reveals an important problem in the situation being analyzed.

Scale for degree of concern for this work system:
1 = No problem; 2 = Work system does not conform; 3 = A significant problem

Work system principle	*Degree of concern*	*Comment or explanation*
Customers		
#1: Please the customers.	3	Several groups of customers have concerns. Top management is concerned about inadequate quality of the loan portfolio, while also feeling pressure to increase short-term profitability. Some clients are displeased when their loans are denied or when fees and interest rates seem too high. Loan officers are frustrated when loan denials reduce their bonuses.
#2: Balance priorities of different customers.	2	Management and loan officers are caught in contradictions between short-term profitability and long-term quality of the loan portfolio. Loan officers want to maximize the dollar volume of approved loans, because that is the basis of their bonuses. Management also wants to maximize the dollar volume of approved loans, which would generate higher short-term profitability. However, they realize that a push to maximize loan volume could reduce the quality of the loan portfolio even further.
Products & Services		
#3: Match process flexibility with product variability.	1	This principle is a reminder that different types of situations might call for variations in the process. The current work system already handles larger and more complex loans differently from the way it handles small loans. (It might be able to go further in handling different types of loans differently, but principle #3 seems to be satisfied in the current system.)
Work Practices (Major Activities or Processes)		
#4: Perform the work efficiently.	3	Aspects of the work are not performed efficiently. The loan officers might be more efficient if they could use a loan evaluation model to help establish realistic expectations. Credit analysts use disjointed tools that exacerbate the disjointed nature of their work. Senior credit officers and loan committees might be more efficient if they could receive better guidance than is available from the current loan evaluation spreadsheet.
#5: Encourage appropriate use of judgment.	2	The poor state of the current loan portfolio results partly from poor decisions in the past. Moving to a more controlled decision process would constrain judgment, but should not preclude the use of judgment because no known procedure takes into account all possible factors and contingencies related to the client and the current competition for the client's business.

#6: Control problems at their source.	2	Some of the market and financial projections provided by clients are questionable. Identifying the most questionable information of this type might help reducing approval errors. Also, clients sometimes feel misled by preliminary discussions. This problem might be reduced by providing some form of preliminary loan evaluation model that could be used to set realistic expectations.
#7: Monitor the quality of both inputs and outputs.	2	Over-optimistic market information and sales forecasts provided by clients with the help of loan officers have contributed to a number of inappropriate loan decisions. Subjective information of this type should be evaluated with some care, especially for new clients. On the output side, the decisions should be rated and tracked.
#8: Boundaries between process steps should facilitate control.	1	This principle is a reminder that there might be some cases in which a subsequent step is begun before a previous step is completed. This does not seem applicable in this situation.
#9: Match the work practices with the participants.	1	Matching the work practices with the participants is a problem in some work systems, but does not appear to be a problem in this work system because the various work system participants within the bank seem well suited for their roles in this work system.
		Participants
#10: Serve the participants.	2	Some clients believe they are not well served when initial discussions with loan officers establish unrealistic expectations. Currently the senior credit officers are overloaded because two of them quit recently. Credit analysts are frustrated by a combination of awkward technology, inexperience, frequent interruptions, and many starts and stops in their work.
#11: Align participant incentives with system goals.	2	Loan officers have strong incentives to maximize loan approvals. In some cases, these incentives lead them to cut corners, exaggerating clients' strengths, and downplaying their weaknesses. Giving senior credit officers and loan committees decision power is a counterbalance. However, the incentives for the loan committees are somewhat unclear because they are torn between supporting short-term profitability versus long-term quality of loan portfolio.
#12: Operate with clear roles and responsibilities	1	Lack of clear roles and responsibilities does not appear to be a problem in this particular work system.
		Information
#13: Provide information where it will affect action.	1	Information is available where it will affect action. There is some question about the quality of some of the information that is used, especially market and financial projections.
#14: Protect information from inappropriate use.	2	Last year there were two examples in which a loan officer improperly discussed confidential client information with individuals in the bank who were not involved in loan approvals and therefore had no reason to be told about confidential client information.
		Technologies
#15: Use cost/effective technology	1	Cost/effectiveness of current technology seems not to be a major problem. The major problem is that the current loan evaluation spreadsheet does not provide adequate guidance.

#16: Minimize effort consumed by technology.	1	The use of technology absorbs a lot of effort in some work systems. This appears not to be major issue for the current work system
Infrastructure		
#17: Take full advantage of infrastructure.	1	Infrastructure appears to be adequate in this situation.
Environment		
#18: Minimize unnecessary conflict with the external environment	2	TRBG's competitive environment includes a number of competitor banks that are taking business from TRBG. Several of those banks have strong loan portfolios and therefore are in a position to take on a bit more risk than TRBG can afford. The regulatory environment is quite relevant because the FDIC may take action if the quality of TRBG's loan portfolio slips further.
Strategies		
#19: Support the firm's strategy	3	The current work system does not support TRBG's strategy of increasing profit while also improving the quality of its loan portfolio. On the other hand, the bank itself does not have a clear strategy for how to accomplish these partially contradictory goals.
Work System as a Whole		
#20: Maintain compatibility and coordination with other work systems.	2	The current work system operates largely in isolation from other work systems. There are no computerized links between the loan approval system and other systems related to monitoring loan payments and other aspects of the client relationship. On the other hand, nothing about the current work system causes direct conflicts with other work systems.
#21: Control the system using goals, measurement, evaluation, and feedback.	2	The current work system focuses on generating and evaluating loan applications. It monitors the number and dollar volume of loans, but does not have formal feedback mechanisms that might improve the quality of decisions. Such mechanisms would collect information about clients, loan decisions, payment histories, and other relevant information to help in developing indicators that might help improve loan decisions in the future.
#22: Minimize unnecessary risks.	2	Loan decisions are inherently risky, but the work system itself has several unnecessary sources of risk. Weaknesses in the bank's loan portfolio imply that current methods, including a willingness to override guidance from the loan evaluation model, may have led to accepting unnecessary risks. Also, the tools used by credit analysts are disjointed and error prone.
#23: Maintain balance between work system elements.	1	The only significant imbalance is that the loan models (part of the technology) provide inadequate guidance and should be upgraded or replaced.
#24: Maintain the ability to adapt, change, and grow.	1	Nothing in the work system prevents it from adapting, changing, and growing.

Work System Scorecard

It is almost a cliché that it is hard to manage something without measuring it. It is worthwhile to consider a range of metrics related to different issues because work systems have many facets. The Work System Scorecard identifies different types of performance indicators for a work system and leaves space for identifying related metrics that are useful in the situation that is being analyzed. In most situations, some types of performance indicators listed in the scorecard template are important and others are unimportant.

Note: For preliminary analysis, the current value and desired value of performance indicators can be estimates that should be verified through data gathering later in the analysis. Performance indicators can be quantitative (such as 5.2 units per day) or qualitative (such as high or low). Quantitative, objectively measured performance indicators are usually more convincing for management decisions and action.

Type of performance indicator	*Metric in this situation (or N/A if not applicable)*	*Current value*	*Desired value*
	Customers		
Customer satisfaction	Overall satisfaction rating (on 1-7 scale) from a newly instituted annual client survey	5.2	6.0
	Among clients whose applications are denied, percentage who complain about the decision	30%	15%
Customer retention	Percentage of clients from previous year who are still clients at yearend and have not moved their primary banking relationships. (Only partially related to this work system.)	90%	90%
	Products & Services		
Cost to the customer	Ratio of average loan fee to average loan fees reported in industry surveys	.8	.8
	Average number of client hours required to obtain a simple loan (estimated)	7	5
	Percentage of customers who believe TRBG's loan interest rates are competitive.	50%	70%
	Percent of applicants who believe loan interest rates or fees are too high	30%	20%
Quality perceived by the customer	Percent of applicants who believe the explanation for a denial or for stringent covenants was inadequate	25%	10%
	Adequacy of the loan decisions produced by the current work system	produces too many poor decisions	will produce fewer poor decisions
	Excess percentage of nonperforming loans compared to competitor banks	50%	0%

Responsiveness to customer needs	Average delay between submission of loan application and receipt of approval or denial decision	4 weeks	3 weeks
Reliability	Guidance provided by the loan evaluation spreadsheet	Inadequate - unclear	clear, based on past experience
Conformance to standards or regulations	Conformance to FDIC regulations	100%	100%
Intangibles	Feeling that loan officers and TRBG as a whole treats customers professionally	not measured	not measured
Work Practices (Major Activities or Processes)			
Metrics commonly used for business processes:			
Activity rate	Number of loan applications processed from new clients per year	2100	2500
	Number of loan denials appealed every year by loan officers	85	45
Output rate	Yield percent of loan applications (signed loans/ loan applications)	725/2100 = 35%	45%
	Average number of loan decisions per committee per week by a loan committee	7.2	At least 7.2
	Average number of write-ups per week by a credit analyst	3.1	4
Consistency	Percentage of loan decisions that conform to procedures and do not contain exceptions to policies related to completeness of information, creditworthiness, and other issues	90%	98%
Speed	Average delay between submission of loan application and receipt of approval or denial decision	4 weeks	3 weeks
Efficiency	Number of credit analyst hours devoted to an uncomplicated loan	7	5
	Average dollar cost of processing a loan application through approval	$1300	$1100
Error Rate	Percent of loan documents produced with significant errors	.02%	.01%
Rework Rate	Average percentage of loan applications that are returned to the loan officer and client for clarifications, revisions of the desired loan amount, or other changes.	23%	15%

Value Added	Not applicable (more relevant to manufacturing situations)	--	--
Uptime	Not measured. (even though illness and other matters some cause delays)	--	--
Vulnerability	Annual rate of incidents related to information security and confidentiality	2 incidents last year	none
	Metrics for communication:		
Clarity of messages	Average scores for annual feedback to loan officers about their loan presentations (scale of 1-7)	5.4	6.5
Absorption of messages	Not applicable	--	--
Completeness of understanding	Percentage of loan applications that are returned to loan officers because important information is not understood.	5%	1%
Efficiency: Value compared to amount of information	Not measured. (even though some loan committee members believe that the loan officer presentations are inefficient)	--	--
	Metrics for decision making:		
Quality of decisions	Percentage of loan approvals falling outside of TRBG's approval policies Percentage of questionable loan denials	15% 3%	5% 3%
Degree of consensus attained	Percentage of decisions with one or no dissenting votes (out of 5 committee members)	89%	95%
Range of viewpoints considered	(Not measured)	--	--
Satisfaction of different legitimate interests	(Not measured)	--	--
Justifiability of decisions	Not measured currently, but might be measured by comparing the actual decisions to the recommendations from better models.	--	--
	Participants		
Individual or group output rate	Decisions per committee per week	7.2	7.2 (assuming committee size does not change)
Individual or group error rate	Not measured. (although some loan committee members believe loan applications sponsored by certain loan officers contain more serious errors than applications from other loan officers.)	--	--

Training time to achieve proficiency	Average months to proficiency for credit analysts	6 months	4 months
Job satisfaction	Loan officer job satisfaction from an annual employee survey (scale from 1 to 7)	5.1	6.0
	Credit analyst job satisfaction (1 to 7)	5.3	6.0
	Senior credit officer job satisfaction (1 to 7)	3.2	6.0
	Loan committee member's satisfaction with work on loan committee (via annual survey)	6.2	6.0
Turnover rate	Annual turnover rate of credit analysts	35%	15%
	Annual turnover of loan officers	8%	8%
Management attention required	Amount of time per month devoted to loan officers by their direct supervisors (estimate)	3 hours	2 hour
Information			
Accuracy	Percentage of loan applications identified as containing significant errors or misstatements.	8.5%	3%
Precision	Not an issue.	--	--
Age	Not measured, but some of the marketing information and financial projections in the loan applications are up to a year old.	--	--
Believability	Loan applicants have incentives to exaggerate strengths and ignore weaknesses. Some may provide misleading information. Not measured	--	--
Traceability	Not an issue.	--	--
Ease of access	The current combination of a spreadsheet for consolidating information, a loan evaluation model, and a Microsoft Word template makes it awkward to access loan evaluation information while the write-up is being produced.	--	--
Controllability of selection and presentation	(See above.)		
Relevance	Not an issue.	--	--
Timeliness	Amount of time loan committee members have for reviewing loan write-ups before the committee meetings	1 day	Up to 6 days,
Completeness	Market and financial projections for startups are often sketchy. Projections for firms in rapidly changing industries may ignore important factors and issues.	--	--

Appropriateness	Not an issue.	--	--
Conciseness	Not measured, but an average loan write-up is 15 pages long	--	--
Ease of understanding	Not an issue. All participants understand what the information means.	--	--
Vulnerability to inappropriate access or use	Not a significant issue.	--	--
Technologies			
Functional capabilities	Quality of guidance provided by loan evaluation model	fair	good
Ease of use	Months for credit analyst to master the loan evaluation model Months for credit analyst to master the entire loan write-up process, including mechanical procedures and use of judgment.	2 months (estimated) 6 months	1 month 3 months
Uptime	Percentage of uptime for computers and communication technologies	99.6%	99.9%
Reliability	Number of significant errors traced to technology	0	0
Compatibility with complementary technologies	Not an issue.	--	--
Maintainability	Not measured. The current spreadsheet is poorly documented but not very complicated.	--	--
Price/ performance	Not measured. The current model is inexpensive to run, but should be replaced by a more effective model that will provide better guidance.	--	--
Training time to achieve proficiency	Months for credit analyst to master the loan evaluation spreadsheet (or the new model in the future)	2 months (estimated)	1 month
Time absorbed by setup and maintenance	Not measured	--	--
Environment			
Fit with environment	Not measured.	--	--
Infrastructure			
Adequacy of human infrastructure	Not measured, but seems adequate.	--	--
Adequacy of information infrastructure	Not measured, but seems adequate.	--	--

Adequacy of technical infrastructure	Not measured, but seems adequate.	--	--
(Organizational) Strategy			
Fit with organizational strategy	The organization's strategy is unclear, but it is clear that the current work system does not support the organization's goals.	?	?
Relationships with other work systems			
Quality of inputs from other work systems	Not linked to other work systems, except through financial statements from clients.	--	--
Timeliness of inputs from other work systems	Not an issue because the loan approval process is triggered by the submission from the client.	--	--
Effectiveness of communication with other work systems	Not linked to other work systems, except through financial statements from clients.	--	--
Effectiveness of coordination with other work systems	Not measured. Loan committee members devote the majority of their time to other work systems.	--	--

Strategy Decisions Checklist

It is tempting to deal with symptoms by suggesting work system changes before considering big picture issues about the work system's design. This checklist provides reminders about many of these big picture issues, which are called strategic issues for the work system (related to the strategy for the work system, but not necessarily for the entire organization or firm). Each of the issues listed here may represent an important design choice in the situation you are analyzing. On the other hand, many of these issues may be unimportant or irrelevant to the work system that is being analyzed.

Note: The template identifies typical strategy dimensions for each work system element. The user of the checklist summarizes the desired direction for change in the second column. If no change is necessary or if the issue is not applicable, the user should leave the cell blank or should enter "no change" or N/A.

Strategy dimension	***Desirable direction for change***
Customers	
Degree of customer segmentation	There are several customer segments, but no changes are anticipated.
Equality of treatment for customers	The largest clients receive higher priority. No changes anticipated.
Degree of personalization in customer experience	No change. The loan officer should continue trying to establish personal ties with applicants, while TRBG's decision process should continue to appear impersonal.
Products & Services	
Product emphasis	Attempts to make the decision process more consistent will increase product rather than service emphasis during the decision process.
Service emphasis	Use of a better loan evaluation model, including a simplified version for loan officers, might allow loan officers to provide better service to clients.
Information emphasis	No change anticipated.
Emphasis on physical things	No change anticipated.
Commodity vs. customized	Use the existing, somewhat customized approach for working with the client. Enforce greater consistency in making approval decisions.
Degree of controllability by customer	N/A
Degree of adaptability by customer	N/A
Value of by-products	Make greater use of loan write-ups generated during the loan approval process. Archive them and use them for analysis purposes to improve future decisions.
Work Practices (Major Activities or Processes)	
Degree of structure	Increase the degree of structure in the approval process by enforcing procedures and standards.
Range of involvement	No change anticipated.

Degree of integration	Produce loan write-ups through a more integrated process that uses an integrated software tool. Provide a better way to link past loan applications with the performance of past loans, thereby providing better support for making future decisions.
Complexity	Reduce the complexity of producing loan write-ups by providing better technology. Reduce the complexity of the decisions made by the loan committees by providing better guidance through a more effective, better-calibrated loan evaluation model.
Variety of work	No change anticipated.
Degree of automation	The new tool for credit analysts should automate some of the manual copying that is currently done. A new evaluation model should increase automation in the decision process, while leaving room for judgment.
Rhythm – frequency	No change anticipated.
Rhythm – regularity	No change anticipated. Each loan committee will still meet once a week on a particular day.
Time pressure	Providing loan write-ups more than one day in advance whenever possible should reduce the time pressure to make decisions during a meeting despite inadequately digested information.
Amount of interruption	Use of a better tool may increase accuracy and therefore reduce interruptions of credit analysts related to errors and confusion.
Degree of attention to planning and control	Almost no workload planning occurs. Due to seasonal fluctuations in application levels (more in the spring and fall, fewer in the summer) historical workload fluctuations might be used to plan committee time during projected workload peaks.
Error-proneness	A better method for producing the loan write-ups, including a better loan evaluation model, should reduce errors by credit analysts and might make decision making process less error prone.
Formality in exception handling	The new decision process should enforce greater consistency. Exceptions to policies and decisions contrary to model recommendations should be recorded and their effects analyzed.
Participants	
Reliance on personal knowledge and skills	Better tools should make it easier for inexperienced credit analysts to succeed more quickly.
Personal autonomy	Stronger enforcement of procedures and standards should reduce the autonomy of the loan committees.
Personal challenge	Too much of the challenge for credit analysts is about working with awkward tools. The main challenge for senior credit analysts is overload, which must be reduced in order to keep them from leaving.

Information	
Quality assurance for information	Although some of the information is inaccurate or biased, no changes are anticipated in quality assurance for information.
Awareness of information quality	Better tools for producing loan write-ups should flag likely errors or statistically questionable information.
Ease of use	A better loan write-up tool should make information easier to use. A better way to link past loan applications with the performance of past loans should provide better support for making future decisions.
Information security	No changes anticipated
Technologies	
Functionality	The current loan evaluation model provides inadequate guidance and should be replaced. In combination, the three tools for producing loan write-ups are awkward.
Ease of use	The new model should have a better user interface and should provide more effective indicators for guiding decisions.
Technical support for technology usage	A commercial loan evaluation model will be better documented and will have backup support from the vendor.
Maintenance	Maintenance on the existing spreadsheet and model is sporadic and disorganized. A vendor should provide much better maintenance for its products.
Infrastructure	
Degree of reliance on human infrastructure	No changes anticipated
Degree of reliance on information infrastructure	No changes anticipated
Degree of reliance on technical infrastructure	No changes anticipated
Environment	
Alignment with culture	A loan evaluation model that provides better guidance should reduce some of the autonomy in the current culture.
Alignment with policies and procedures	No changes anticipated in relation to policies and procedures elsewhere the bank.
Work System as a Whole	
Degree of centralization	No changes anticipated.
Capacity	The work system should be able to process a larger number of loan applications with the existing work system participants (after replacing the two senior credit officers who left.)
Resilience	No changes anticipated.
Scalability	The work system will be more scalable because the tools will be easier to learn.
Agility	No changes anticipated.
Transparency	Enforcing procedures and recording exceptions should make the loan approval process more transparent.

Possibilities Checklist

This checklist identifies many common types of changes to work systems. It asks you to consider the possibility that each type of change might be worthwhile in the situation you are analyzing. If a type of change might be worthwhile, briefly describe how that type of change might be beneficial. The checklist is designed as a starting point for identifying possible changes, but some possible changes may be inconsistent with other possible changes. The recommendation should ignore the possibilities that are listed here but are not to be pursued. Also, although many common types of changes are listed in the first column, please recognize that many other types of changes are not mentioned.

The following checklist is filled out as it might be filled out in the middle of the analysis, when many different possibilities are being considered, but it is recognized that only some of them will be recommended.

Relevance: 1 = Not applicable; 2 = might be worthwhile; 3 = probably a good idea

Common types of work system changes	***Relevance (1, 2, or 3)***	***Summary of specifics for this situation***
	Customers	
Add or eliminate customer groups	1	
Change customer expectations	2	Some customers have unrealistic expectations. Need to set more realistic expectations.
Change the nature of the customer relationship	1	
Change the customer experience	1	
	Products & Services	
Change information content	3	Provide better guidance using a better loan analysis model. Might want to generate information that can be used to audit the approval process, such as information about whether and why the loan evaluation model was overridden.
Change physical content	1	
Change service content	1	
Increase or decrease customization	1	
Make products and services more controllable by the customer	1	
Make products and services more adaptable by the customer	1	
Provide better intangibles	1	
Change by-products	2	(See above.)

Work Practices (Major Activities or Processes)		
Change roles and division of labor	2	With better tools it might be possible to give credit analysts more responsibility.
Improve business process by adding, combining, or eliminating steps, changing the sequence of steps, or changing methods used within steps	2	A new loan evaluation model might be obtained, and an integrated interface and tool for credit analysts might be provided.
Change business rules and policies that govern work practices	3	Need to do something to improve the loan portfolio. This probably involves controlling or reducing the overrides.
Eliminate built-in obstacles and delays	3	The tools used by credit analysts are an obstacle. Immediate availability of completed loan write-ups will eliminate a delay.
Add new functions that are not currently performed	3	Develop or acquire a better loan evaluation model.
Improve coordination between steps	1	
Improve decision making practices	3	Need to improve decision making, probably with a better model and with better procedures and feedback.
Improve communication practices	1	
Improve the processing of information, including capture, transmission, retrieval, storage, manipulation, and display	2	Need to find better ways to obtain projected market and financial information, instead of relying so much on what the client says about the future.
Change practices related to physical things (creation, movement, storage, modification, usage, protection, etc.)	1	
Participants		
Change the participants	3	Need to replace the two senior credit officers who quit.
Provide training on details of work	2	Going with a loan evaluation model from a vendor might provide much better training material.
Assure that participants understand the meaning and significance of their work	1	
Provide resources needed for doing work	3	Provide a well-designed, integrated tool for credit analysts instead of an awkward combination of disjointed tools.
Change incentives	2	Loan officers have a moral conflict between earning larger bonuses and maintaining the quality of the bank's portfolio.
Change organizational structure	2	Might change the size or structure of the loan committees.
Change the social relations within the work system	2	Might be necessary to do something about relationships that lead to the appearance of favoritism for some loan officers.

Change the degree of interdependence in doing work	2	Overworked senior credit officers are highly involved in work by less experienced credit analysts.
Change the amount of pressure felt by participants	3	Senior credit officers are under extreme pressure because two of them quit.
	Information	
Provide different information	3	Need to do something to digest information and focus it on decision issues. This might require better models.
Codify currently uncodified information	3	Need more clarity about the evaluation criteria for approving loans.
Eliminate some information	1	
Organize information so it can be used more effectively	2	The new integrated tool should organize information to make its use more effective.
Improve information quality	2	Some of the information supplied by clients and loan officers is surely biased. It would be good to have ways to identify any excessive bias.
Make it easier to manipulate information	2	Better models might perform the required manipulation of the loan-related information.
Make it easier to display information effectively	1	
Protect information more effectively	1	
Provide different codified knowledge	1	
Assure understanding of details of tasks and use of appropriate information and knowledge in doing work	1	
Provide access to knowledgeable people	1	The senior credit officers often wish they weren't quite as available.
	Technologies	
Upgrade software and/or hardware to a newer version	3	The current models are old and seem not to provide useful guidance.
Incorporate a new type of technology	1	
Reconfigure existing software and/or hardware	1	
Make technology easier to use	3	The existing combination of the spreadsheet, the loan evaluation model, and MS Word is awkward to use.
Improve maintenance of software and/or hardware	2	The existing loan evaluation spreadsheet is poorly maintained.

Improve uptime of software and/or hardware	1	
Reduce the cost of ownership of technology	1	
Infrastructure		
Make better use of human infrastructure	1	
Make better use of information infrastructure	1	
Make better use of technical infrastructure	1	
Environment		
Change the work system's fit with organizational policies and procedures (related to confidentiality, privacy, working conditions, worker's rights, use of company resources, etc.)	1	
Change the work system's fit with organizational culture	2	TRBG's culture is quite autonomous. It might be a good idea to establish more procedural guidelines.
Respond to expectations and support from executives	1	
Change the work system's fit with organizational politics	2	TRBG's autonomous culture affects internal politics. It might be a good idea to establish more procedural guidelines.
Respond to competitive pressures	3	Probably need greater clarity about TRBG's strategy regarding short term profitability versus long term health of the portfolio.
Improve conformance to regulatory requirements and industry standards	3	Need to make sure that the portfolio stays within regulatory bounds.
Strategy		
Improve alignment with the organization's strategy	1	
Change the work system's overall strategy	1	
Improve strategies related to specific work system elements (See strategies checklist)	2	Might change to a higher degree of automation in making the decision.

Risk Factors and Stumbling Blocks Checklist

This checklist provides a reminder of common risk factors and stumbling blocks that affect the performance of work systems. It asks you to decide whether each is important in the situation you are analyzing. Your recommendation should address risk factors and stumbling blocks that are within control of the people improving or managing the work system.

Degree of threat: 1 = Low threat or not applicable; 2 = might matter; 3 = needs positive action

Risk factors and stumbling blocks, organized by work system element	***Degree of threat (1, 2, or 3)***	***Comment related to current or proposed work system***
	Customers	
Unrealistic expectations	2	Some customers have unrealistic expectations that inadequately justified loans will be approved.
Unmet customer needs or concerns	1	
Disagreement about customer requirements or expectations	1	
Customer segments with contradictory requirements or needs	1	
Unsatisfying customer experience	2	Some customers are dissatisfied with their experience.
Lack of customers or customer interest	3	Part of the bank's profitability problem may reflect an insufficient number of prospects.
Lack of customer feedback	1	
	Products & Services	
Difficulty using or adapting the work system's products and services	1	
Unfamiliar products or services	1	
High cost of ownership	1	
Complex product, stringent requirements	1	
Incompatibility with significant aspects of the customer's environment	1	
	Work Practices (Major Activities or Processes)	
Inadequate quality control	3	TRBG's poorly performing loan portfolio is partly a result of inadequate quality control.
Uncertainty about work methods	2	Each time a new credit analyst starts there is difficulty related to poorly defined work methods.
Excessive variability in work practices	2	Policies and procedures exist, but are bypassed or overridden in some situations.
Frequent changes in work practices	1	
Over-structured work practices	1	To the contrary, loan approval decisions may bypass policies and procedures too often.
Excessive interruptions	2	Senior credit officers find their jobs quite stressful, with frequent interruptions from loan officers, credit analysts, and management.
Excessive complexity	3	Credit analysis is complex, but not excessively complex in relation to its goal. Awkward tools create excessive complexity for credit analysts.
Inadequate security	1	

Inadequate methods for planning the work	2	Workload bunches up occasionally.
Omission of important functions	3	There is no feedback function that evaluates past loan decisions based on loan performance and other indicators, thereby providing useful lessons for improving future loan decisions.
Built-in delays	2	
Unnecessary hand-offs or authorizations	1	.
Steps that don't add value	1	All of the steps seem to add value.
Unnecessary constraints	1	
Unclear or poorly explained business rules and policies	3	Past inconsistencies in processing some loans is related to unclear policies for approving loans.
Low value variations in methods or tools	1	
Large fluctuations in workload	2	The loan committees have higher workloads in the fall and spring, and lower workloads in the summer.
Participants		
Inadequate skills, knowledge, or experience	2	New credit analysts and loan officers often take several months to come up to speed.
Inadequate understanding of reasons for using current methods	1	
Multiple, inconsistent incentives	3	The key conflict is between the need for short-term profitability and long-term quality of the bank's loan portfolio. Loan officers also feel conflict between their own drive to receive bonuses and their responsibilities toward the bank.
Unclear goals and priorities	3	The bank has not established clear priorities related to the inconsistent incentives faced by the loan committees.
Responsibility without authority	1	
Inadequate role definitions	1	
Lack of accountability	2	Because the approval or denial for large loans is a committee decision, several managers believe there is a lack of accountability.
Inadequate management or leadership	2	Mixed messages and lack of enforcement related to loan riskiness may be an indication of inadequate management.
Unnecessary layers of management	1	
Inconsistency between the organization chart and actual work patterns.	1	
Poor morale	3	Senior credit officers are overstretched.
Disgruntled individuals	3	Two senior credit officers resigned last month, causing more pressure on those remaining.
Lack of motivation and engagement	1	
Ineffective teamwork	1	
Turnover of participants	2	Turnover among credit analysts and senior credit officers causes inefficiency.
Inattention	2	Several top managers believe that loan officers pay too little attention to the quality of some of the loan applications they submit.
Excessive job pressures	3	The loan officers feel substantial pressure to bring in new clients and to avoid losing existing clients to other banks that are offering extremely aggressive loan deals. Senior credit officers feel excessive job pressures.

Failure to follow procedures	2	In the last five years there were a number of cases in which the loan committees made decisions before receiving a complete loan write-up, and without considering all relevant information about the borrower. In some cases, competitive pressures led to loans for marginally qualified borrowers.
Departmental rivalries and politics	2	The three loan committees have existed for five years and feel some rivalry related to the quantity and quality of loans each of them approved.
Information		
Use of obsolete or inaccurate information	2	Although the financial data that is provided is rather standard, in some cases the market analysis for new clients is based on outdated information.
Difficulty accessing information	3	Loan write-ups produced by credit analysts are a necessary first step, but they don't provide enough guidance on borderline loan applications.
Misuse of information developed for a different purpose	2	Inexperienced loan officers and credit analysts sometimes misuse accounting information and projections that were originally developed for a different purpose.
Misinterpretation of information	2	(See above.)
Multiple versions of the same information	2	In the past, loan applications that were resubmitted after clarifications or changes were sometimes confused with the original applications. A careful numbering scheme eliminated the problem.
Inconsistent coding of information	2	Inconsistent coding of risk ratings occasionally causes problems.
Manual re-entry of previously computerized information	2	Most of the information in the loan application exists in the client's financial information systems, but credit analysts re-enter this information into the loan evaluation spreadsheet.
Inadequate control of information access and modification	1	
Unauthorized access	2	There have been instances of unauthorized discussions of confidential information.
Poorly articulated knowledge about work practices	2	Training material for new loan officers and credit analysts is not adequate.
Technologies		
Use of inadequate technology	3	The loan evaluation spreadsheet provides inadequate guidance for loan committees.
Undocumented technology	2	The loan evaluation spreadsheet is poorly documented.
Inadequately maintained technology	2	The loan evaluation spreadsheet is poorly maintained.
Technology incompatibilities	1	
Technology complex or difficult to understand	2	The loan evaluation spreadsheet is poorly maintained.
Non- user friendly technology	2	The user interface of the loan evaluation model is somewhat confusing.
Equipment and software downtime	1	
New or unproven technology	1	
Mismatch with needs of work practices	3	The loan evaluation spreadsheet provides inadequate guidance for loan committees.
Unauthorized usage	1	

Software bugs	1	
Unauthorized changes to software	1	
Infrastructure		
Poor fit with or use of human infrastructure	1	
Poor fit with or use of information infrastructure	2	TRBG's history of loan approvals and denials has not been analyzed to identify warning signs of making the same mistakes that were made in the past.
Poor fit with or use of technical infrastructure	1	Not an evident problem, but the bank's technical infrastructure may be underutilized in obtaining and consolidating information from clients.
Environment		
Lack of management support and attention	1	
Poor fit with organizational policies and procedures	1	
Poor fit with organizational culture	1	
Negative impacts of recent organizational initiatives	2	An attempt to upgrade TRBG's internal accounting system encountered many problems that may discourage people from getting involved with a new system-related project.
Poor fit with organizational and competitive pressures	3	The current work system is not meeting TRBG's goals in a highly competitive environment.
Noncomformance to regulations and industry standards	3	The FDIC expressed concerns related to the quality of the bank's current loan portfolio.
High level of turmoil and distractions.	1	
Strategy		
Misalignment with the organization's strategy	1	
Poorly articulated corporate strategy	3	TRBG's relative priority for short-term profit versus long-term quality of the loan portfolio has been unclear for many years.
Work System as a Whole		
Inadequate resources or capacity	1	
Inadequate management	1	
Inadequate security	1	
Inadequate measurement of success	3	The tradeoff between short-term profit versus long-term portfolio quality has been unclear for years.

Level Three – Diagrams and Templates

Explanations of work practices and information in Chapters 10 and 12 showed several of the types of diagrams that can be used for specifying processes and information more precisely and in more depth. These specific diagrams were:

Figure 10.1: Flowchart
Figure 10.2: Swimlane Diagram
Figure 10.3: Data Flow Diagram
Figure 12.1: Entity-Relationship Diagram

These and other techniques from software development and quality management are explained in many books and web sites and will not be repeated here.

Shown below are a number of templates that can be used in the RJ phase. Templates that were presented earlier in this Appendix are not repeated. These templates are presented to identify some of the issues that should be considered. Many companies with well-established project analysis methods have their own templates that address these issues.

RJ7: How can the recommendations be implemented?

Templates RJ7-1, RJ7-2, and RJ7-3 provide a starting point for thinking about whether the recommendation is practical to implement. They are not filled out here because of the constructed nature of the example. Because a complete project plan would involve substantially more information, many companies use project planning templates that are more detailed.

Ownership and Management (RJ7-1)

Role	*Responsible individual*
Work system owner	
Project manager	
Source of funding	

Tentative Project Plan (RJ7-2)

Step in project	*Responsible individual*	*Start date*	*End date*
Initiation phase			
Vision for the new or improved work system			
Operational goals			
Feasibility study			

Development phase			
Detailed requirements for the new work system (including information system requirements)			
Software production, modification, or acquisition and configuration			
Hardware installation			
Documentation and training materials			
Debugging and testing of hardware, software, and documentation			
Implementation phase			
Implementation approach and plan			
Change management efforts about rationale and positive or negative impacts of changes			
Training on details of the new or revised information system and work system			
Conversion to the new or revised information system and work system			
Rework on software, documentation, and training materials (if required)			
Acceptance testing			

Resource Requirements (RJ7-3)

Step in project	***Person days***	***Other resources***	***Budget***	***Key people***
Initiation phase				
Vision for the new or improved work system				
Operational goals				
Feasibility study				
Development phase				
Detailed requirements for the new work system (including information system requirements)				
Software production, modification, or acquisition and configuration				
Hardware installation				
Documentation and training materials				
Debugging and testing of hardware, software, and documentation				
Implementation phase				
Implementation approach and plan				
Change management efforts about rationale and positive or negative impacts of changes				
Training on details of the new or revised information system and work system				
Conversion to the new or revised information system and work system				
Rework on software, documentation, and training materials (if required)				
Acceptance testing				

RJ9: Are the recommended changes justified in terms of costs, benefits, and risks?

RJ9-1 through RJ9-5 illustrate templates that can be used for cost/benefit/risk justifications. Many organizations use templates such as these. The complexity templates used in industry depends on the purpose of the analysis and the organization's experience with system-related projects. The templates shown below are not filled out because a great deal of additional explanation would be required to provide enough context to make the numbers meaningful.

Summary of Project Justification (RJ9-1)

Aspect of the justification	*How the recommendation addresses this issue*
Direct costs by major category	* * *
Tangible benefits by category (in monetary terms)	* * *
Intangible benefits	* * *
Indirect and hidden costs	* * *
Benefits compared to costs	Net present value: Internal rate of return:
Payback period	*
Value of options opened by the project	* * *
Business and strategic priority	* * *
Significant risks	* * *
Comparison with other possible uses of resources	* * *

Direct Costs of the Project (RJ9-2)

Project phase/ category of direct costs	*Estimated cost*	*Comment*
Initiation phase		
Salary and overhead for IT staff		
Costs related to IT contractors and consultants		
Cost of communication and travel related to the project		
Other direct costs related to initiation		
Development phase		
Salary and overhead for IT staff		
Costs related to IT contractors and consultants		
Equipment purchase and installation costs		
Purchase (if any) of system or application software		
Site modifications such as wiring offices		
Other direct costs related to development		
Implementation phase		
Salary and overhead for IT staff and trainers		
Costs related to IT contractors and consultants		
Cost of communication and travel related to the project		
Other direct costs related to implementation		
Operation and maintenance phase		
Salary and overhead for IT staff		
Costs related to IT contractors		
New direct costs related to salaries and overhead for other employees or contractors		
Software license fees (if any)		
Hardware rental or depreciation		
Ongoing facilities costs		
Other direct costs related to operation and maintenance		

Indirect Costs of the Project (RJ9-3)

Project phase/ category of indirect costs	*Estimated cost*	*Comment*
Initiation phase		
Salary and overhead of work system participants and management involved in the analysis		
Other work that is displaced in favor of work on the project		
Other indirect costs		
Development phase		
Salary and overhead of work system participants and management involved in the analysis		
Additional site modifications not included in direct costs such as wiring offices		
Other indirect costs		
Implementation phase		
Salary and overhead of work system participants and management involved in the analysis		
Disruption of work during implementation process		
Effect of internal disruptions on customer service and customer satisfaction		
Other indirect costs		
Operation and maintenance phase		
Salary and overhead of work system participants and management doing extra work related to the new system		
Other indirect costs		

General risks applicable to this project (RJ9-4)

Template RJ9-4 is not shown here because it uses the Work System Risk Factors and Stumbling Blocks Checklist that was illustrated earlier in this Appendix. That checklist was applied to a work system that is being improved. The same checklist can be applied to the project because the project is a work system on its own right.

Project Risks Applicable to Specific Phases of a Project (RJ9-5)

RJ9-4 focuses on risks that apply to any work system, and hence, any project or phase of a project (since a project or a subproject is also a work system on its own right). RJ9-5 adds risks that do not apply to work systems in general, but rather, apply specifically to the initiation, development, or implementation phases of a project.

Degree of threat: 1 = Low threat or not applicable; 2 = might matter; 3 = needs positive action

Project risks applicable to specific phases	*Severity of risk*	*Comment related to risks for the designated phase of this project*
Initiation phase		
Lack of consensus on the need for the project or the approach used		
Inadequate representation of stakeholder interests		
Key stakeholders unwilling or unable to participate		
Social and organizational issues downplayed or ignored		
Unwarranted assumptions about technology		
Functional requirements misdirected or over-ambitious		
Preliminary project plan over-ambitious		
Inadequate consideration of environmental factors		
Inadequate consideration of the organization's current or future strategy		
Development phase		
Error prone development process		
Overly costly or overly complex development process		
Analysis paralysis		
Excessive fixation on the schedule		
Excessive fixation on the method rather than the results.		
Inadequate experimentation and proof of concept.		
Inadequate attention to documentation and training materials		
Incomplete debugging and testing		
Preliminary project plan over-ambitious		
Inadequate consideration of environmental factors		
Inadequate consideration of the organization's current or future strategy		
Participants not fully able to perform the work [e.g., user representatives not senior enough to make judgments about what might work)		
Non-engaged participants do slipshod work (e.g., programmers who don't care very much about long term quality issues)		
Project participants not suited to work practices chosen for the project		
Participants have too little experience with the technology used in the development phase		
Imbalance between business and IT professionals		

Skepticism about whether the project can be done within the allotted time and resources		
Inadequate availability of subject matter experts		
Fear that the new work system changes will lead to staff reductions and de-skilling		
Beginning development from unclear or otherwise inadequate goals and functional requirements		
Information for detailed requirements analysis is inaccurate or incomplete		
Management unwillingness to allocate necessary resources		
Culture of ineffective cooperation on projects		
Lack of consensus about project governance		
Lack of consensus on the need for the project		
Development strategy mismatched to the situation. (e.g., should have done a prototype or should have purchased more of the software)		
Implementation phase		
Inappropriate implementation approach		
Disillusionment and other problems due to inadequate training and training materials		
Training happens too early and many work system participants forget the training by the time the conversion takes place		
Backup procedures prove inadequate when the initial attempt to convert encounters problems		
Unexpected resistance to the implementation process		
Implementation process quashes or ignores resistance that could have been useful feedback		
Implementation involves excessive amounts of time, effort, and pain		
Eforts at change management are inadequate or inappropriate		
Work system participants have difficulty switching or resist switching to a new way to do their work		
Implementers lack interpersonal skills, empathy, and abilities related to change management		
Fear that the new work system changes will lead to staff reductions and de-skilling		
Denial of new indications that the new system will not be effective or will fail totally		
Inadequate attention to resistance and other signals that the implementation is in trouble		
Unrealistic expectations about what work system changes are supposed to accomplish		
Strategy of the implementation is unrealistic or otherwise flawed. [e.g., should have done a phased implementation but did a big bang implementation)		
Urgency in the timetable for the implementation does not match the urgency required by the surrounding strategies		

Notes

Chapter 1: Why Are So Many Systems Such a Mess?

[1] Michael Schrage, "IT's Hardest Puzzle," *CIO Magazine*, July 15, 2005 http://www.cio.com/archive/071505/leadership.html

[2] CRM stands for customer relationship management. CRM is a category of commercial software that focuses on customer-facing activities such as sales cycles, order entry, and customer service. Different CRM packages address different aspects of these topics, implying that a statement such as "we are installing CRM" is quite unclear without an explanation of which specific functions are being addressed.

[3] The term *work system* appeared in two articles in the first volume of *MIS Quarterly* (Bostrom and Heinen, 1979a; 1979b). Mumford and Weir (1979, p. 3) spoke of "the design and implementation of a new work system." Davis and Taylor (1979, p. xv) mentioned "attempts at comprehensive work systems design, including the social systems within which the work systems are embedded." Trist (1981) said that "primary work systems are the systems which carry out the set of activities involved in an identifiable and bounded subsystem of a whole organization - such as a line department or service unit." (p. 11) and "The primary work system ... may include more than one face-to-face group along with others in matrix and network clusters." (p. 35) More recently, Mumford (2000) summarized sociotechnical insights cited by Pasmore (1985), such as "The work system should be seen as a set of activities contributing to an integrated whole and not as a set of individual jobs" and "The work system should be regulated by its members, not by external supervisors." Land (2000) said "socio-technical methods focus on design of work systems to improve the welfare of employees. The prime aim of redesigning work systems is the improvement of the quality of working life." Other IS researchers such as Sumner and Ryan (1994) and Mitchell and Zmud (1999) also used the term. The term work system appeared in the title of Jasperson, Carter, and Zmud (2005). In addition, the term *high performance work system* has appeared occasionally in the popular business press and in some the consulting circles to describe organizations with high degrees of participation and self-management.

- Robert, P. Bostrom and J. Stephen Heinen, "MIS Problems and Failures: A Socio-Technical Perspective. Part I: The Causes." *MIS Quarterly*, (1)3, December 1977, pp. 17-32.
- Robert, P. Bostrom and J. Stephen Heinen, "MIS Problems and Failures: A Socio-Technical Perspective. PART II: The Application of Socio-Technical Theory." *MIS Quarterly*, (1)4, December 1977, pp. 11-28.
- Lou E. Davis and James C. Taylor *eds*., *Design of Jobs*, 2nd ed., Santa Monica, CA: Goodyear Publishing Company, 1979
- Jon Jasperson, Pamela E. Carter, and Robert W. Zmud,"A Comprehensive Conceptualization of Post-Adoptive Behaviors Associated with Information Technology Enabled Work Systems," *MIS Quarterly*, 29(3), Sept. 2005, pp. 525-557.
- Frank Land, "Evaluation in a Socio-Technical Context," *Proceedings of IFIP W.G.8.2 Working Conference 2000, IS2000: The Social and Organizational Perspective on Research and Practice in Information Systems,* Aalberg, Denmark, June 2000
- Victoria L. Mitchell and Robert W. Zmud, "The Effects of Coupling IT and Work Process Strategy in Redesign Projects," *Organization Science*, (10)4, 1999, pp. 424-438.
- Enid Mumford, "Socio-technical Design: An Unfulfilled Promise?" *Proceedings of IFIP W.G.8.2 Working Conference 2000, IS2000: The Social and Organizational Perspective on Research and Practice in Information Systems,* Aalberg, Denmark, June 2000.

- Enid Mumford and Mary Weir, *Computer Systems in Work Design – the ETHICS method*, New York: John Wiley & Sons, 1979
- William A. Pasmore, "Social Science Transformer: the Socio-technical Perspective." *Human Relations,* (48)1 January 1985. pp. 1-22
- Mary Sumner and Terry Ryan, "The Impact of CASE: Can It Achieve Critical Success Factors?" *Journal of Systems Management*, (45)6, 1994, pp.16-21.
- Eric Trist, "The Evolution of Socio-Technical Systems: A Conceptual Framework and an Action Research Program." in Van de Ven and Joyce, *Perspectives on Organizational Design and Behavior*, NY: Wiley Interscience, 1981.

[4] Judy E. Scott and Iris Vessey, "Managing Risks in Enterprise Systems Implementations," *Communications of the ACM*, Vol. 45, No. 4, April, 2002, pp. 74-81.

[5] Alison Bass, "Cigna's self-inflicted wounds," *CIO Magazine*, March 17, 2003. http://www.cio.com/archive/031503/cigna.html

[6] Evan Perez and Rick Brooks, "For Big Vendor of Personal Data, A Theft Lays Bare the Downside," *Wall Street Journal*, May 3, 2005, p. A1.

[7] Shelley Branch. "Hershey Foods Says It Expects to Miss Lowered 4th Quarter Earnings Target," *Wall Street Journal*, Dec. 29, 1999.

[8] Marc L.Songini, "Nike says profit woes IT-based," *Computerworld*, March 5, 2001. http://www.computerworld.com/industrytopics/manufacturing/story/0,10801,58330,00.html For an update see: Christopher Koch, "Nike Rebounds: How (and Why) Nike Recovered from its Supply Chain Disaster," *CIO Magazine*, June 15, 2004. http://www.cio.com/archive/061504/nike.html

[9] Mel Duvall, "In the Pipeline," *Baseline*, Nov. 2005, pp. 58-63.

[10] *Software Magazine*, "Standish: Project Success Rate Improved Over 10 Years," Jan. 15, 2004, http://www.softwaremag.com/L.cfm?Doc=newsletter/2004-01-15/Standish

[11] Byron Reimus, "The IT System that Couldn't Deliver," *Harvard Business Review*, May-June 1997, pp. 22-25.

[12] M. Lynne Markus and Robert Benjamin, "The Magic Bullet Theory of IT-Enabled Transformation," *Sloan Management Review*, Winter 1997, pp. 55-68.

[13] Michael J. Gallivan, "Meaning to Change: How Diverse Stakeholders Interpret Organizational Communication About Change Initiatives," *IEEE Transactions on Professional Communication*, 44(4) Dec. 2001, pp. 243-266.

Chapter 2: Work Systems - The Source of Business Results

[14] For example, Churchman's book *The Systems Approach* says that five basic considerations should be "kept in mind when thinking about the meaning of a system:

- the total system objectives (and performance measures),
- the system's environment (the fixed constraints),
- the system's resources,
- the components, their activities, goals and measures of performance,
- the management of the system.

A number of somewhat similar frameworks have been proposed and used, especially in Europe. Checkland's soft systems methodology is well known in England. It looks at systems in a somewhat more general way, and seems more directed at using systems concepts to develop business or organizational strategies rather than at improving specific work systems that are supported by information technology. Its framework contains six elements described by the acronym CATWOE: customers, actors, transformation process, weltanschauung (world view), organization, and environmental constraints. Other European approaches have a conscious ideological bent, such as Enid Mumford's ETHICS approach for improving sociotechnical systems by separately looking at the technical system and the social system. Scandinavian researchers have also worked extensively on system development techniques designed to give heavy weight to the worker's interests. See:

- C. West Churchman, *The Systems Approach*. Revised and Updated, New York, NY: Dell Publishing, 1979, p. 29.
- Peter Checkland, *Systems Thinking, Systems Practice (Includes a 30-year retrospective)*, Chichester, UK: John Wiley & Sons, 1999.
- Enid Mumford and Mary Weir, *Computer Systems in Work Design – the ETHICS method*, New York: John Wiley & Sons, 1979

[15] For example, see Roberta Lamb and Rob Kling, "Reconceptualizing Users as Social Actors in Information Systems Research," *MIS Quarterly*, 27(2), June 2003, pp. 197-235.

[16] The distinction between an assembly line and case manager approach was one of the important examples used in Hammer and Champy's original book on reengineering: Michael Hammer and James Champy, *Reengineering the Corporation: A Manifesto for Change*, New York: HarperBusiness, 1993.

Chapter 3: Overview of the Work System Method

[17] The first step in the work system method is to identify the problem and opportunity. Peter Checkland's Soft System Methodology addresses a similar goal but at a more philosophical level. The first step in soft system methodology is to produce a "rich picture" that illustrates the main activities, actors, and issues on a single page. A rich picture might be used in conjunction with a work system snapshot at the beginning of an analysis. (See Checkland, 1999)

[18] An influence diagram uses words and arrows to illustrate the flow of impacts from initial causes through intermediate effects to results. Drawing an influence diagram at the outset to identify initial beliefs may be useful, although analysis will often reveal that some of the initial beliefs are not correct, and that the influence diagram should be corrected.

[19] For example, a quick look at the web sites of major academic publishers such as Pearson Prentice Hall, McGraw-Hill, John Wiley and Sons, and Course Technologies would find many good books on systems analysis and design for IT majors. Similarly, web sites for software firms and professional consortiums would find highly usable summaries of topics such as unified modeling language (UML), BPMN (business process modeling notation), and SCOR (the supply chain reference model).

[20] For example, Silver et al (1995) cite Ackoff (1993) as stating that "it is not possible to understand a system by *analyzing* it alone - that is, by simply decomposing it into its constituent parts. One must first *synthesize* it - determine its function in the supersystem, the next higher level system of which it is a part. In the case of an information system, this supersystem is the organizational (or inter-organizational) system to which it belongs." Similarly, Checkland notes, "whenever one system serves or supports another, it is a very basic principle of systems thinking that the necessary features of the system which serves can be worked out only on the basis of a *prior* account of the system served." (Checkland (1997), cited in Rose and Meldrum (1999)).

- Russell L. Ackoff, Presentation at the Systems Thinking In Action Conference, Cambridge, MA, 1993 (cited by Silver et al)
- Mark Silver, M. Lynne Markus, and Cynthia M. Beath, "The Information Technology Interaction Model: A Foundation for the MBA Core Course," *MIS Quarterly* 19(3), September 1995, pp. 361-390.
- Peter Checkland, *Information, Systems, and Information Systems,* Chichester, UK: Wiley, 1997.

Jeremy Rose, and Mary Meldrum, "Requirements generation for web site developments using SSM and the ICDT model," Proceedings of the 9th Business Information Technology Conference, 1999, Ray Hackney and Dennis Dunn, eds., http://www.cs.auc.dk/~jeremy/pdf%20files/BIT%201999.pdf

[21] Punched cards are vivid in the memories of people who worked in businesses in the 1960s and 1970s but the idea of punching data on physical, 80-column cards may seem unbelievably primitive to people who grew up with PCs. Around 2003, I explained to an MBA class that my first computing job required typing programs on punched cards, and that the history of card-based data processing goes back to the card tabulation machines developed by Herman Hollerith for the 1890 census. One student raised his hand and basically said, "You really punched programs onto cards? Surely you're kidding."

[22] Some systems experts might argue that a goal is a necessary part of a work system. Although it is usually desirable to have explicit goals, if you think about work systems that you are familiar with, you will probably be able to identify some that do not have explicit goals. Many work systems lack explicit goals and even if they have explicit goals those goals may change at any time, especially if a new manager arrives. Accordingly, the work system method assumes a work system's goal is not an inherent part of the system.

[23] See Suchman's study of air traffic controllers. Lucy Suchman. "Supporting Articulation Work," Rob Kling, ed., *Computerization and Controversy.* 2nd ed., San Diego, CA: Academic Press, 1996, pp. 407-423.

[24] For example, see Kevin Crowston, Joseph Rubleske, and James Howison, "Coordination Theory: A Ten Year Retrospective," working paper, December,

2004, to appear in Ping Zhang and Dennis Galletta, (eds.) *Human-Computer Interaction in Management Information Systems*. M.E. Sharpe.

[25] For example, see the discussion of thinklets for supporting different thinking patterns in: Robert O. Briggs, Gert-Jan de Vreede, and Jay F. Nunamaker, Jr., "Collaboration Engineering with ThinkLets to Pursue Sustained Success with Group Support Systems." *Journal of Management Information Systems*, 19(4), pp. 31–64.

[26] This definition fits a relational database. The distinction between relational databases, multi-dimensional databases, and other types of pre-defined databases is beyond the scope of this discussion.

[27] Robert L. Mitchell, "Beyond Paper," *Computerworld*, June 13, 2005. http://www.computerworld.com.au/pp.php?id=1047268187&fp=512&fpid=1267916587

[28] For an overview of important issues in knowledge management and reuse, see M. Lynne Markus, "Toward a Theory of Knowledge Reuse: Types of Knowledge Reuse Situations and Factors in Reuse Success," *Journal of Management Information Systems*, 18(1), Summer 2001, pp. 57-93

[29] For example, see Wanda J. Orlikowski, "Learning from Notes: Organizational Issues in Groupware Development," *The Information Society*, 9(3), July-Sept, 1993, pp. 237-250.

[30] John Seeley Brown and Paul Duguid, *The Social Life of Information*, Boston, MA: Harvard Business School Press, 2000.

[31] Peter Weill and Jeanne W. Ross, *IT Governance: How Top Performers Manage IT Decision Rights for Superior Results*, Boston, MA: Harvard Business School Press, 2004, p. 35.

Chapter 5: Identifying Issues and Possible Improvements

[32] See Robert S. Kaplan and David P. Norton, "Using the Balanced Scorecard as a Strategic Management System." *Harvard Business Review*, Jan/Feb 1996, pp. 75-85

[33] The distinction between an assembly line and case manager approach was one of the important examples used in Hammer and Champy's original book on reengineering: Michael Hammer and James Champy, "Reengineering the Corporation: A Manifesto for Change," New York: HarperBusiness, 1993.

[34] Philip Evans and Bob Wolf, "Collaboration Rules," *Harvard Business Review*, July-Sept 2005, pp. 96-104.

[35] Figure 5.4 is based on two companion articles:

- Susan A. Sherer and Steven Alter, "A General but Readily Adaptable Model of Information System Risk," *Communications of the AIS*, 14(1), July 2004, pp. 1-28.
- Steven Alter and Susan A. Sherer, "Information System Risks and Risk Factors: Are They Mostly about Information Systems?" *Communications of the AIS*, 14(2), July 2004, pp. 29-64.

[36] The trial and error development of the principles is described in Alter (2004). The closest models for the principles presented here are Cherns' principles of sociotechnical systems, Deming's 14 principles, and principles for manufacturing systems developed by Majchrzak and Borys. Other attempts to restate or extend Cherns principles include Berniker (1992) and Clegg (2000)

- Steven Alter, "Making Work System Principles Visible and Usable in Systems Analysis and Design," *Proceedings of AMCIS 2004, the Americas Conference on Information Systems*, New York, NY, Aug. 8-10, 2004, pp. 1604-1611.
- Albert Cherns, "Principles of Socio-technical Design", *Human Relations*, 2(9), 1976, pp. 783-792.
- W. Edwards Deming, *Out of Crisis*, Cambridge, MA: MIT Press, 2000
- Pat Oliphant, "Pat Oliphant's Illustrations of Deming's 14 Points," viewed on April 21, 2006, http://www.managementwisdom.com/freilofdem14.html
- Ann Majchrzak and Bryan Borys, "Generating testable socio-technical systems theory," *Engineering Technology Management*, 1105, 2001 pp. 1-22.
- Eli Berniker, "Some Principles of Sociotechnical Systems Analysis and Design," Pacific Lutheran University, Oct. 1992, viewed on April 21, 2006,http://www.plu.edu/~bernike/SocioTech.htm
- Chris W. Clegg, "Sociotechnical principles for systems design," *Applied Ergonomics*, 31, 2000, pp. 463-477.

[37] The work system principles that are mentioned were compiled from a Western viewpoint. Chinese students in a leading MBA program in Beijing noted that principles such as "do the work efficiently" might be appropriate in a typical Western company, but do not fit traditional Chinese state owned enterprises, where maintaining employment of large numbers of people was often more important than efficiency. The Chinese students were very

interested in discussing what would happen as state owned enterprises transformed themselves to fit today's globalized economy.

[38] Principles #3, 5, 6, 7, 8, 11, 13, and 15 are restatements of sociotechnical principles proposed by Cherns (1976).

[39] See Deming (2000)

[40] See Raymond Panko, "What We Know about Speadsheet Errors," 2005, viewed on Mar. 27, 2006 at http://panko.cba.hawaii.edu/ssr/

[41] For example, Brown and Duguid (2000) quote a Paul Strassmann, former vice president of Xerox's Information Products Group and onetime chief information technologist for the Department of Defense, as saying that "most businesses that are well endowed with computer technology lose about $5,000 a year per workstation on 'stealth spending.' Of this, he argues, "22% [is] for peer support and 30% is for 'time users spend in a befuddled state while clearing up unexplained happenings [and] overcoming the confusion and panic when computers produce enigmatic messages that stop work."

Paul A. Strassmann, *The Squandered Computer: Evaluating the Business Alignment of Information Technologies*, New Canaan, CT: Information Economics Press, 1997, p. 77.

Chapter 6: Justifying a Recommendation

[42] Techniques related to the "voice of the customer" such as Kano analysis and quality function deployment are not discussed here but are covered in books on Six Sigma and articles such as: John Hauser, "How Puritan-Bennett Used the House of Quality," *Sloan Management Review*, Spring 1993, Vol. 34, No. 3 pp. 61-70.

[43] Sociologists and people who study organizational behavior have developed a number of methods for describing interactions between people in groups or organizations. Use of these techniques may be useful in situations where interactions between people or roles are important. The article in Wikipedia on Social Networking provides an introduction. http://en.wikipedia.org/wiki/Social_networking

[44] Erik Brynjolfsson, "The IT Productivity Gap," Issue 21, July 2003, http://www.optimizemag.com/issue/021/roi.htm

[45] Russell L. Ackoff, " Ends Planning, " Chapter Ten in *Ackoff's Best: His Classic Writings on Management*, New York: John Wiley & Sons, 1999, p. 122. (Reprinted from *Creating the Corporate Future*, John Wiley & Sons, 1981.)

[46] Complications such as these are one of the reasons for the emerging field of requirements traceability. For example, see Balasubramaniam Ramesh, "Factors Influencing Requirements Traceability Practice," *Communications of the ACM*, 41(12) December 1998, pp. 37-44

[47] One such story is cited in Shoshana Zuboff, "New Worlds of Computer-Mediated Work." *Harvard Business Review*, Sept-Oct 1982, pp. 142-152.

[48] Jacques Ellul, "The Technological Order." *Technology and Culture.* Fall 1962, p. 394. Quoted in Wilson P. Dizard, Jr., *The Coming Information Age*. New York: Longman, 1982.

[49] Shoshana Zuboff, *In the Age of the Smart Machine: The Future of Work and Power*, New York, NY: Basic Books, 1988. The claims processing example is on pp. 133-136.

[50] Net present value is the discounted value of the total benefits expressed in dollars (by period) minus the total costs (by period). The discounting takes into account the time value of money. For example, a $100 benefit occurring one year from now might be discounted 10% to reflect the time value of money and to indicate that benefits a year from now are not as valuable as benefits today. Internal rate of return is the interest rate that would give an equivalent monetary return on the money and other resources invested in the project.

Chapter 7: The Work System Life Cycle

[51] The work system life cycle model first appeared in Steven Alter, "Which Life Cycle -- Work System, Information System, or Software?" *Communications of the AIS*, 7(17), October 2001.

[52] Explanations of agile programming methods and extreme programming are readily available on the web. See, for example, http://en.wikipedia.org/wiki/Agile_Software_Development and http://en.wikipedia.org/wiki/Extreme_programming

[53] Geoffrey Darnton and Sergio Giacoletto. *Information in the Enterprise.* Digital Press, 1992, p. i.

[54] Peter G. W. Keen, "Information Systems and Organizational Change." *Communications of the ACM,* Vol. 24(1), Jan. 1981, pp. 24–33.

Chapter 8: Example Illustrating the Work System Method

[55] Timothy W. Koch and S. Scott MacDonald, "Chapter 16: Evaluating Commercial Loan Requests," pp. 589-635 in *Bank Management*, 5th ed., Mason, OH: Thompson/ South-Western, 2003, pp.

[56] For example, see the Small Business Administration's summary of the 5 C's of Credit at United States Small Business Administration, "Applying for a Loan," http://www.sba.gov/financing/basics/applyloan.html

[57] Koch and McDonald, p. 571.

[58] Federal Reserve Bank of San Francisco, "Using CAMELS Ratings to Monitor Bank Conditions," FRBSF Economic Letter 99-19, June 11, 1999.

[59] For example, as explained in Koch and McDonald, p. 599, current ratio is a measure of liquidity; quick ratio is a related measure that emphasizes assets that can be converted to cash quickly. Activity ratios that measure the efficiency of a firm include sales to asset ratio, days accounts receivables collection period, days inventory on hand, and inventory turnover.

- Current ratio = current assets / current liabilities.
- Quick ratio = (cash+accounts receivable)/current liabilities
- Days accounts receivables collection period = accounts receivables/ average daily credit sales
- Days inventory on hand = inventory/ average daily purchases
- Inventory turnover = cost of goods sold/ inventory.

Other ratios measure the firm's ability to meet loan payments with current income.

- Times interest earned = EBIT/interest expense (EBIT = earnings before interest and taxes)
- Fixed charge coverage ratio = (EBIT + lease payments) /(interest expense + lease payments)

Chapter 9: Customers and Products & Services

[60] Davis, S. and Meyer, C. *Blur: the speed of change in the connected economy.* Addison-Wesley, Reading, MA, 1998.

[61] There are many similar models of the stages of a customer experience, such as:

Blake Ives and Gerald Learmonth, "The information system as a competitive weapon," *Communications of the ACM,* 27(12), December 1984, pp. 1193–1201.

[62] The difference between encounters and relationships is explained in Barbara A. Gutek, *The Dynamics of Service: Reflections on the Changing Nature of Customer/Provider Interactions*, San Francisco, CA: Jossey-Bass Publishers, 1995, pp. 7-8.

[63] This definition of quality expresses the customer's viewpoint. An alternative definition emphasizes conformance to manufacturing specifications and expresses a producer's viewpoint that is measured using metrics such as defect rate and rework rate.

Chapter 10: Work Practices

[64] Michael Hammer and James Champy. *Reengineering the Corporation: A Manifesto for Business Revolution*. New York: Harper Business, 1993.

[65] Extensive information about UML is available on the Object Management Group's UML Resource Page at http://www.uml.org/

[66] For example, an IBM flowcharting manual from 1969 identifies a large number of flow chart symbols that are directly related to the data processing of the time.

IBM, *Data Processing Techniques: Flowcharting Techniques*, GC20-8152-1, 1969, revised 1970. http://www.fh-jena.de/~kleine/history/software/IBM-FlowchartingTechniques-GC20-8152-1.pdf

[67] Source: Steven Spear and H. Kent Bowen, "Decoding the DNA of the Toyota Production System," *Harvard Business Review*, Sept – Oct 1999, pp. 97-106.

[68] Steven Spear, "The Health Factory," *New York Times*, Aug. 29, 2005, p. A19.

[69] Robert Slater, *The New GE: How Jack Welch Revived an American Institution*, Homewood, IL: Irwin, 1993, p. 216.

[70] For example, some decision support systems provide relevant information and tools for analyzing that information, but do not provide models that help in evaluating potential decisions. Similarly, model-based decision support systems may not provide current information but may help in identifying problems to solve.

[71] For example, a 2005 commission report to President Bush said that the FBI and CIA have increased joint operations, but that "clashes have become all too common as well, particularly in the context of intelligence gathered in the United States. When sources provide information to both agencies, the FBI complains that conflicting or duplicative reports go up the chain, causing circular or otherwise misleading streams of reporting. In response, CIA claims that FBI headquarters is more concerned about credit for intelligence production than the quality of its reporting. If the agencies' fight were limited to disputes about who gets credit for intelligence reports, it would be far less alarming. Unfortunately, it extends beyond headquarters and into the field, where lives are at stake." (p. 468)

Source: Lawrence H. Silberman, Charles S. Robb, et al, *Report to the President of the United States by the Commission on the Intelligence Capabilities of the United States Regarding Weapons of Mass Destruction*, March 31, 2005. www.wmd.gov/report/wmd_report.pdf

[72] Many examples appear in Steven Alter, "Goals and Tactics on the Dark Side of Knowledge Management, *Proceedings of HICSS-39, Hawaii International Conference on Systems Sciences*, Poipu, Kauai, HI, Jan.4-7, 2006.

[73] The common distinction between data and information is not relevant here because the information within a work system is by definition relevant to the work system. Otherwise it wouldn't be included in the work system.

[74] Charles Perrow, *Normal Accidents: Living with High-Risk Technologies*. New York: Basic Books, 1984.

[75] Tom Herman and Robert Guy Matthews, "Tax Preparers Come Under Fire," *Wall Street Journal*, Apr. 4, 2006

[76] Alan Borning, "Computer System Reliability and Nuclear War," in Anatoly Gromyko and Martin Hellman, *Breakthrough: Emerging New Thinking, Soviet and Western Scholars Issue a Challenge to Build a World Beyond War*, New York: Walker and Company, 1988, published online in 2001, pp. 31-38. http://www-ee.stanford.edu/~hellman/Breakthrough/book/contents.html

Also see: Tony Forester and Perry Morrison. *Computer Ethics: Cautionary Tales and Ethical Dilemmas in Computing*, Cambridge, MA: MIT Press, 1994, p. 108.

[77] Kasra Ferdows, Michael A. Lewis, and Jose A.D. Machuca, "Rapid-Fire Fulfillment," *Harvard Business Review*, Nov 2004, pp. 104-110.

[78] Martha L Maznevski and Katherine M. Chuboda, "Bridging Space over Time: Virtual Team Dynamics and Effectiveness," *Organization Science*, 11(5), Sept-Oct., 2000, pp. 473-492.

[79] Ina Fried, "Driven to distraction by technology," c/net News.com, July 21, 2005 http://news.com.com/Driven+to+distraction+by+technology/2100-1022_3-5797028.html

[80] Don Phillips, "Cockpit Confusion Found in Crash of Cypriot Plane," *International Herald Tribune*, Sept. 7, 2005.

[81] Susanne Craig, "Wall Street Puts Its Finger on the Latest Trading Glitch," *Wall Street Journal,* Sept. 3, 2004, p. C3

[82] Martin Fackler, "Tokyo Exchange Struggles with Snarls in Electronics," *New York Times*, Dec. 13, 2005.

Three years earlier, a Lehman Brothers trader executed a large, erroneous sell order at the end of the trading day, when very few buyers were in the market. The order to sell a basket of securities worth 300 million pounds instead of 3 million pounds, as intended, led to a 130 point drop in the British FTSE index, and substantial losses by many investors. Source: BBC News, "FTSE collapse remains a mystery," *BBC Business News*, May 16, 2001,

http://news.bbc.co.uk/1/hi/business/1333405.stm

[83] Jared Sandberg, "Never a Safe Feature, 'Reply to All' Falls into the Wrong Hands," *Wall Street Journal*, Oct. 25, 2005.

[84] Charles Perrow, *Normal Accidents: Living with High-Risk Technologies*. New York: Basic Books, 1984.

[85] Ed Joyce, "Software Bugs: A Matter of Life and Liability." *Datamation,* May 15, 1987, pp. 88–92.

[86] Philip E. Ross, "The day the software crashed." *Forbes*, Apr. 25, 1994, pp. 142-156.

Chapter 11: Participants

[87] For an illustrative, but artificially constructed example, plus comments by four experts, see Byron Reimus, "The IT System that Couldn't Deliver," *Harvard Business Review*, May-June 1997, pp. 22-35

[88] V.G. Narayanan and Ananth Raman, "Aligning Incentives in Supply Chains," *Harvard Business Review*, Nov 2004, pp. 94-102.

[89] A famous paper about this topic is Steven Kerr, "On the Following of Rewarding A when Hoping for B," *Academy of Management Journal*, 18, 1975, pp. 769-783. http://pages.stern.nyu.edu/~wstarbuc/mob/kerrab.html

[90] Eileen C. Shapiro, *Fad Surfing in the Board Room: Managing in the Age of Instant Answers*, Reading, MA: Addison-Wesley, 1996, p. 57

[91] Harold Salzman and Stephen R. Rosenthal. *Software by Design: Shaping Technology and the Workplace*. New York: Oxford University Press, 1994, p. 134.

[92] Rachel Zimmerman, "Why Emergency Rooms Rarely Test Trauma Patients for Alcohol, Drugs," *Wall Street Journal*, Feb. 24, 2003.

[93] David Pitches, Amanda Burls, and Ann Fry-Smith, A., "How to make a silk purse from a sow's ear--a comprehensive review of strategies to optimise data for corrupt managers and incompetent clinicians," *British Medical Journal*, 327, December 20-27, 2003, pp. 1436-1439.

[94] Carol A. Derby et al, "Possible effect of DRGs on the classification of stroke subtypes: Implications for surveillance." *Stroke*, 32, 2001, pp. 1487-1491.

[95] Marc Santora, "Cardiologists Say Rankings Sway Choices on Surgery," *New York Times*, Jan. 11, 2005.

[96] Louis V. Gerstner, Jr. "Our Schools Are Failing: Do We Care?" *New York Times*, May 27, 1994, p. A15.

[97] Barbara Garson, *The Electronic Sweatshop*. New York: Penguin Books, 1989.

[98] Daniel J. Power, "Randy Fields Interview: Automating 'Administrivia' Decisions", *DSSResources.COM*, April 9, 2004.

[99] Steven Alter, "Equitable Life: A Computer-Assisted Underwriting System." in *Decision Support Systems: Current Practice and Continuing Challenges,* Reading, MA: Addison-Wesley, 1980.

[100] Kurt H. Fleming, Curt Coffman, and James K. Harter, "Manage Your Human Sigma," *Harvard Business Review*, July-August 2005, pp. 107-114.

[101] Shoshana Zuboff, *In the Age of the Smart Machine,* New York: Basic Books, 1988, pp. 127, 135.

[102] "Healthy Lives: A New View of Stress." *University of California, Berkeley Wellness Letter,* June 1990, pp. 4–5.

[103] Robert Karasek and Tores Theorell. *Healthy Work*. New York: Basic Books, 1990.

[104] Anna Wilde Mathews, "New Gadgets Trace Truckers' Every Move," *Wall Street Journal*, July14, 1997.

[105] Dean Foust, "The Ground War at FedEx,' *BusinessWeek*, Nov. 28, 2005, pp. 42-43.

[106] Working by the book is also a tactic used in labor disputes. For example, a reference to pilots working by the book appears in *Chicago Tribune*, "Summer of hell exacts heavy toll," Third of four parts, July 15, 2003, http://www.chicagotribune.com/chi-uniteday3-story,0,4838025.story?page=4&coll=chi-unitednavover-misc

[107] Christina Elnora Garza, "Studying the Natives on the Shop Floor," *BusinessWeek*, Sept. 20, 1991, pp. 74-78.

Chapter 12: Information

[108] Charles Forelle, "As Need for Data Storage Grows, A Dull Industry Gets an Upgrade," *Wall Street Journal*, Oct. 14, 2005, p. A1.

[109] The decision to recommend going ahead with the launch despite warnings from engineers is widely cited in discussions of organizational communication. For a summary and references to more detailed descriptions see: Department of Philosophy and Department of Mechanical Engineering, Texas A&M University, "Engineering Ethics: The Space Shuttle Challenger Disaster," http://ethics.tamu.edu/ethics/shuttle/shuttle1.htm

[110] Terrorist threats prior to 9/11 are mentioned in news articles such as:

* David Johnston, "Pre-Attack Memo Cited Bin Laden," *New York Times*, May 15, 2002.

* Associated Press, "U.S. Was Aware of bin Laden Threat before Sept. 11 Attacks," *New York Times*, Sept. 19, 2002.

* Frank Bass and Randy Herschaft, "AP: U.S. Foresaw Terror Threats in 1970s, *Boston.com News*, Jan. 23, 2005. http://www.boston.com/news/nation/washington/articles/2005/01/23/ap_us_foresaw_terror_threats_in_1970s/

[111] A doctor describes this situation in: Abigail Zuger, "When the Computers Crash, All That's Left Are the Patients," *New York Times*, Sept. 14, 2004.

[112] A good example is the Northwind database that Microsoft distributes with its database product Microsoft Access.

[113] George Houdeshel and Hugh J. Watson, "The Management Information and Decision Support System (MIDS) at Lockheed-Georgia," *MIS Quarterly*, 11(1), March 1987, pp. 127-140.

[114] Thomas H. Davenport, "Saving IT's Soul: Human-Centered Information Management," *Harvard Business Review*, Mar-Apr 1994, pp. 119-131.

[115] Paul R. Spickard, "The Illogic of American Racial Categories," *Frontline*, Public Broadcasting System, http://www.pbs.org/wgbh/pages/frontline/shows/jefferson/mixed/spickard.html

[116] U.S. Census Bureau, *Population Division, Special Population Staff.* "Racial and Ethnic Classifications Used in the 2000 Census and Beyond," April, 12, 2000, http://www.census.gov/population/www/socdemo/race/racefactcb.html

[117] Steven A. Holmes, "The Politics of Race and the Census," *New York Times*, Mar. 19, 2000.

[118] For example, see, Marc L. Songini, "ETL: Quick Study," *Computerworld*, Feb. 2, 2004. http://www.computerworld.com/databasetopics/businessintelligence/datawarehouse/story/0,10801,89534,00.html

[119] James M. Dzierzanowski et al, "The Authorizer's Assistant: A Knowledge-Based Credit Authorization System for American Express." In Herbert Schorr and Alain Rappoport, *Proceedings of the First Annual Conference on Innovative Applications of Artificial Intelligence*. Menlo Park, CA: American Association for Artificial Intelligence, 1989, pp. 168-172.

Dorothy Leonard-Barton, and John J. Sviokla. "Putting Expert Systems to Work," *Harvard Business Review*, Mar-Apr 1988, pp. 91-98.

[120] Hugh Sidey, "The Lessons John Kennedy Learned from the Bay of Pigs," *Time Online Edition*, Apr. 16, 2001. http://www.time.com/time/nation/article/0,8599,106537,00.html

[121] Amos Tversky and Daniel Kahneman, "The Framing of Decisions and the Psychology of Choice," *Science*, Vol. 2111, Jan. 30, 1981, pp. 453-458.

[122] John Rubin, "The Dangers of Overconfidence," *Technology Review*, July 1989, pp. 11-12.

[123] The first four categories are from data quality categories discussed by Diane M. Strong, Yang W. Lee, and Richard Y. Wang, "Data Quality in Context," *Communications of the ACM*, May 1997, 40(5), pp. 103-110. They included access security as a subcategory within accessibility. In Figure 12.1, the accessibility dimension from Strong *et al* is divided into two dimensions: accessibility of information (whether authorized users access what they need) and security and control (whether access is denied to unauthorized users). The two aspects of accessibility are treated as separate dimensions because they are different in nature and because problems related to information security and control receive so much attention today.

Although most of the dimensions within the categories in Figure 12.1 match those proposed by Strong *et al*, some are renamed slightly, appear under different categories, or simply are different. Readers interested in the source of the dimensions should see Strong *et al* or the more detailed article, Richard Wang and Diane M. Strong, "Beyond Accuracy: What Data Quality Means to Data Consumers," *Journal of Management Information Systems*, 12(4), Spring 1996, pp. 5-33.

[124] Barbara D. Klein, Dale L. Goodhue, and Gordon B. Davis, "Can Humans Detect Errors in Data? Impact of Base Rates, Incentives, and Goals," *MIS Quarterly*, 21(2), June 1997, pp. 169-194.

A similar estimate of 1% to 5% appears in Thomas C. Redman, "The Impact of Poor Data Quality on the Typical Enterprise," *Communications of the ACM*, 41(2), Feb. 1998, pp. 79-82.

[125] Kenneth C. Laudon, "Data quality and due process in large interorganizational record systems," *Communications of the ACM*, 29(1), Jan. 1986, pp. 4-11.

[126] Debra D'Agostino, "Data Management: Getting Clean," *CIO Insight*, Aug. 1, 2004.

[127] Deborah Gage and John McCormick, "ChoicePoint: Blur," *Baseline*, June 14, 2005, http://www.baselinemag.com/article2/0,1540,1826200,00.asp

[128] *Ibid.*

[129] Barbara Whitaker, "How to Mend a Credit Report That's Not Really Broken," *New York Times* Aug. 1, 2004.

[130] Timothy W. Koch and S. Scott MacDonald, *Bank Management*, 5th ed., Mason, OH: Thompson/ South-Western, 2003, p. 650.

[131] Tom Herman, "IRS Beefs Up Audits, Enforcement, Even as Customer Service Improves," *Wall Street Journal*, Nov. 30, 2005. The data cited is from a GAO report: Government Accountability Office, "Tax Administration: IRS Improved Some Filing Season Services, but Long-Term Goals Would Help Manage Strategic Trade-offs, GAO 06-51, November 2005. http://www.gao.gov/new.items/d0651.pdf

[132] Frederick F. Reichhold, "The One Number You Need to Grow," *Harvard Business Review*, Dec. 2003, pp. 46-54

[133] Commission on the Intelligence Capabilities of the United States Regarding Weapons of Mass Destruction, *Report to the President of the United States*, March 31, 2005. http://www.wmd.gov/report/wmd_report.pdf

[134] For example, see, Ramesh Balasubramaniam, "Factors Influencing Requirements Traceability Practice," *Communications of the ACM*, 41(12), Dec. 1998, pp. 37-44.

[135] See the web site of the Equal Employment Opportunity Commission, http://www.eeoc.gov/

[136] Charles Forelle, "IBM Policy Bars Use of Genetic Data in Employment," *Wall Street Journal*, Oct. 11, 2005, p. B2.

[137] Robin Cooper and Robert S. Kaplan, "The Promise – and Peril – of Integrated Cost Systems," *Harvard Business Review*, Jul-Aug 1998, pp. 109-119.

[138] Robin Lloyd, "Metric mishap caused loss of NASA orbiter," *CNN.com*, Sept. 30, 1999, http://www.cnn.com/TECH/space/9909/30/mars.metric.02/

[139] Christopher Mayer, "The Fed as Oracle," Ludwig von Mises Institute, Sept. 23, 2003, http://www.mises.org/fullstory.aspx?control=1335&id=78

[140] Evan Perez and Rich Brooks, "For a Big Vendor of Personal Data, A Theft Lays Bare the Downside," *Wall Street Journal*, May 3, 2005.

[141] Charles Perrow, *Normal Accidents: Living with High-Risk Technologies*, New York: Basic Books, 1984.

[142] Matthew L. Wald, "Colombians Attribute Cali Crash to Pilot Error," *New York Times*, Sept. 26, 1996, p. 12.

[143] Dow Jones Newswires (2004) "Ex-Teledata Worker Enters Guilty Plea in Identity Theft," *Wall Street Journal*, Sept. 15, 2004.

Chapter 13: Technology and Infrastructure

[144] Mel Duvall and John McCormick, "Triumphs & Trip-Ups In 2004," *Baseline*, Dec. 20, 2004. http://www.baselinemag.com/article2/0,1397,1744725,00.asp

[145] Hal R. Varian, "The Information Economy. How much will two bits be worth in the digital marketplace?" *Scientific American*, Sept. 1995, pp. 200-201.

[146] For example, if you want to learn how RFID technology operates, you can receive a number of explanations by typing something like "RFID technology" or "How does RFID operate?" in the query box on Google.com, Yahoo.com, MSN.com, AOL.com, Answers.com, Wikipedia.org, and other general purpose sources of information.

[147] Arthur C. Clarke, *Profiles of the Future*, revised edition, New York: Holt, Rinehart, and Winston, 1984.

[148] See, for example, M. Lynne Markus and Robert I. Benjamin, "The Magic Bullet Theory of IT-Enabled Transformation," *Sloan Management Review*, Winter 1997, pp. 55-68.

[149] Zeller, Tom, Jr., "Beware Your Trail of Digital Fingerprints," *New York Times*, Nov. 7, 2005.

[150] Electronic Privacy Information Center, "The Drivers Privacy Protection Act (DPPA) and the Privacy of Your State Motor Vehicle Record," updated Aug. 28, 2005. http://www.epic.org/privacy/drivers/

[151] Nicholas Carr, "IT Doesn't Matter," *Harvard Business Review*, May 2003, pp. 41-49.

[152] Timothy L. Smunt and Candace L. Sutcliffe, "There's Gold in Them Bills," *Harvard Business Review*, September 2004, pp. 24-25.

[153] First described in the 1960s, packet switching is a technique for transmitting data from one computer to another. With packet switching, which is used by the Internet, a message is subdivided into packets that are individually transmitted from the source to the destination and then reassembled at the destination. Each packet carries not only the information it conveys, but also the Internet address of its source and destination.

[154] Lee Smith, "New Ideas from the Army." *Fortune*, Sept. 19, 1994, pp. 203-212.

[155] Researchers have found positive correlations between IT spending and productivity across matched firms, but they note many reasons why IT expenditures may not generate positive results and why expenditures on IT are only the tip of the iceberg when trying to produce better corporate results through initiatives that involve IT. See Erik Brynjolfsson, "The IT Productivity Gap," *Optimize Magazine*, Issue 21, July 2003, http://www.optimizemag.com/showArticle.jhtml?articleID=17700941

Furthermore, Brynjolfsson says that companies in one study averaged "$10 of organizational capital associated with every $1 of technology capital." The additional costs include "direct costs, such as hiring consultants, the time spent by internal and external software developers to rewrite the code and adapt the code to work for the organization, the training of the personnel at all levels, whether it be the secretaries or the line workers or the sales people or the middle managers, all the way up. Then there's the management time in thinking about which processes need to change and how." Mark Kindery, "Hidden Assets" *CIO Insight*, Oct. 2001. http://www.cioinsight.com/article/0,3658,s=301&a=16867,00.asp

[156] William M. Bulkeley, "The Data Trap: How PC Users Waste Time." *Wall Street Journal*, Jan. 4, 1993, p. B2.

[157] Richard Heygate, "Technophobes, Don't Run Away Yet." *Wall Street Journal,* Aug. 15, 1994, p. A8.

[158] For detailed information about the development of floppy disks, enter the term *floppy disk* into the Wikipedia search box in http://en.wikipedia.org/wiki/Main_Page. Wikipedia provides similarly detailed historical information related to technologies such as the CD-ROM, DVD, hard disk, and microprocessor.

[159] Clayton M. Christensen and Michael E. Raynor, *The Innovator's Solution: Creating and Sustaining Successful Growth*, Boston, MA: Harvard Business School Press.

[160] David Pogue, "An iPod Worth Keeping an Eye On," *New York Times*, Oct. 18, 2005.

[161] See discussion of Moore's Law in Wikipedia, http://en.wikipedia.org/wiki/Moore%27s_Law, and on the Intel web page on the 40th anniversary of Moore's Law, http://www.intel.com/technology/mooreslaw/index.htm

[162] The five generations are mechanical computing devices, electro-mechanical (relay based) computers, vacuum-tube computers, discrete transistor computers, and integrated circuit computers. Ray Kurzweil, *The Age of Spiritual Machines*, New York: Viking Penguin, 1999. For an illustration, see http://en.wikipedia.org/wiki/Image:PPTMooresLawai.jpg

[163] Intel Corporation, "Microprocessor Transistor Count Chart," downloaded from a site called "Moore's Law," http://www.intel.com/museum/archives/history_docs/mooreslaw.htm

[164] Walter A. Mossberg. "Annual Buying Guide: How to Ensure New PC Can Use Windows Vista," *Wall Street Journal*, April. 13, 2006.

[165] Walter A. Mossberg. "A Screen Test for the Video iPod," *Wall Street Journal*, Oct. 19, 2005.

[166] Elizabeth Horwitt, "N.Y. sites unfazed by outage," *Computerworld*, Sept. 23, 1991, p. 1 +.

Also, Gary H. Anthes, "FCC blasts AT&T for New York blowout," *Computerworld*, Nov. 18, 1991, p. 58

[167] Peter Weill, "The Role and Value of Information Technology Infrastructure: Some Empirical Observations." pp. 547-572 in Rajiv Banker, Robert Kauffman, and Mo Adam Mahmood, *Strategic Information Technology Management: Perspectives on Organizational Growth and Competitive Advantage,* Idea Group Publishing, 1993.

[168] John A. Byrne, "The Pain of Downsizing." *BusinessWeek,* May 9, 1994, pp. 60-69.

[169] Lakshmi Mohan, and William K. Holstein. "Marketing Decision Support in Transition," pp. 230-252 in Robert C. Blattberg, Rashi Glazer, and John D. C. Little, *The Marketing Information Revolution.* Boston: The Harvard Business School Press, 1994.

[170] Thomas H. Davenport, "Putting the Enterprise into Enterprise Systems," *Harvard Business Review*, Jul-Aug 1998.

[171] Robert B. McKersie and Richard E. Walton. "Organizational Change," pp. 244-277 in Michael S. Scott Morton, ed., *The Corporation of the 1990s: Information Technology and Organizational Transformation*, New York: Oxford University Press, 1991.

[172] Peter Weill and Jeanne W. Ross, *IT Governance: How Top Performers Manage IT Decision Rights for Superior Results*, Boston, MA: Harvard Business School Press, 2004.

[173] Weill and Ross, *pp. 30.*

[174] Marianne Broadbent and Peter Weill, "Management by Maxim: How Business and IT

Managers Can Create IT Infrastructures," *Sloan Management Review*, Spring 1997, pp. 77-92

[175] Tables 3 and 4 in Broadbent and Weill (1997) identify four to ten business and IT maxims that they found in each category in their research.

[176] Weill and Ross, *pp. 30-31.*

[177] Jeanne W. Ross, "Creating a Strategic IT Architecture Competency: Learning the Stages," *MIS Quarterly Executive*, March 2003, pp. 31-43.

[178] Ross (2003)

[179] Peter Weill, Mani Subramani, and Marianne Broadbent, "Building IT Infrastructure for Strategic Agility," *MIT Sloan Management Review*, Fall 2002, pp. 57-65.

[180] Sinan Aral and Peter Weill, "IT Savvy Pays Off: How Top Performers Match IT Portfolios and Organizational Practices," Center for Information Systems Research, Working Paper No. 353, May 2005.

Chapter 14: Environment and Strategy

[181] Hau L. Lee, "The Triple-A Supply Chain," *Harvard Business Review*, 82(10), Oct. 2004, pp. 102-112.

[182] V.G. Narayanan and Ananth Raman, "Aligning Incentives in Supply Chains," *Harvard Business Review*, 82(11), Nov. 2004, pp. 94-102.

[183] Robert J. Thomas, *What Machines Can't Do.* Berkeley: University of California Press, 1994, pp. 76-87

[184] Edgar Schein, *Organizational Culture and Leadership*, 2nd ed., San Francisco, CA: Jossey-Bass Publishers, 1992

[185] *Ibid.*

[186] Sumner and Ryan say that part of the problem is the failure to integrate CASE software into work systems. Mary Sumner and Terry Ryan "The Impact of CASE: Can it achieve critical success factors?" *Journal of Systems Management*, 45(6), 1994, p. 16, 6 pages.

[187] Richard Farson, *Management of the Absurd*, New York: Simon & Schuster, 1996.

[188] James C. Collins and Jerry I. Porras. *Built to Last: Successful Habits of Visionary Companies.* New York: Harper Business, 1994, p. 117.

[189] Harold Salzman and Stephen R. Rosenthal. *Software by Design: Shaping Technology and the Workplace.* New York: Oxford University Press, 1994, p. 109.

[190] Salzman and Rosenthal. pp. 96-97.

[191] Youichi Ito and Takaaki Hattori, "Mass Media Ethics in Japan," in Thomas W. Cooper, *Communications Ethics and Global Change*, White Plains, NY: Longman, 1989, pp. 69-84.

[192] David Bjerklie, "E-mail: The Boss Is Watching." *Technology Review.* Apr. 1993, p. 14.

[193] Bruce Caldwell, "Consultant, Heal Thyself." *Information Week*, June 21, 1993, pp. 12-13.

[194] James Brian, Quinn, Philip Anderson, and Sydney Finkelstein. "Managing Professional Intellect: Making the Most of the Best," *Harvard Business Review*, Mar-Apr, 1996, pp. 71-80

[195] Jeff Moad, "The Second Wave." *Datamation*, Feb. 1, 1989, pp. 14-20.

[196] John Hazard, "Nearly Half May Not Make Second Sarbanes-Oxley Deadline,"*CIO Insight*, Sept. 27, 2005. http://www.cioinsight.com/article2/0,1540,1867018,00.asp

[197] Jerry Luftman, Raymond Papp, and Tom Brier, "Enablers and Inhibitors of Business-IT Alignment," *Communications of the Association for Information Systems*, 1(11), March 1999.

[198] Jerry Luftman, "Key Issues for IT Executives for 2004," *MIS Quarterly Executive*, June 2005, pp. 269-285.

Chapter 15: Work System Ideas in a Broader Context

[199] Six Sigma is a term that originally had one meaning and eventually took on a very different meaning. The original meaning is related to the field of statistics. Six Sigma came from the notion that a process generates results according to a "normal distribution" The Greek letter sigma refers to the statistical concept of standard deviation. In a symmetrical bell shaped distribution called a normal distribution, approximately 68% of the values are within one standard deviation (one sigma) of the average value. As part of its total quality management program in the 1990s, Motorola decided to aim for Six Sigma in highly repetitive processes. Six Sigma refers to

no more than 3 defects per million. Eventually the term Six Sigma took on a life of its own as statistically oriented quality management approaches adopted the name Six Sigma.

[200] Abigail J. Sellen and Richard Harper, *The Myth of the Paperless Office*. Cambridge, MA: MIT Press, 2002

[201] Raju Narisetti, "IBM to Launch Global Web-Business Blitz," *Wall Street Journal*, Oct. 6, 1997, p. B11.

[202] Daniel Amor, *The E-business revolution*. Upper Saddle River, NJ: Prentice Hall PTR, 2000, p.7.

[203] Ravi Kalakota and Marcia Robinson, *E-Business: Roadmap for Success*. Reading, MA: Addison-Wesley, 1999, p. xvi.

[204] Sid L. Huff, Michael Wade, Michael Parent, Scott Schneberger, and Peter Newson. *Cases in Electronic Commerce*. Boston: Irwin-McGraw Hill, 2000, p. 4.

[205] For research related to spreadsheet errors see Raymond Panko's spreadsheet research site at http://panko.cba.hawaii.edu/ssr/

[206] "On 14 October 2005, Wal-Mart announced the results of research showing that customers found items they wanted in stock more often in stores that use RFID technology with embedded electronic product codes (EPCs) than in stores that do not. Researchers at the University of Arkansas found a 16 percent drop in out-of-stock (OOS) merchandise at RFID-equipped stores due to better in-store stock management. OOS items with EPCs were replenished three times faster than comparable items using standard bar-code technology." Source: Jeff Woods, "Wal-Mart's RFID Project Promising for Retail, CPG Industries" Gartner Research, Oct. 19, 2005, http://www.gartner.com/DisplayDocument?doc_cd=133493

[207] Linear programming, the most common technique used in optimization models, assumes a linear objective function and linear constraints. There are many other types of optimization models. Mathematically inclined readers might be interested in further explanation from web-base sources such as http://en.wikipedia.org/wiki/Linear_programming

[208] Steven Alter, "A Work System View of DSS in its Fourth Decade," *Decision Support Systems*, 38(3), December 2004, pp. 319-327.

[209] Source: MicroStrategy, Inc., "The 5 Styles of Business Intelligence: Industrial-Strength Business Analysis," White Paper, viewed on Jan. 12, 2005 at http://www.microstrategy.com/Download/files/solutions/5_styles/5_styles.pdf

[210] Thomas H. Davenport, and Jeanne G. Harris, "Automated Decision Making Comes of Age," *MIT Sloan Management Review*, Summer 2005, pp. 83-89.

This page is blank

Index